International Cryogenics Monograph Series

Series Editors
J. G. Weisend II, European Spallation Source
Lund, Sweden
Sangkwon Jeong, Department of Mechanical Engineering
KAIST
Daejeon, Korea (Republic of)

The International Cryogenics Monograph Series was established in the early 1960s to present an opportunity for active researchers in various areas associated with cryogenic engineering to cover their area of expertise by thoroughly covering its past development and its present status. These high level reviews assist young researchers to initiate research programs of their own in these key areas of cryogenic engineering without an extensive search of literature.

Ernst Wolfgang Stautner • Kiruba S. Haran
Phillip J. Ansell • Constantinos Minas

Aircraft Cryogenics

Ernst Wolfgang Stautner
GE HealthCare – Technology
and Innovation Center (HTIC),
One Research Circle
Niskayuna, NY, USA

Kiruba S. Haran
Department of Electrical and Computer
Engineering
University of Illinois at Urbana-Champaign
Urbana, IL, USA

Phillip J. Ansell
Department of Aerospace Engineering
University of Illinois Urbana-Champaign
Urbana, IL, USA

Constantinos Minas
GE Research Aviation, One Research Circle
Niskayuna, USA

ISSN 0538-7051 ISSN 2199-3084 (electronic)
International Cryogenics Monograph Series
ISBN 978-3-031-71407-8 ISBN 978-3-031-71408-5 (eBook)
https://doi.org/10.1007/978-3-031-71408-5

© The Editor(s) (if applicable) and The Author(s), under exclusive license to Springer Nature Switzerland AG 2024

This work is subject to copyright. All rights are solely and exclusively licensed by the Publisher, whether the whole or part of the material is concerned, specifically the rights of translation, reprinting, reuse of illustrations, recitation, broadcasting, reproduction on microfilms or in any other physical way, and transmission or information storage and retrieval, electronic adaptation, computer software, or by similar or dissimilar methodology now known or hereafter developed.
The use of general descriptive names, registered names, trademarks, service marks, etc. in this publication does not imply, even in the absence of a specific statement, that such names are exempt from the relevant protective laws and regulations and therefore free for general use.
The publisher, the authors and the editors are safe to assume that the advice and information in this book are believed to be true and accurate at the date of publication. Neither the publisher nor the authors or the editors give a warranty, expressed or implied, with respect to the material contained herein or for any errors or omissions that may have been made. The publisher remains neutral with regard to jurisdictional claims in published maps and institutional affiliations.

This Springer imprint is published by the registered company Springer Nature Switzerland AG
The registered company address is: Gewerbestrasse 11, 6330 Cham, Switzerland

If disposing of this product, please recycle the paper.

*Dedicated to my wife Jeena, my mentor
Albert Hofmann, the KIT and GE folks, and
all aspiring, creative cryogenic engineers.*

Preface

This book is a step-by-step guidance approach on the cryogenic design of cryogenic components for an all-electric aircraft system that is in part also applicable to liquid hydrogen combusting aircraft as well as hybrids.

There is no shortage on publication on hydrogen-fueled aircraft; however, in this book we look through a magnifying glass on what has been done in the past, condensing it into a strategy on what must be done next to enable liquid hydrogen storage in aircraft, given today's technological advancement.

Emphasis is particularly placed on tank manufacturability, safety features, and minimum tank weight, providing a holistic focus on the logistics of hydrogen management of all major components within the cryoaircraft as well as on emerging, superconducting motor architecture.

The intention is further to fully exploit the benefit of an available cold reservoir of liquid hydrogen with relevance to cooling of various superconducting components, e.g., motors and superconducting cables, as well as for heat sinking numerous power electronic devices on fueling the fuel cell stack system. A liquid hydrogen tank hold-time analysis reveals the main governing factors and describes the required efforts for minimizing onboard boil off for aircraft designs with middle to long flight mission duration.

This is followed by an outlook showing where cryotankage technology and cryogenic aircraft architecture may move to within the next 10–20 years, and how basic research will need to play a major supporting role to help realizing these future designs by consequently eliminating whitespace within today's technology landscape in a world, already suffering from effects of a global climate change.

In one way this is an unusual book with contents that refers and points out to anticipated, advanced technology for future systems embedded in a truly green hydrogen infrastructure.

Both do not exist yet at the time of compiling this book. We hope that the many ideas herein will be taken on, validated and re-shaped, and gradually fill the white space. With this book we sometimes tread "no-man's land" expanding our knowledge based on the cryogenic foundation that has been established over the last 100 years.

Niskayuna, NY, USA Ernst Wolfgang Stautner
Urbana, IL, USA Kiruba S. Haran
Phillip J. Ansell
Constantinos Minas

Introduction to Chapters

Chapter 1 begins with thoughts on "why hydrogen."

Chapter 2 explains on how hydrogen is stored, liquefied, and subcooled, with a special focus on hydrogen, as an aircraft fuel.

Chapter 3 looks at the different design steps involved for storing liquid hydrogen in a novel composite tank focusing on various thermal parameters that need to be taken into consideration.

Chapter 4 gives deeper insight on our current cryogenic composite technology landscape for liquid hydrogen tanks from the mechanical perspective.

Chapter 5 ponders on which way composite tanks can be integrated into an airframe, given the freedom of changing fuselage to accommodate new tank shapes with special focus on CHEETA.

Chapter 6 talks about interfacing the tank with various piping, the integration of components and instrumentation within aircraft cryo-circuit, that could be seen as a survey on the technology that is available now and from the engineering perspective, where we need to focus on.

Chapter 7 reviews pumps used with liquid cryogens, followed by an excursion on possible high temperature superconducting pump designs. Furthermore, we gauge the present status of hydrogen pumps and where we need components to improve this technology to make them aircraft worthy, followed by general tank safety concerns.

Chapter 8 connects to the previous chapter, but here we give a high-level review on current hydrogen leak detection methods based on the hydrogen automotive industry.

Chapter 9 extensively reviews cooling system technologies for DC superconducting motors with their field windings, followed by analyzing AC loss densities on the armature winding.

Chapter 10 features an excursion to investigate how rotating heat transfer can be possible in a vacuum environment, which may lead to novel methods impacting other engineering modalities.

Chapter 11 circles back to Chap. 1, emphasizing the necessary effort on current and future hydrogen infrastructure required for aircraft fueling at the airport, followed by an overview on safety and available standards.

Chapter 12 gives a summary and outlook on what needs to be done to get cryogenics and superconductivity "on board" and outlines major technology gaps that need almost immediately be closed and highlights engineering white space, where we tread on uncharted territory, that asks for further research with universities, aid from government organizations and industry partners.

Appendix 1 gives an overview on material selection for tanks, cryogenic components, like heat exchangers or liners, for composite tanks.

Since a hydrogen-powered aircraft will not "fly" with heavy base material, the focus here is on material embrittlement and material selection, with supply of rare data made available for engineers and researchers.

Appendix 2 complements the material side with necessary hydrogen properties in all its states. Those rarely found data have been made available with approval by NBS/NIST. Most of the thermophysical data, including those for heat transfer give the engineer the opportunity to arrive at novel solutions for accommodating liquid hydrogen in an aircraft.

Appendix 3 provides information on various safety aspects and resources to tap into.

Appendix 4 covers most non aerospace liquid hydrogen storage methods.

Appendix 5 covers the Airbus activities in cryogenics and superconductivity, as publicly available of 2024.

Appendix 6 is meant for the new generation of students or researchers who would like to learn more about cryogenics by presenting a list on cryogenic reference books, whereas as the main focal point.

Appendix 7 highlights the relevant key resources provided by NBS/NIST for future use of hydrogen, to create and accelerate the build of the required hydrogen infrastructure for aircraft and other technology fields, like off-shore superconducting generators with superconducting cables for energy and liquid hydrogen transfer as well as for other superconducting applications that use high temperature superconductors and maritime engineering.

Appendix 8 points out critical vacuum sealing technology for cryogenic applications.

Acknowledgment

Support from the Center for High-Efficiency Electrical Technologies for Aircraft (CHEETA) by NASA under award number 80NSSC19M0125 is gratefully acknowledged. Special thanks to the NHA (National Hydrogen Association), Professor Sandy Smith of the University of Manchester, and the research centers of GE Aerospace, GE HealthCare, and GE Vernova for their generous support.

This work was supported by NASA under award number 80NSSC19M0125 as part of the Center for High-Efficiency Electrical Technologies for Aircraft (CHEETA)

Contents

1 **Introduction** 1
   1.1 Introduction 1
   1.2 Element 1 3
   1.3 Scope 4
   References.. 5

2 **Hydrogen Storage Technology: Options and Outlook** 7
   2.1 Hydrogen Storage Technology—Options and Outlook 7
       2.1.1 Liquid Hydrogen—Properties 8
       2.1.2 Liquid Hydrogen as an Aircraft Fuel 12
       2.1.3 Liquid Hydrogen for all Electric Aircraft—The CHEETA
              Project 15
   References.. 17

3 **Cryogenic Liquid Hydrogen Tank Design Aspects: General
Overview** 19
   3.1 Cryogenic Liquid Hydrogen Tank Design Aspects—General
       Overview 19
       3.1.1 Tank Volume Determination 23
       3.1.2 Hydrogen Infrastructure in an all-Electric Aircraft 24
   3.2 Cryotankage—Heat Sources 25
       3.2.1 Selecting the Appropriate Thermal Protection Scheme 25
       3.2.2 Mitigating Thermal Radiation 29
       3.2.3 Implementing Insulation—Structural Example 32
   3.3 Cryotankage—Simplified Design Example 34
   References.. 61

4 **Cryotankage: Structural Thoughts** 63
   4.1 Cryotankage—Structural Thoughts 63
       4.1.1 Early Liquid Hydrogen Tank Designs 63
       4.1.2 Tank Wall—Composites 64
       4.1.3 Optimized Liquid Tank Structure 87

4.2　Cryotankage—Design for Tank Faults/Operating Conditions 87
　　　　　4.2.1　Creating a Composite Pressure Vessel Shell 88
　　　　　4.2.2　CHEETA Stress Analysis Example 89
　　　4.3　Minimum Tank Mass Weight Fraction 96
　　　References 100

5　Cryotankage—Tank Shapes and Airframe Integration 105
　　　5.1　Non-integral vs. Integral Tank Designs 106
　　　　　5.1.1　Dual Tank—Separate Dual Tank Walls 107
　　　　　5.1.2　Tank Structural Designs 108
　　　　　5.1.3　Tank Accommodation Options 110
　　　　　5.1.4　Tank Fuselage Changes 110
　　　　　5.1.5　A321 XLR Hybrid-Electric Aircraft
　　　　　　　　　with Superconducting Propulsors 115
　　　　　5.1.6　CHEETA All-Electric Aircraft with Superconducting
　　　　　　　　　Propulsors and Superconducting Components 118
　　　References 121

**6　Hydrogen Tank—Cryocircuit, Integration of Components,
　　Instrumentation** 123
　　　6.1　Design of Cryogenic Penetrations for Liquid Bulk Tanks 123
　　　　　6.1.1　Inclined Tubes 126
　　　　　6.1.2　Metal/Composite Bonding 131
　　　6.2　Typical Process Flow Circuit for a Liquid Hydrogen Storage
　　　　　Tank with Instrumentation 131
　　　6.3　Tank Instrumentation 133
　　　　　6.3.1　Fill Level Sensors for Aircraft 133
　　　6.4　Cryogenic Transfer Lines 134
　　　　　6.4.1　Vacuum Jacketed Lines (VJ) 134
　　　　　6.4.2　Jacketed Lines Without Vacuum 137
　　　　　6.4.3　Cable Cryostats (CC) 137
　　　6.5　Valves, Safety Valves, and Special Features 141
　　　6.6　Tank Chilldown /Refueling 143
　　　6.7　Other Designs 145
　　　6.8　Defueling 146
　　　6.9　Liquid Hydrogen Transfer Process 147
　　　　　6.9.1　Logistics of Single-Tank Operation 147
　　　　　6.9.2　Logistics of Multitank Operation 148
　　　References 149

7　Liquid Hydrogen Pump Overview/Tank Safety 151
　　　7.1　Cycling Pumping 155
　　　　　7.1.1　Cryogenic Magnetic Bearing Technology 156
　　　7.2　Liquid Hydrogen Pumps—Current Status 159
　　　　　7.2.1　Phase Separator 164

Contents   xv

	7.3	Tank Safety 166
		7.3.1 Compromised Vacuum Conditions 166
		7.3.2 Sudden Tank Vacuum Failure 166
		7.3.3 Inner Tank Rupture 168
		7.3.4 Ice Formation on Outer Surface of Inner Tank—Refer to Section 168
		7.3.5 Lightning Protection 168
		7.3.6 Stratification, Explosive Boil-Off 168
		7.3.7 Example: The X-33 Project 169
	References 170

8   Hydrogen Detection, Leak Detection, Zero Emission Vehicles (ZEV), Sensor Types 173
    8.1   Leak Detection on Hydrogen Aircraft 173
        8.1.1   Introduction 173
        8.1.2   H2 Detection on ZEV 173
        8.1.3   H2 Detection at H2 Infrastructure Sites 175
        8.1.4   Types of H2 Sensors 176
        8.1.5   H2 Detection on H2 Aircraft 176
        8.1.6   Example of a H2 Aircraft Detection System 178
    References 180

9   Cooling System Technologies on Superconducting Rotating Machines 181
Uijong Bong and Kiruba S. Haran
    9.1   Introduction 181
    9.2   Review on Cooling System Technologies for Rotary Superconducting DC Field Winding 183
        9.2.1   Closed-Loop Cooling System 183
        9.2.2   Open-Loop Cooling System 186
        9.2.3   Comparison of Cooling Technologies 187
    9.3   Considerations on Cooling System Technologies for Stationary Superconducting AC Armature Winding 187
        9.3.1   Challenge with Closed-Loop Cooling System 188
        9.3.2   Opportunity in Open-Loop Cooling System 196
    9.4   Summary 197
    References 197

10 Rotating Vacuum Heat Transfer, Rotating Cryocoolers, Slip Rings, Rotating Bearings, Ball Bearings 201
    10.1   Excursion: Rotating Heat Transfer for Motors 201
        10.1.1   Rotating Cryocoolers 202
        10.1.2   Thermal Slip Ring/Brush Design 203
        10.1.3   Bearings and Their Derivatives 205
    10.2   Conclusion 210
    10.3   Other Alternatives 210
    References 211

11	Airport Infrastructure Requirements for Liquid Hydrogen Supply and Distribution	213
	11.1 Introduction	213
	11.2 Current Aviation Fuel and Energy Infrastructure	213
	11.3 Hydrogen Production and Distribution Networks	215
	11.3.1 Hydrogen Production Methods	215
	11.4 Hydrogen Infrastructure at Airports	217
	11.4.1 Airport Hydrogen Supply Chain	218
	11.4.2 Airport Local Distribution	219
	11.4.3 Refueling with LH_2	224
	11.4.4 Safety and Standards	229
	References	231
12	Technology Gaps, Cryogenic Power Electronics, Cryogenic Current Leads, Transfer Lines, Liquid Hydrogen Pumps	233
	12.1 Summary and Outlook	233
	12.2 Technology Gaps	234
	12.2.1 Cryogenic Power Modules: Knowledge Gap	234
	12.2.2 Cryogenic Current Leads: Knowledge Gap	236
	12.2.3 Flight Verification	236
	References	237

Appendices ... 243

References ... 321

Index ... 325

Introduction

> *Why should I care about future generations – what have they ever done for me?*
>
> Groucho Marx

1.1 Introduction

At the recent 26th United Nations Climate Change Conference last year, over 140 countries pledged to achieve net-zero emissions to combat climate change. And in a dramatic appeal to attain sustainability in the skies, Europe's Flightpath 2050 [1] initiated a bold effort to reduce CO_2 emissions worldwide by 75%, NOx emissions by 90%, and the noise footprint by 60% by the mid-century mark.

The implications and effects of climate change have been contemplated for decades. Just consider the following excerpt from the Global 2000 Report [2] commissioned by President Carter in 1980 that postulated the impact of rising concentrations of carbon dioxide on future generations:

> The full effects of rising concentrations of carbon dioxide, depletion of stratospheric ozone, deterioration of soils, increasing introduction of complex persistent toxic chemicals into the environment, and massive extinction of species may not occur until well after 2000. Yet once such global environmental problems are in motion they are very difficult to reverse. In fact, few if any of the problems addressed in the Global 2000 Study are amenable to quick technological or policy fixes; rather, they are inextricably mixed with the world's most perplexing social and economic problems.

Even before the Global 2000 report, the effect of environmental pollution had been addressed in one of the first publications of Scientific American in 1973. The call was made then by D.P. Gregory [3] for new energy regimes when the energy crisis hit the world, and specifically for a hydrogen economy (Fig. 1.1).

As early as 1973 it was believed that most of the domestic oil supply was already depleted, and gas would have exhausted well even earlier with 90% of the world's oil supply being depleted by 2032 [4].

At this time, the aircraft industry made a serious effort to replace Jet-A fuel with other fuel types, e.g., hydrogen, or fuel mixes, resulting in the first attempt of "going green" of the aircraft industry. In 1975 NBS released its bibliography on "Hydrogen Future Fuel-A" that gives deep insight into the effort it needs to get us there [5].

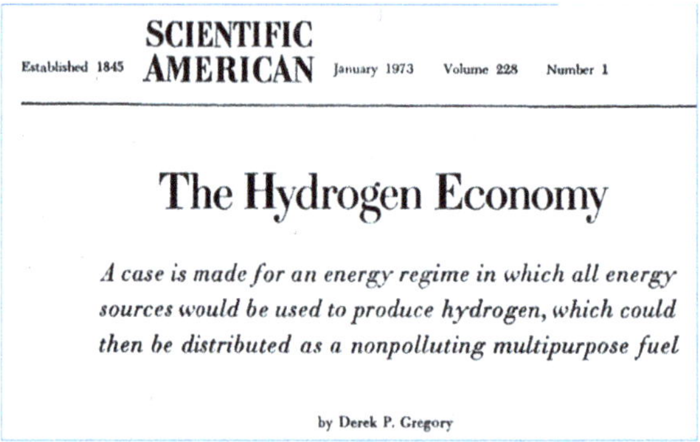

Fig. 1.1 Hydrogen economy

> **SCIENTIFIC AMERICAN**
> Established 1845   January 1973   Volume 228   Number 1
>
> ## The Hydrogen Economy
>
> *A case is made for an energy regime in which all energy sources would be used to produce hydrogen, which could then be distributed as a nonpolluting multipurpose fuel*
>
> by Derek P. Gregory

However, it needs to be understood that unleashing all of these new applications with hydrogen to effectively combat climate change will require a concerted action to establish a comprehensive, integrated hydrogen economy. For example, even though hydrogen-powered aircraft systems were possible 50 years ago, there was no infrastructure available to commercially fill liquid hydrogen into fuel tanks at airports. In addition, we didn't have dedicated hydrogen storage systems, apart from hydrogen trucks or means to produce liquid hydrogen at the point of source where needed.

This infrastructure for producing and distributing hydrogen had to be built and was thought to come from steam reforming of natural gas, coal gasification, water electrolysis, a continuous steam-iron process, by thermochemical decomposition of water (electrolyzers), solar-derived hydrogen iron processes, and liquefying hydrogen using magnetic refrigeration.

Alden Armagnac [6] suggested to run electrolyzers during off-peak hours from nuclear power plants and even make good use of any produced oxygen as well.

Finally, in 2021, we are beginning to see real action by several European countries that is setting the stage for an integrated hydrogen economy that far surpasses what we have seen in the last fifty years. This action includes a promising new generation of high-tech electrolyzers, fuel cells, and other important components such as liquid hydrogen pumps. The list of green energy initiatives is long, backed by more than 30 European countries. A great example is "The HyDeal Ambition" [7].

Of course, the different ways hydrogen is created are characterized by different "colors," including green, blue, grey, brown, or black, turquoise, and still others. In this article, we will only refer to "green hydrogen," produced entirely from clean, zero-carbon energy.

We should however not deny the fact that the realization of the hydrogen economy is still struggling, but even conventional fuel-supplying companies seem to realize the need for change [8].

To achieve our climate change goals, hydrogen indeed will have a major role in high energy-consuming industries (steel) as well as for individual and mass transportation, future power generation and production, and even long-distance space travel. The applications in mass transportation are numerous, ranging from hydrogen-powered cars and trains to ships and aircraft systems. Hydrogen also has future applications in healthcare as well, including superconducting magnet applications operating at 20 K (see Appendix 4).

To be clear, hydrogen is not being used as a propulsor fuel in the traditional sense with all electric aircraft. Nevertheless combusting hydrogen would still result in the production of contrails and cause NOx emission but half as with Jet-A fuel. Instead, its primary uses would be for cooling superconducting aircraft components and for feeding stacks of fuel cells that generate only water as an unwanted by-product.

As we shall see, these tasks will require a working, sustainable cryogenic environment with element 1 that we briefly recall in the following.

1.2 Element 1

Since the dawn of time hydrogen and helium have both emerged as the most useful cryogens. Those elements that astronomers told us have been created by the early universe, approx. 380,000 years after the Big Bang [9]. Both elements are still today the most important ones for cooling cryogenic components as well for keeping superconducting magnets or components at their superconducting temperature (Fig. 1.2).

Hydrogen was first discovered by Cavendish in 1776 which at that time was called "inflammable air." He observed that "air" given off when metals react with acids is distinctly different from the air we breathe. He also found that 423 measures of this inflammable air are very closely sufficient to phlogisticate 1000 of common air (a near 2:1 ratio). Nearly a century later, in 1871, Mendeleyev first had the

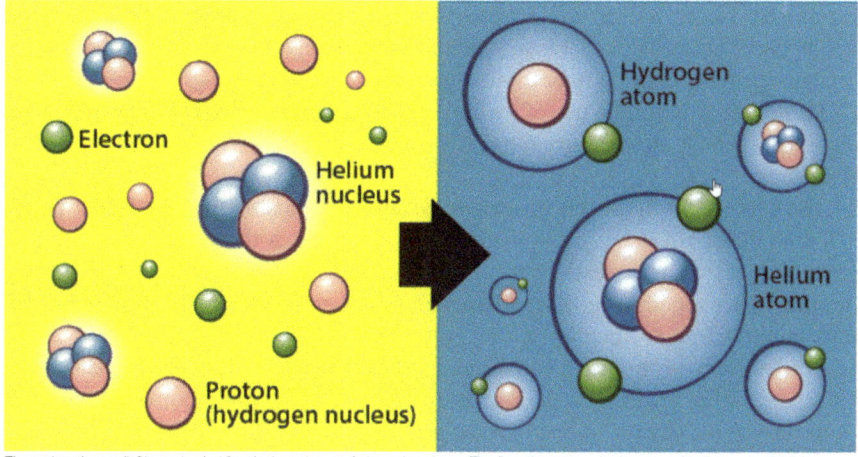

The early universe (left) was too hot for electrons to remain bound to atoms. The first elements — hydrogen and helium — couldn't form until the universe had cooled enough to allow their nuclei to capture electrons (right), about 380,000 years after the Big Bang.

Fig. 1.2 Early universe

courage among other researchers to publish his early version of the table of elements by placing them corresponding to their weight with hydrogen at the top [10].

In the following, we show the role the cryogen hydrogen will play for all-electric aircraft systems and how a cryogenic and superconducting infrastructure is integrated into an all-electric hydrogen-powered aircraft.

It needs to be understood that some of the items discussed will also apply for other modalities yet to come or need to be explored, that as of today we may not even know about yet.

1.3 Scope

The following book provides a step-by-step approach on the cryogenic design aspects of liquid hydrogen tanks for all-electric aircraft and hydrogen-fueled systems. Emphasis is kept on tank manufacturability, safety, and minimum weight. In this book, we also keep the complete logistics of hydrogen management in mind for all the components within the aircraft. The main intention is to further exploit the full benefit of a cold reservoir of liquid hydrogen with respect to cooling of various superconducting components, e.g., motors and power transfer cables, as well as the power electronics and the cooling and fueling fuel cell stacks. A liquid hydrogen hold-time analysis reveals the main governing factors and describes the required efforts for minimizing onboard boil-off for some aircraft designs with different flight mission duration.

This is then followed by an outlook showing where the technology would move to within the next 20 years and how basic research would help us realize these future designs by consequently eliminating the whitespace within the technology landscape that is, unfortunately, still present as of today.

Here we also draw the circle (span the scope) toward all superconducting applications for future use of liquid hydrogen. We draw from previous experience in neighboring fields like general storage of hydrogen, hydrogen transport, as well as on hydrogen use in the automotive and renewables industry [11].

Pacing the way for the use of hydrogen in all-electric aircraft systems requires a significant boost in mission range by optimizing those liquid hydrogen tank structures.

The safe use of hydrogen in aircraft systems foremost requires a deeper understanding of the underlying cryogenic effort. Cryogenics thus needs to encompass structural tank design questions for given operating conditions and for a long-range aircraft power topology with its components, as well as options for embedding the tank in the airframe structure.

Furthermore, with the CFM-Airbus RISE announcement as of February 2022, CFM and Airbus decided on pioneering hydrogen combustion technology with LH_2 as the cryogenic storage method for the prototype A380 [12] and recently presented a very early design of a liquid hydrogen tank, stating that if one were to store the same energy of 4600 kg of Jet fuel, one would need 1600 kg of LH_2. As we shall see, the drawback is an LH_2 volume that is four times larger than that for Jet fuel.

Thus, with the given current uncertainties in providing fuel resources for industry and domestic applications, together with climate warming reduction efforts, demand for production of "green hydrogen" will have to quickly increase to an unprecedented level.

References

1. Flightpath 2050, https://op.europa.eu/en/publication-detail/-/publication/296a9bd7-fef9-4ae8-82c4-a21ff48be673
2. https://en.wikipedia.org/wiki/The_Global_2000_Report_to_the_President
3. D. P. Gregory, The Hydrogen Economy, Scientific American, 1973, vol. 228, no. 1
4. Daily Press January 4th, 1973, Virginia, http://dailypress.newspapers.com
5. Olien N A Schiffmacher S A Hydrogen-Future "Fuel-A" (with emphasis on cryogenic technology) NBS 664 1975
6. J.P. Penland, "Liquid hydrogen fueled Boeing 737-200", Penland diaries, from popular Science 1973
7. Green hydrogen initiative https://finance.yahoo.com/m/2c876bdf-4db1-3249-aefe-e84d2c38c187/green-hydrogen-initiative.html
8. Ball B Basile A Nejat Veziroglu T Chapter 11 'The hydrogen economy - vision or reality?' in Compendium of Hydrogen Energy Volume 4: Hydrogen Use, Safety and the Hydrogen Economy, Elsevier 2015
9. First element: https://astronomy.com/magazine/ask-astro/2018/12/the-first-element
10. J. Gribbin, The Scientists, Random house, 2002
11. R. Kottenstette, J. Cotrell "Hydrogen storage in wind turbine towers" International Journal of Hydrogen Energy 29 pp 1277–1288 2004
12. https://www.airbus.com/en/newsroom/press-releases/2022-02-airbus-and-cfm-international-to-pioneer-hydrogen-combustion

Hydrogen Storage Technology: Options and Outlook

2.1 Hydrogen Storage Technology—Options and Outlook

To store a cryogen at light weight, the storage density is the important factor for aircraft. Figure 2.1, taken from the first liquid hydrogen-fueled car [1] (BMW Hydrogen 7, see Appendix 4), compares different storage densities at various temperatures and pressures. To achieve a storage density of approx. 80 g/l, gaseous hydrogen is compressed to 300 bar working pressure at 38 K, which means we need to work with cryo-compressed hydrogen. For room temperature storage of compressed hydrogen at 350 bar, the storage density drops to 25 g/l, and to 40 g/l at a working pressure of 700 bar. Basically, a factor of 2 less than what can be achieved with cryo-compressed hydrogen. Figure 2.2 on the left hints on the wall thickness of 12 mm that would be required for 350 bar operating pressure with a GEN 4 type tank of diameter 200 mm.

As can be seen from Fig. 2.1, for aviation cryo-compressed gas storage will be too heavy and bulky, constraining available space.

This leaves liquid hydrogen storage as the only possible option, with respect to minimum pressure vessel weight and achievable storage densities of 70 g/l at 1 bar, which can be used to support a superconducting motor operating temperature of 20 K.

Efforts are underway worldwide to find additional cryogenic and non-cryogenic storage options for hydrogen. But even today the same holds true what has long been suspected, even promising metal hydrides are still too heavy and bulky to be considered as an option [2]: Metal hydrides ($MgNiH_2$) for example can store roughly the same amount of liquid volume as a traditional tank at a fuel weight that is at least 20 times higher, including the hydride bed [3]. A further pending question is on how gaseous storage can be distributed along the transfer lines fast enough to a propulsor or for fuel cell supply. Figures 2.2 and 2.3 show the presently available compressed hydrogen tanks [4].

2 Hydrogen Storage Technology: Options and Outlook

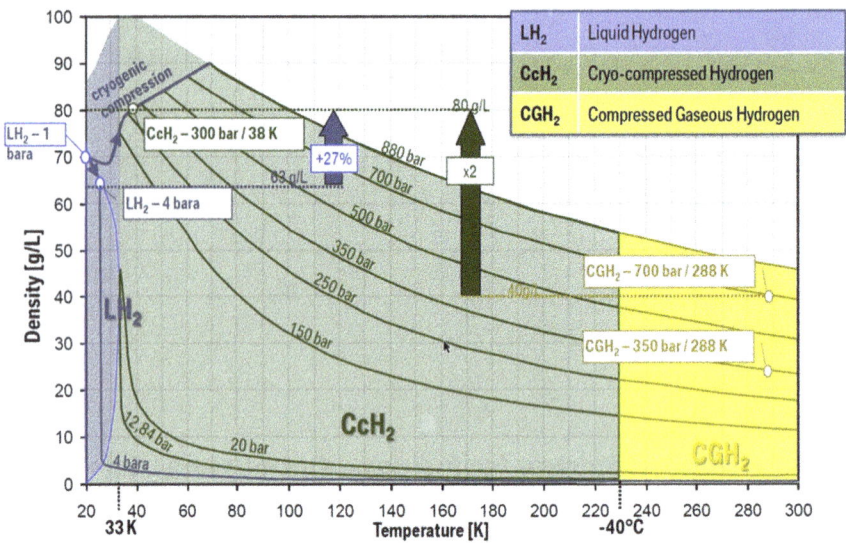

Fig. 2.1 Hydrogen storage options

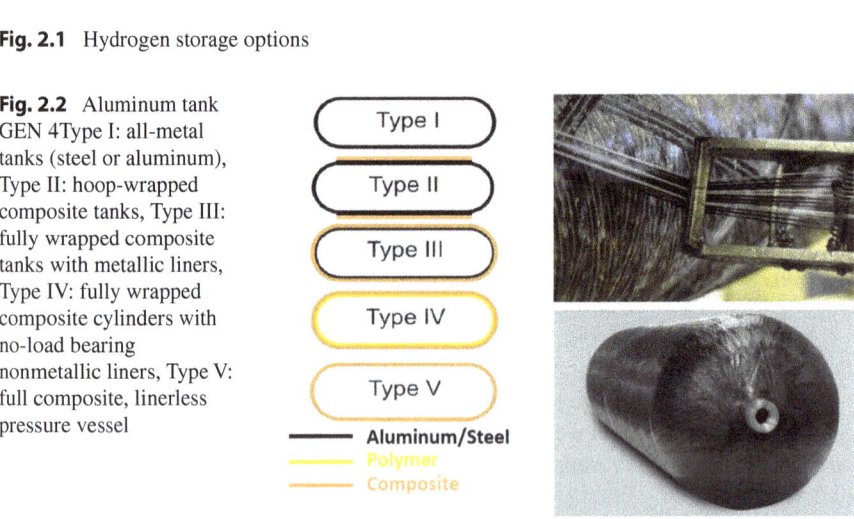

Fig. 2.2 Aluminum tank GEN 4Type I: all-metal tanks (steel or aluminum), Type II: hoop-wrapped composite tanks, Type III: fully wrapped composite tanks with metallic liners, Type IV: fully wrapped composite cylinders with no-load bearing nonmetallic liners, Type V: full composite, linerless pressure vessel

Figure 2.3 shows a state-of-the-art high-pressure carbon fiber composite (CFC) reinforced aluminum tank for a typical operating pressure of 70 MPa [5].

2.1.1 Liquid Hydrogen—Properties

It was first Olszewski, in 1895, who applied the adiabatic expansion method to the dynamic state liquefaction of hydrogen [6]. In that, he compressed hydrogen gas to 190 atmospheres and pre-cooled it with liquid oxygen boiling at reduced pressure (−211 °C). When the pressure was released, he noticed a fog of liquid hydrogen

Fig. 2.3 High-pressure CFC reinforced aluminum tank

Table 2.1 Properties of common cryogens

Cryogen	Latent heat of vaporization (kJ/kg)	Volume boiling off from an applied heat load to liquid of 1 W (ml/h)	Gas flow at standard conditions (STP 0 °C, 1 at), of this boil-off rate of 1 W (l/min)	Enthalpy change at 1 at (ideal heat gas absorption) at T-range (J/g)
Helium	20.91	1377	16.05	87 (4.2–20 K) 384 (4.2–77 K) 1542 (4.2–300 K)
Hydrogen[a]	448.3	114.5	1.5	590 (22–77 K) 3490 (20–300 K)
Neon	87.2	34.7	0.77	283.5 (27–300 K
Nitrogen	199.1	22.5	0.2423	233.5 (77–300 K)

[a]A comprehensive list of physical hydrogen properties is given in Appendix 2

drops. Soon after, the first commercial liquefaction of hydrogen started in 1898. It was the liquefaction of hydrogen, which enabled Onnes to liquefy helium in 1908 and introduced the world to the phenomenon of superconductivity [7].

Table 2.1 explains why hydrogen is found suitable for many applications as compared to other cryogens.

With respect to costly helium, liquid hydrogen boil-off is significantly lowered per W of applied heat, by a factor of almost 12, which makes liquid hydrogen a suitable cooling medium for many applications. In its gaseous form hydrogen as a flowing gas can absorb twice as much heat as helium and is also 12 times more efficient in that process than gaseous nitrogen considering the same temperature range

(77–300 K), thus making it one of the most useful cryogens for cold mass cooldown, only inferior to costly neon, even beating helium vapor by a factor of 3.5.

This well-known fact was exploited in the early days of cryogenics and magnet technology for designing high-field magnet systems that used ultra-high purity aluminum wire maintained at liquid hydrogen temperatures and revealed an interesting alternative concept compared to superconducting magnets that were to be operated at costly liquid helium temperatures. In fact, in the early days and in the absence of any superconducting wire, high-purity aluminum wire was used to wind high field magnets, so-called cryogenic coils with iron return yokes and a central field of 2.6 Tesla, that were maintained and operated at liquid hydrogen temperature [8].

With respect to liquid hydrogen boil-off compared with helium as per Table 2.1, higher heat loads to the tanks are acceptable, especially for a dormant aircraft idling at the tarmac. Liquefaction most likely will not be required if the tank structure is optimized for low parasitic losses. If it is however found necessary for some aircraft types, the recondensing power needs to be 12 times higher as compared to helium (typical GE MRI magnet systems recondensing liquid helium use 1 W cryocoolers). Since the cooling power of cryocoolers at 20 K is in the range of 20 to 30 W this is still a good option. Hydrogen ZBO designs have been analyzed and designed by GE [9] and with NASA Glenn, by Plachta [10]. A typical test implementation is shown in Fig. 2.4 for a spherical tank, intended for space missions and microgravity.

In summary, liquid tanks tolerate higher heat loads, but that also means a bigger recondensing surface on the cryocooler cold finger is required.

2.1.1.1 Slush Hydrogen

At 13.957 K liquid hydrogen turns into so-called slush hydrogen, a very interesting phase not well researched yet. Slush hydrogen is a mixture of liquid hydrogen with

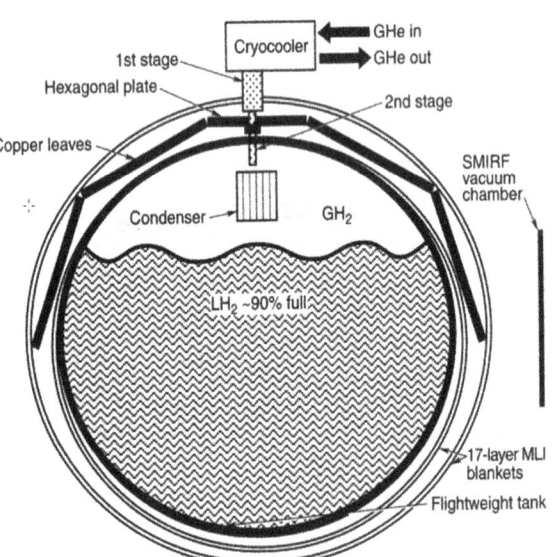

Fig. 2.4 Liquefying hydrogen

solid hydrogen in the form of many small particles, resembling crushed ice. Interestingly as Keller [11] notices, solid hydrogen ice particles age and change shape and size over time with the particles becoming more spherical in shape and denser. Most advantageous method to generate slush is by removing hydrogen vapor by pumping, extracting the heat of vaporization as the liquid boils off and keeps the remaining liquid. With a liquid temperature of 20.39 K the temperature difference of 6.433 K to slush hydrogen, unlike neon, with a temperature difference of only 2.5 K, makes the slush phase control less difficult.

Slush or supercooled hydrogen (sLH$_2$) is increasingly relevant for large-scale trucks, e.g., by Daimler. One truck with approx. 80 kg of LH$_2$ enables a cruising range of 1000 km with a tank fill time of 15 min. A subcooled hydrogen-filling station has recently been opened in Germany, Wörth am Rein.

Therefore, slush hydrogen may obviously seem to be an attractive option for the space and automotive industry as well as for aircraft whenever one needs to store liquid at a higher density. For some liquid tank designs this could result in a higher "payload."

The benefit for space applications was recognized already very early in 1968 [12] and is summarized in Table 2.2.

More recently Verstraete [13] notices: "In the Cryoplane project, the consequences of the adoption of slush hydrogen on aircraft design, performance and economics were assessed based on a practical slush of 50% solid in terms of mass at a temperature of −260 °C [14]. Due to the higher density, a 20% smaller tank volume was possible compared to LH$_2$ while the higher heat capacity of the slush also allows reduction of the insulation thickness and weight. However, due to the higher production costs (in the order of 8–17%) and the correspondingly higher fuel cost, a direct operating cost reduction compared to LH$_2$ was not clearly identifiable.

Table 2.2 Comparing slush with saturated liquid hydrogen for space applications

Design parameter	Atmospheric saturated liquid hydrogen	Subcooled liquid or slush hydrogen
Tank loading technique	Fill and top	Fill and recirculate
Loading parameters	Liquid level, pressure, temperature	Liquid level, pressure, temperature, flow rate, solid fraction
Ground venting	Standard practice	In emergency only
Ground pressurization	Optional	Mandatory
Engine and pump status	Qualified	Qualification required[a]
Hydrogen temperature at inlet	20.55–22.22 K Nominal	13.88 K or more (depending on storage time)
Optimum insulation thickness	Nominal	Less than nominal
Space storability without venting	Nominal	Better than nominal
Pressure requirement	Nominal	Larger than nominal

[a]for slush-fed systems only

It was therefore concluded that the use of slush hydrogen does not offer a convincing advantage to justify major dedicated efforts."

A trade-off study on slush hydrogen was also recently executed by the MagnaSteyr and the Airbus D Team [15]. Also in a recent study, Park [16] gives a comprehensive review on slush properties. As for pumping slush hydrogen it was concluded that net positive suction head (NPSH) requirements, pump efficiency, and cavitation constant are the same for slush hydrogen and triple-point hydrogen, showing no performance difference.

One further advantage of slush hydrogen, however, is the ability to efficiently transfer slush through cryogenic transfer lines (see also Section on Cryogenic transfer lines in Chap. 6).

For further analysis on tank filling strategies, please see Chap. 9.

For a summary of the characterization study of slush hydrogen please see Sindt [17].

For slush hydrogen physical properties, see Appendix 2.

2.1.2 Liquid Hydrogen as an Aircraft Fuel

Short History of Hydrogen Flights—Liquid Hydrogen as a Fuel.

Using liquid hydrogen for aerospace is nearly as old as storing liquid hydrogen in cryostats and in trailers (see Appendix 4), approximately starting around the year 1950. Most of those early designs were initiated by NASA [18, 19].

The first in flight use of liquid hydrogen was on an U.S. Air Force B-57 twin-engine bomber, flown in 1956, within the scope of a test program to determine the combustion characteristics of hydrogen in an engine at high altitude.

Very early on in 1988, the science community also recognized the importance of hydrogen as a powerful means to reduce harmful emissions reporting that both environmental and energy supply concerns for domestic air transport should be the major factors responsible for initiation of a hydrogen program. It was recognized that in principle, "**hydrogen can be obtained and liquefied on any point of the globe**" and that "**this is the best fuel as regards ecology**." It was hoped that LH$_2$ and LNG-fueled commercial transport aircraft designs can be launched [20]. For example, one Trijet Tupolev T-155 transport aircraft was in air for 21 min and burned liquid hydrogen in one of its turbofan engines.

GE also very early recognized the potential of liquid hydrogen as an aircraft fuel.

Brewer [20] with Lockheed, especially emphasized that use, saying (Fig. 2.5):

Fig. 2.5 Brewer's comment

> It is my belief that liquid hydrogen will be used as fuel in aircraft of the future
>
> (Brewer 1991)

2.1 Hydrogen Storage Technology—Options and Outlook

In which way now aircraft fuel compares with hydrogen with respect to combustion.

First of all, the fuel type for aircraft is kerosene. Jet A-1 has a flash point higher than 38 °C and a freezing point of −47 °C. Jet A is a similar kerosene fuel type that is normally available only in the United States. After refining, aviation fuel is mixed with extremely small amounts of several additives. Considering the same energy content, kerosene weighs 2.8 times of liquid hydrogen, which means higher payload and range for a fixed aircraft empty weight. The downside is that liquid hydrogen needs more than 4 times the volume than Jet-A, which most likely requires a new aircraft fuselage design, as we shall see in the Cryotankage Section on Tank shapes and Airframe integration. For further combustion properties, see Appendix 3 (Tables 2.3 and 2.4).

In an effort to move to use sustainable fuels the following feedstocks are currently being researched that may be able to produce sustainable alternative fuels (SAFs), since an estimated 1 billion dry tons of biomass is currently being collected per year in the United States according to general IATA statements (Table 2.5).

Bicer and Dincer [21] et al. also looked at the toxicity of different aircraft fuels revealing the benefit of liquid hydrogen (see Fig. 2.6).

2.1.2.1 The Early Work at Boeing

The first flight using liquid hydrogen as a fuel [20] took place on June 20th, 1988, by William H. Conrad with a Grumman-American "Cheetah," powered by a Lycoming 0320-E2D engine, see Fig. 2.7.

The other recourse was to design and build a tank specifically for the space available in the airplane, insulated to contain the cryogenic hydrogen, and capable of operating at a working pressure of 400 psi.

As fabricated, the tank consisted of an inner vessel 20 inch in diameter by 33.25 inch long, constructed of Type 304 stainless steel. The end caps were of ellipsoidal shape. The inner vessel, which contains the liquid hydrogen, is supported inside an outer vessel which is 26 inch in diameter and 41.25 inch long and has flat ends. The outer vessel is aluminum. The inner tank is supported at multiple points inside the aluminum vessel by low-conductivity spacers. The annulus between the vessels is evacuated and contains about 25 layers of aluminized mylar which serve as radiation shields for improved thermal insulation [20].

Table 2.3 Jet-A fuel properties as compared to liquid hydrogen

Property	Liquid hydrogen	Jet-A (kerosene)	Ratio Jet-A/LH$_2$
Liquid density (kg/m^3)	70.96	810.53	11.42
Heat of combustion (MJ/kg)	120.0	42.80	0.3567
Energy density (MJ/l)	8.515	34.689	4.074

Table 2.4 Comparison with other alternative aviation fuel

Property	Liquid methane	Methanol	Ethanol	Biodiesel	Liquid ammonia
Liquid density (kg/m³)	424.0	796.0	794.0	870.0	730.0
Heat of combustion (MJ/kg)	50.0	19.9	27.2	38.9	18.6
Energy density (MJ/l)	21.2	15.9	21.6	33.9	13.6

Table 2.5 Further Jet-A alternative fuels

Typical SAFs
Corn grain
Oil seeds
Algae
Other fats, oils, and greases
Agricultural residues
Forestry residues
Wood mill waste
Municipal solid waste streams
Wet wastes (manures, wastewater treatment sludge)
Dedicated energy crops

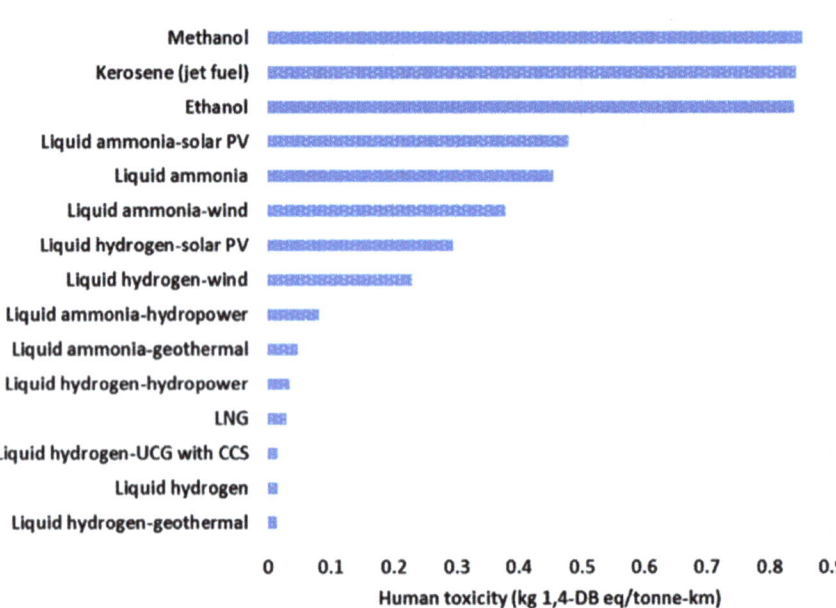

Fig. 2.6 Human toxicity values of various fueled aircrafts per traveled tons-km

2.1 Hydrogen Storage Technology—Options and Outlook

Fig. 2.7 LH$_2$-fueled Cheetah

2.1.3 Liquid Hydrogen for all Electric Aircraft—The CHEETA Project

Phillip J. Ansell Center for High-Efficiency Electrical Technologies for Aircraft (CHEETA); https://cheeta.illinois.edu/

As of 2018, the global CO$_2$ emissions produced by aviation exceeded 1 billion metric tons per year [21]. While the efficiency of aircraft and associated propulsion systems have dramatically improved over the past century, these improvements have been overtaken by the substantial increase in air traffic, resulting in a consistent net growth of environmental impact associated with civil air transportation. The environmental impact of aircraft emissions produced at high altitudes is difficult to quantify and characterize, though it is widely accepted that the long-lasting impacts of CO$_2$ deposition into the troposphere, the short-term and long-term products created from NO$_X$ emissions, and the formation of cirrus contrail clouds all contribute to an equivalent radiative forcing that encourages warming behavior. With the strong potential for further growth in air traffic in decades to come, it is anticipated that aviation will pose a growing influence in the overall environmental impact of transportation.

In order to facilitate improvements in the environmental sustainability of aviation in the future, several academic, industry, and government research groups have set ambitious goals and put forward new ultra-efficient aircraft concepts. In 2019, the Center for High-Efficiency Electrical Technologies for Aircraft (CHEETA) was founded under a NASA-supported University Leadership Initiative (ULI) program, composed of a multi-disciplinary consortium of experts in areas of aeronautics,

Table 2.6 Comparison of emissions Jet-A vs LH_2

Emission product	Emission Index ($kg_{emission}/kg_{fuel}$)	
	Jet A	LH_2
CO_2	3.16	0
NOx	0.0152	0.0108
H_2O	1.23	9

electrical systems, material sciences, and cryogenics. This group has put forward a visionary concept for aircraft systems that feature a fundamental shift away from hydrocarbon fuels and thermal engines, instead relying on liquid hydrogen energy storage and electrochemical devices for electrical power generation. This approach was selected as a means to completely eliminate CO_2 emissions produced by aircraft, as the energy carrier itself is absent carbon in composition, as well as all NOX emissions, which would otherwise be created in an airbreathing thermal engine system. Instead, the only emission produced by this system is water vapor, which is known to have little effective radiative forcing impact [22] so long as the production of nucleation sites for contrail formation is avoided. Hydrogen also has the potential of being produced in a fully renewable fashion. While many methods of hydrogen production exist, electrolysis or reverse fuel cell operation can be used to create hydrogen and oxygen from water in a clean fashion if the electricity used to drive this process is renewably sourced.

Given the ultra-high specific energy of hydrogen, at 120 MJ/kg (lower heating value), it serves as a promising energy carrier for aircraft applications. However, the energy density of hydrogen is significantly lower than conventional kerosene-based jet fuels, which poses a challenge to meeting the energy storage requirements for large aircraft. For this reason, hydrogen used for commercial aircraft is typically envisaged to be stored in a cryogenic liquid form. Given the low-temperature boiling point of hydrogen, at 20.28 K, the use of this energy storage medium as a cryogen is also a way to enable the use of superconducting electrical power transmission and propulsion systems. Doing so allows for high-power electrical transmission to be achieved with no ohmic losses, and superconducting electrical machines to be utilized that feature extremely high efficiencies and specific power capabilities. Furthermore, the design of an aircraft utilizing stored hydrogen energy can leverage a strategic integration of the storage tanks into the design of the airframe in order to reduce many of the drag penalties introduced by the high-volume storage system.

Thus, hydrogen is envisaged as an enabling and viable energy carrier for a future zero-carbon, environmentally sustainable aviation industry. Certainly, there are many challenges associated with making a hydrogen aircraft ecosystem realizable but investing in new technologies and capabilities that permit safe, scalable growth of aircraft hydrogen storage and distribution systems will serve as a crucial contribution to these efforts. As such, the following focus on the cryogenics of liquid hydrogen fuel tanks is anticipated to be an important contribution to this vision.

Table 2.6 gives a first estimate on emission levels, for a Mach 0.85 aircraft.

For further information on general liquid hydrogen storage see Brewer [23].

References

1. K. Kunze, O. Kircher, "Cryo-compressed hydrogen storage", Cryogenic Cluster Day, Oxford, 2012
2. L. Klebanoff, Hydrogen Storage Technology, CRC Press, 2012
3. S. J. Canfer, D. Evans, "Properties of materials for use in liquid hydrogen containment vessels", Advances in Cryogenic Engineering (Materials), Vol. 44, 1998, pp. 253-259
4. Source: fcto_cold_cryo_h2_storage_wkshp_1_doe file
5. NHA, National Hydrogen Conference, 2010
6. Kapitza https://societypublishing.org/doi/10.1098/rspa.1934.0214
7. https://link.springer.com/chapter/10.1007%2F978-94-009-2079-8_7, Onnes: Developed in view of the statical liquefaction of hydrogen and the obtaining of a permanent bath of liquid hydrogen (Comm. N°. 94() at which I was working then [PDF,PU00013525].
8. H. Brechna, Superconducting Magnet Systems, "Technische Physik in Einzeldarstellungen", Band 18, Springer Verlag, 1973, page 484
9. W. Stautner, M. Xu, S. Mine and K. Amm, "Hydrogen Cooling Options for MgB_2-based Superconducting Systems", AIP Conf. Proc., vol. 82, 2014, pp 82–90 and US20130104570A1, patent pending
10. D.W. Plachta "Hybrid Thermal Control Testing of a Cryogenic Propellant tank" AIP Conf. Proc., vol. 45A, 1999, pp. 468
11. Keller, Lockheed, Space / Aeronautics, 1968, p 108
12. Space / Aeronautics, 1968, p 110
13. D. Verstraete, PhD thesis, "The Potential of Liquid Hydrogen for long range aircraft propulsion", p 40
14. A. Westenberger, H2 Technology for Commercial Aircraft, Airbus Deutschland GmbH, 2007 andreas.westenberger@airbus.com
15. A. Westenberger, MagnaSteyr and Airbus D Team
16. Y. M. Park, Literature research on the production, loading, flow, and heat transfer of slush hydrogen, International Journal of Hydrogen Energy, 35, 2010, pp. 12993–13003
17. C. Sindt, "A Summary of the characterization study of slush hydrogen" Cryogenics vol. 10, issue 5 pp. 372–380 1970
18. A. Silverstein, E.W. Hall, "Liquid Hydrogen as a Jet Fuel for High-Altitude Aircraft," Research Memorandum, Lewis Flight Propulsion Laboratory, April 1955
19. D.B. Fenn, L.W. Acker, J.S. Algranti, "Flight Operation of a Pump-Fed Liquid Hydrogen Fuel System," Technical Memorandum X-252, NASA Lewis Research Center, 1960
20. G.D. Brewer, Hydrogen Aircraft Technology, CRC Press, 1991
21. Y. Bicer, I. Dincer, Life cycle evaluation of hydrogen and other potential fuels for aircrafts, International Journal of Hydrogen Energy, 42, 2017 10722-10738
22. D.S. Lee, D.W. Fahey, A. Skowron, M.R. Allen et al, "The contribution of global aviation to anthropogenic climate forcing for 2000 to 2018," *Atmospheric Environment*, Vol. 244, 2021, 117834
23. G.D Brewer, R.E Morris, R.H. Lange, J.W. Moore, Report no. NASCAR-132559, Study of the application of hydrogen fuel to long range subsonic transport aircraft vol. 2 1975

Cryogenic Liquid Hydrogen Tank Design Aspects: General Overview

3.1 Cryogenic Liquid Hydrogen Tank Design Aspects—General Overview

In the following, we give a brief overview on certain important tank design aspects that need to be considered. Depending on the size of aircraft and mission profile parameters those may change, or parameters need to be added, or may come with different weightings.

At this point, we recall Brewer's cautious comments on the difficulty of designing liquid hydrogen tanks for aircraft systems (Fig. 3.1).

Followed by the assessment of Boll [1] 30 years later, nothing much seems to have changed with respect to the cryogenic challenges that liquid hydrogen tanks pose for aircraft. Within the framework of this book, we can highlight only a few of those items that are most important. In fact, the below parameters are generally considered and applied for many different cryogenic modalities:

Design parameters:

- Integral wet wing/non-integral/modular.
- Materials.
- Insulation selection.
- Tank shape and size.
- Tank weight mass fraction (tank/LH$_2$).
- Tank wall stack-up.
- Slosh, bulk-head baffles.
- Vibration dampers.

Loads:

- Operating pressure/External/Internal/Test.
- Static/dynamic weight of liquid cryogen.

20   3 Cryogenic Liquid Hydrogen Tank Design Aspects: General Overview

> The design and development of the fuel tanks and their thermal protection systems to contain the cryogenic LH$_2$ in a satisfactory manner is regarded as one of the crucial technical challenges confronting use of LH$_2$ in operational aircraft
>
> (Brewer 1991)

> Integration of the hydrogen tank in actual aircraft is challenging, novel concepts might be required
>
> (Boll 2021)

Fig. 3.1 Brewer's and Boll's comments

- Vacuum conditions.
- Thermal stress/strain.
- Fatigue over time (composites and other materials).
- Cooldown.
- Thermal loads: Steady state/transient.
- Sloshing/parasitic loads.
- Vibration.
- Altitude changes.
- Mechanical: g-loads (tbd.).

Design boundary conditions:

- Mission profile - Cruising.
- Limit design (load factors).
- Ultimate design.
- Fail-safe design.
- Emergency.
- Proof test.
- Burst test.
- Leak test.
- Thermal test.

Design for manufacturing:

- Tank/fiber wrap/high-strength aluminum structure.
- Fuel transfer lines/(liquid/gas).

Design for inspection and maintenance:

- Bulk heads need man-holes, etc.

Safety aspects:

- Closed tank, safety valves etc.

Process control from tank to fuel cells:

3.1 Cryogenic Liquid Hydrogen Tank Design Aspects—General Overview

- Pumps etc., valves, etc.

Design boundary conditions:

- Mission profile—Cruising.
- Limit design (load factors).
- Ultimate design.
- Fail-safe design.
- Emergency.
- Proof test.
- Burst test.
- Leak test.
- Thermal tests.

Design for manufacturing:

- Tank/fiber wrap/high-strength aluminum structure.
- Fuel transfer lines/(liquid/gas).

Design for inspection and maintenance:

- Bulk heads, access with manholes, etc.

Safety aspects:

- Closed tank, safety valves, etc.

Process control from tank to fuel cells:

- Pumps, valves, etc.

One of the most important design goals for an aircraft is minimizing tank weight. Most design steps need to take that goal quite seriously, for long range as well as short range, and even hypersonic flight models. Liquid hydrogen tanks thus play a very important role in dictating the viability of hydrogen for the future of zero-carbon aviation.

Figure 3.2 shows the structural process for arriving at a first design for liquid hydrogen tanks, keeping in mind the aforementioned list of parameters.

It is understood that this is not a straightforward, downward approach as shown on the left, but rather a process that iteratively loops back all stages until the specific design objectives are refined. In the following, the required steps will be discussed in detail in each section.

Explanation Any liquid hydrogen storage volume is determined by a number of factors, for example, the flight profile (see Fig. 3.42 Mission profile, flight duration

22 3 Cryogenic Liquid Hydrogen Tank Design Aspects: General Overview

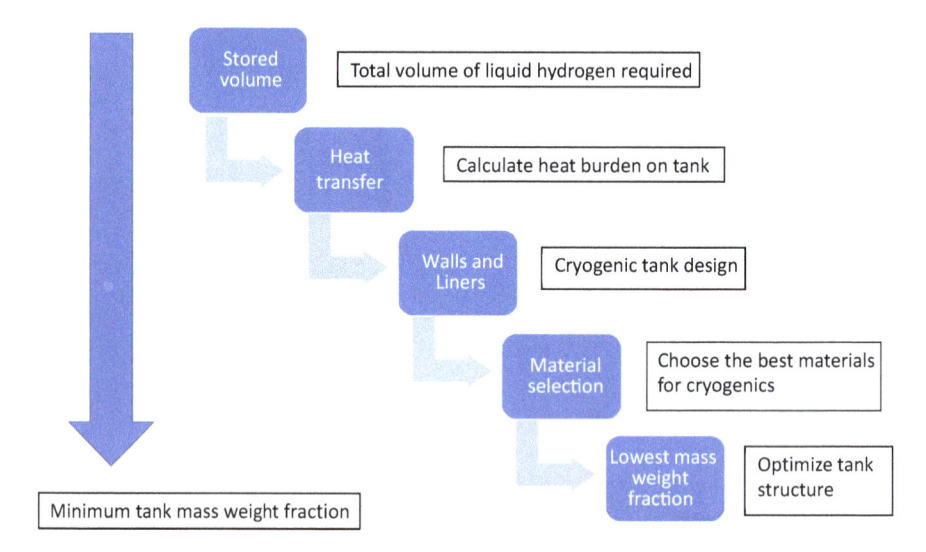

Fig. 3.2 Liquid hydrogen tank design steps

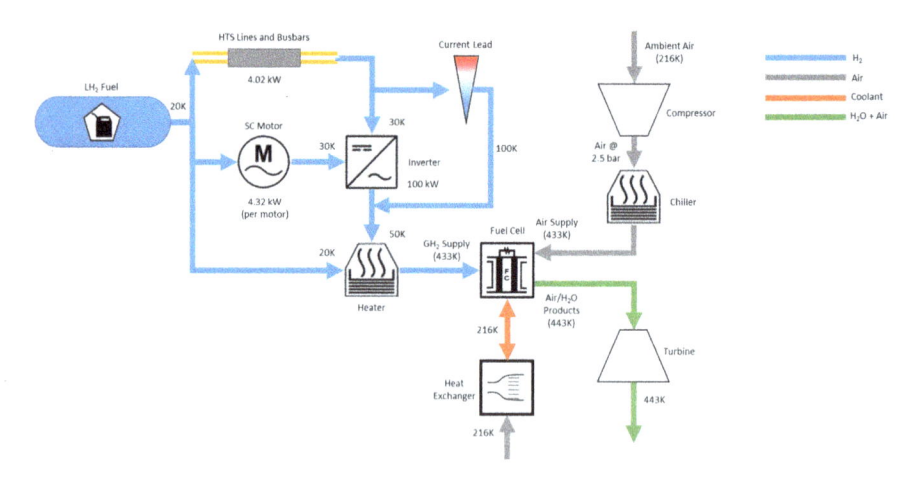

Fig. 3.3 Cryogenic flow circuit—Cryo aircraft

and Table 3.4), the all-electric aircraft components (see Fig. 3.3 and Section on Hydrogen Infrastructure) that consists of the cooling effort of any superconducting motor, superconducting power lines, the hydrogen consumption of the fuel cell stacks as well as other non-cryogenic components, that need to be cooled. To a certain extent, this also applies to hydrogen-fueled aircraft.

1. Once the required total inventory is established, including fill margin and ullage space, the heat burden on the tank can be calculated depending on the tolerable heat leak to the liquid hydrogen tank for a given flight profile. For this study, we

followed the most technically feasible and simple cryogenic approach for the tank structure. Based on prior work on thermal insulation studies for the boil-off of liquid hydrogen, we can safely disregard the use of recondensing coolers for this type of aircraft.
2. Now that a preliminary cryogenic envelop is established, we look at the overall design of the tank and the tank walls and how to preliminarily accommodate the tank within the fuselage. Furthermore, we need to add cryogenic components that deliver the mass flow through cryogenic and non-cryogenic consumers of the aircraft. We then need to agree on the comprehensive operating conditions the tank will experience for the life of the aircraft, including aircraft engine safety aspects, in particular again the flight profile, on-ground fueling, as well as on aircraft liquid hydrogen tank maintenance.
3. This finally leads us to a key point in the tank design, namely the choice of material. Material selection needs to consider the hydrogen specifics [see also Appendices 1 to 3]. See also section on material selection.
4. Now we are in a position to down-select the best materials and able to optimize the tank structure itself. We may also want to divide up the tanks further to fit them more readily into the airframe and change tank shapes as required.

3.1.1 Tank Volume Determination

For that we first need to keep the complete aircraft fuselage in mind and assume we know where to place the components and whether those require cooling or/and whether those need to be fueled.

With all-electric aircrafts or hybrids, fuel cells or batteries may need cooling depending on the type of component or even consume hydrogen itself.

Figure 3.3 shows a principal flow circuit for all-electric aircraft system where the superconducting motor is powered by the fuel cells output.

Once the complete flow circuit is known in detail, we can then arrive at a first guess on the liquid volume requirement.

Note that for cryogenic engineering we always assume a serial flow path for hydrogen, that is to say, flow out of the hydrogen tank will satisfy the thermal requirements of all the components whereas the flow of hydrogen will need to increase in temperature, until it meets the fuel cells high temperature end.

In general, it is good engineering practice to utilize the latent heat of hydrogen to its maximum. Unfortunately, this does not entirely work in the presence of a fuel cell, and we need to rethink and re-optimize the flow circuit.

Depending on the aircraft range (long-range) the fuel cell requirements mainly determine the liquid volume and hence the size of the hydrogen tanks. In addition, component cooling requires a share of the tank volume as well.

The superconducting engine (motor) with its field winding and armature needs to be kept cold at optimum superconducting temperature. This also applies to any power transmission lines, cold plates, and cooling efforts for batteries as well as for any inverters required.

3.1.2 Hydrogen Infrastructure in an all-Electric Aircraft

Following the blue arrows in Fig. 3.3, this simplified flowchart shows the liquid fuel distribution across the aircraft, from the storage tank to the propulsors. Most of the liquid is used for cooling the fully superconducting motor [2–5], the high temperature superconductor (HTS) lines, busbars, and current leads. The superconducting engine (motor) with its field winding and armature needs to be kept superconducting at an optimum temperature. This is achieved by making use of the high heat sinking capability of hydrogen using forced flow boiling through the helical motor windings to fully exploit the latent heat of vaporization.

Superconducting power transmission lines must be maintained below their critical temperature as well. Liquid hydrogen also resides in cryogenic transfer lines and is directed to flow over cold plates for heat sinking cryogenic power electronics and for cooling batteries.

From the motor, we drive the liquid flow to a cryogenic inverter and many other, different cryogenic power electronic components. Those are efficiently heat-sinked to run at their nominal optimal operating temperature, before entering a further heat exchanger. The current design requires a liquid cooling flow through the superconducting motor of only max. 0.1 kg/s, which would account for an approximate 30% drain of the liquid hydrogen tank. A bypass line also leads the liquid hydrogen flow directly to a heat exchanger to generate additional gaseous flow for maintaining the fuel cells. As the fuel cell technology advances, the hydrogen gas flow that accounts for 70% of the liquid hydrogen tank drain will be reduced. Besides, there may be components in an aircraft that can benefit from operating at a cryogenic temperature that we are not aware of yet. Many components tend to work and respond much faster due to the low electrical resistance of their metallic components. This leads us to the question of the overall fuel storage efficiency of an all-electric aircraft.

For comparison, and for future different all-electric aircraft designs, an efficiency factor could be thought of being introduced, that includes the mass flow required for different flow branches in this circuit, e.g.:

m_t = totally stored mass in all liquid hydrogen tanks

$m_f = m_{required} + x1^*$ fuel cells $+ x2^*$ motor cooling $+ x3^*$ component cooling (heat sinking) $+ x4^*$ system losses

whereas: $\eta = m_f / (m_f + m_t)$.

with:

x1: x2: x3: x4 may be % values.

3.2 Cryotankage—Heat Sources

Heat Burden on Tank

Any cryotankage design requires a thorough analysis of the thermal burden that is to be expected. Based on this, one has then to decide on whether the resulting boil-off is acceptable for all given operating conditions, or whether there is a need to reduce that further and whether recondensation needs to be considered, especially during long-term aircraft coasting or idling in particular to avoid pressure build up or to reduce the amount of hydrogen that needs to flushed off the aircraft. For comparison, for the BMW Hydrogen 7 car liquid storage, a boil-off or loss of liquid inventory of 2%/day is envisaged [6], similar to what a liquid helium tank boils off generally.

Below are the main losses that need to be considered to avoid any rapid pressure build up. In general, one maps the complete thermal loads imposed on the storage tank and refines the design until this meets the expectations:

1. Minimize thermal radiation down to the cryo environment.
2. Maintaining a vacuum environment in all operating conditions.
3. Optimize any tank suspensions and openings, feedthroughs, or sleeves for minimum thermal load down into the cryogenic environment.
4. Ensure any outgassing from composites or metal surfaces and other components is well controlled and tolerable.
5. Calculate or at least estimate the sloshing effect thermal loads: compare with/without baffles, optimize baffle structures.
6. Understand the thermal load on the tank induced by vibrations that are being transmitted to the tank by the airframe structure.
7. It is understood that the exothermic conversion of ortho to para hydrogen to over 99% has already taken place and mainly para hydrogen (para-H_2) is fueled into the aircraft tanks from the fueling station [7]. The fueling station can be local at the airport (e.g., with installed electrolyzers) or with liquid hydrogen trucks (see also Chap. 11).
8. Cryogenic liquid boil-off can increase as the aircraft undergoes altitude changes.

3.2.1 Selecting the Appropriate Thermal Protection Scheme

Overview

One might wonder why any thermal analysis is already being considered as a first step after the tank volume is guesstimated.

The reason is that we need to take some decisions in the way we intend to store or contain the liquid volume. This again depends on the flight mission profile, the total size of the aircraft, and whether this is for a short-range or long-range flight. Here we consider, for example, a long-range Boeing type 737, or a similar Airbus engine.

We also need to consider that a liquid hydrogen tank has to be working in a safe environment, protected from any environmental oxygen, see also Appendix 3 on Safety. The tank certainly needs a good choice of insulation although the boil-off rate of 114 ml/h per Watt heat loss at 20 K is rather low for a given tank size of e.g., 6000 liters. Still, the temperature gradient that needs to be kept in mind is at least 380 K, assuming the engine can warm up to 110 degrees (F) in hot countries during taxiing, depending on taxi speed.

We also need to bear in mind some peculiarities that any use of hydrogen imposes on the design, that is the ability of hydrogen to diffuse through certain materials, and the rate of which hydrogen might even chemically react with a confinement wall. Furthermore, we really should keep in mind that any insulation that we would like to apply on a surface should not fall off and work for as long as the aircraft remains in service.

There are lessons to be learned from space flights, e.g., the Ariane launch space center [8]:

> A fallen chunk of foam insulation from the rocket's fuel tank scuppered the latest launch attempt. During a final inspection of the Ariane-5 launch rocket at 0100 GMT on Friday, a 10 by 15-centimeter size like chunk of foam was found on the movable launch "table" that supports the rocket.
> The 30-metre-high cylindrical main stage of the rocket is covered with foam to keep its cryogenically cooled liquid oxygen and hydrogen fuel from heating to the tropical temperatures of the launch site in Kourou, French Guiana.
> If the launch had proceeded, rocket operators say water vapor in the air could have cooled to form a 100-gram block of ice on the exposed surface. During flight, the ice could "separate and cause bad damage", says Jean-Yves Le Gall, chief executive officer of Arianespace, the rocket's manufacturer.

NASA has experience with space liquid hydrogen fuel tanks, but their operational characteristic is somewhat different from aircraft engines. We need to be innovative and adapt the thermal protection scheme to aircraft requirement.

For aircraft systems this is even more important since any loss in insulation would cause hot spots on the tank surface that then in turn start a convective motion of the liquid within the tank and increase the thermal losses of this containment without any opportunity to locate and repair the missing insulation.

As mentioned, the careful choice of selecting the thermal protection scheme is one of the most important design decisions. Brewer [9] and Kaganer [10] give some earlier thoughts on those as well which is summarized in Table 3.1. Liquid hydrogen tanks are being used for some space missions—but those are designed for a very short lifetime and consequently tolerate a high boil-off rate.

Table 3.1 gives insight into early space designs and lessons learned on liquid hydrogen tank insulations with different insulation concepts confirming the need for careful insulation validation.

3.2 Cryotankage—Heat Sources

Table 3.1 Thermal protection scheme

No.	Insulation system concept	Reusable design?	Demo on	Comments
1	He purged	Yes (space shuttle application technology)	NAS 8–27,419, 2.2 m (7.2 ft) tank	Purge jacket is epoxy glass, Teflon® coated; insulation is multilayers; 100 space shuttle flight cycles demonstrated with LH_2
2	He/N_2 double purge	No (orbital application technology)	NAS 3–4199, 2.1 m (6.9 ft) tank	Used Helium purged fiberglass substrate, nitrogen-filled multilayers; simulated one ground hold, launch, orbit flight cycle with LH_2, thickness of He to N_2 layers must be controlled accurately to prevent N_2 liquefaction
3	External polyurethane foam. Nonintegral tank	No (Apollo flight program)	Saturn S-II stage 10 m (33 ft) diameter	Polyurethane foam sprayed on, machined, covered with polyurethane sealer; conductivity rises with time due to displacement of blowing gas with air; flight demonstrated
4	External polyurethane foam; integral tank			
5	Internal polyurethane foam	No (Apollo flight program)	Saturn S-IVB stage 6.7 m (22 ft) diameter	Glass fiber reinforced foam tiles, individually bonded, fiberglass polyurethane resin liquid harrier (GH_2 filled); 135 thermal cycles
6	PPO internal foam/ polyurethane external foam		This combination has not been demonstrated; see comments on systems 3 and 7	
7	PPO internal open cell foam	Yes (space shuttle technology)	NAS 9–10,960, 1.75 m (5.8 ft) tank	Individual tiles bonded to wall; conductivity higher than GH_2, varies with orientation; 100 space shuttle flight cycles demonstrated with LH_2

(continued)

Table 3.1 (continued)

No.	Insulation system concept	Reusable design?	Demo on	Comments
8	Honeycomb gas layer barrier	Yes (SST methane tank technology; space shuttle technology)	NAS 3–12,425 NAS 8–25,974	GH_2 filled insulation
9	Rigid vacuum cell	Yes (space shuttle system technology)	NAS 3–14,369, 2.6 m (8.7 ft) diameter tank	Aluminum honeycomb rigid vacuum shell with aluminum face sheets; shell collapsed after cycling 29 times due to peeling of inner face sheet; external face sheet should be made as a vacuum seal to prevent this; problems making system vacuum tight to 10^{-5} torr; the presence of the closed cell foam, would create difficulty in maintaining the prescribed level of vacuum due to outgassing
10	Microspheres with external flexible metal jacket	Yes (space tug system technology)	NAS 3–17,817, 1.2 m (3.9 ft)	Stainless steel jacket, 0.008 cm (0.003 in.) thick, has demonstrated vacuum integrity to 10^{-06} torr; none of 23.2 m (76 ft.) of resistance seam welds leaked; test program demonstrated 13 flight cycles using LN_2 with no change in thermal performance; microspheres have been loaded compressively in a flat plate 100 times with no change in thermal performance
11	Microspheres with internal liner	Yes (LH_2 aircraft application technology)	This design modification to system 10 has not been demonstrated	
12	Silica insulation with internal liner	Yes (space shuttle high temperature insulation)	Properties of insulation have been determined; liner has not been demonstrated	
13	Self-evacuating shingles	No (orbital application technology)	NAS 3–6289, 0.8 m (2.5 ft) calorimeter tank	Leak tight shingles were not obtained; sealing strips opened upon thermal cycling; this system did not perform as designed; requires further development

(continued)

Table 3.1 (continued)

No.	Insulation system concept	Reusable design?	Demo on	Comments
14	Self-evacuating honeycomb/foam		This combination has not been demonstrated; see comments on systems 3 and 15	
15	Self-evacuating honeycomb/N_2 purge	No (orbital application technology)	NAS 8–11,747, 0.8 m (2.5 ft)	Conductivity of honeycomb degraded with number of LH_2 cycles (up to 14) as gas permeated the honeycomb; had problems with nitrogen purge gas liquefying in the multilayers (honeycomb sublayer should have been thicker)

3.2.2 Mitigating Thermal Radiation

One must decide whether a helium warm gas purging system needs to be installed to ensure any wall insulation exposed to the environment is maintained oxygen free and any ice formation is eliminated that could, once it accumulates, create a safety hazard.

The following highlights some points that need to be taken into consideration:

- Type of insulation and combinations.
- Aging of insulation under repeated thermal and mechanical stress.
- Insulation fixtures.
- Insulation behavior during transients, e.g., when not in operation/landing.
- Insulation fatigue.
- Temperature swings.
- Insulation detachment and long-term quality deterioration.
- Frosting indication sensors and other insulation failure detection methods.
- Insulation thickness.

All those design points feed into the total storage losses and boil-off acceptance requirement.

As an important safety aspect any tank surface held at a temperature of 20 K cannot be exposed to air since otherwise liquid air would freeze out creating severe safety concerns, see also Appendix 3 on Safety.

A way to mitigate those concerns is by employing dual walls where the space between adjacent tank walls is evacuated whereas the achievable vacuum quality can be selected and vary. Groundbreaking work on many different types of insulation material, configurations, and respective insulation quality has been published early by Brewer [9] and Kaganer [10], as mentioned.

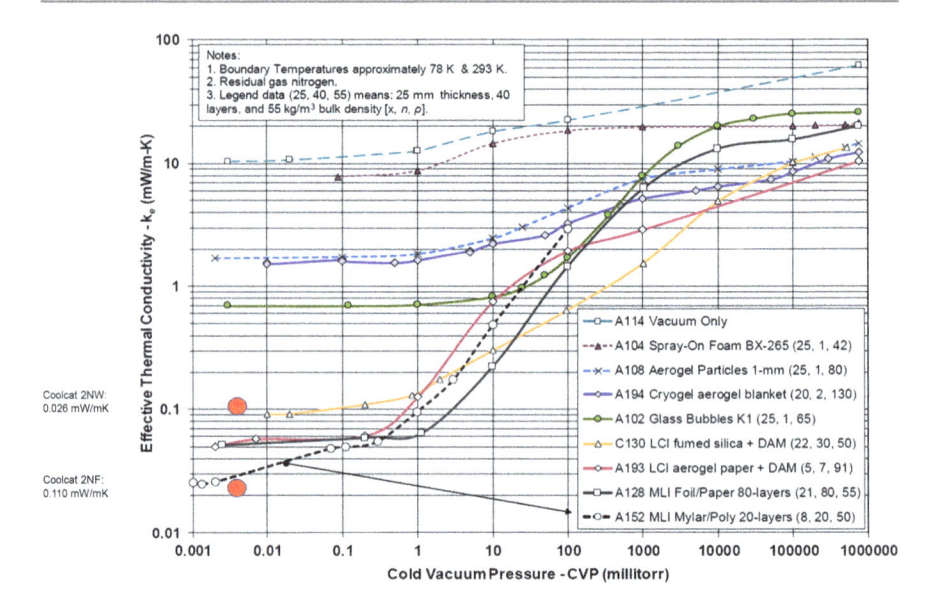

30   3 Cryogenic Liquid Hydrogen Tank Design Aspects: General Overview

Fig. 3.4 Effective thermal conductivity of insulation types

However, the most recent work by Fesmire [11] gives an excellent and extensive update on insulation efforts that are commercially available as of today. Figure 3.4 shows a detailed plot of achievable effective thermal conductivity through the insulation depending on material type and vacuum conditions. This important and well-documented tool lets you decide on what insulation quality to choose from. The effective thermal conductivity is calculated as the perpendicular thermal conductivity through the insulation stack. The vacuum domain can be classified as follows: A rough vacuum condition is maintained at a pressure range of 1000 to 1 mbar, medium range vacuum in 1 to 10^{-03} mbar, high vacuum between 10^{-03} and 10^{-07} mbar and ultra-high vacuum conditions (UHV) in the range 10^{-07} to 10^{-14}. 1 mTorr corresponds to 1.33 mbar.

1. Minimize Thermal Radiation Down to the Cryo Environment.

For taking the thermal heat loads to the absolute achievable minimum, multilayer insulation can be employed. The 2 bullets in Fig. 3.4 show the calculated, most efficient layer combination that results in an effective thermal conductivity of two types of insulation, COOLCAT 2NW and COOLCAT 2NF, supplied by Beyond Gravity [12] as measured at KIT on the THISTA test bay with a test vacuum of 0.0074 mTorr.

Example For a 20-layer COOLCAT 2NW type, the effective thermal conductivity k_e is estimated as 0.026 to 0.030 mW/mK with a 10-layer blanket thickness of 3 mm. This compares to line A152 in Fig. 3.4.

For a 20-layer blanket COOLCAT 2NF (10-layer blanket thickness of 7 mm), k_e increases to 0.118 to 0.196 mW/mK. COOLCAT 2NF is a thick non-flammable foil, hydrogen application blanket, corresponding to line C130.

Figure 3.5 shows a typical insulation blanket (50 m long, 1.9 m width) for a foil thickness of 12 µm. The weight of a blanket consisting of 10 layers is 0.308 kg/m^2, Fig. 3.6 on the right shows a 10 µm thick foil, specifically developed for hydrogen applications with a standard length of 3 to 50 m and a width of 1 m. For 10 foils with a length of 3 m and a width of 1 m the weight is 0.54 kg/m^2 [13].

Appropriate insulation winding techniques need to be developed at the tank end bells. Since a tank also comes with multiple penetrations automatic winding for large-scale production is challenging [14].

Blankets can be fitted with Velcro strips for ease of assembly as shown in Fig. 3.7.

Beyond Gravity gives further details on their insulation type with respect to performance at lower vacuum quality (see Fig. 3.8) [15].

Continuous MLI improvement of suppliers helped to maintain a reasonable thermal radiative heat load to the tank even with degrading vacuum quality up to a pressure of about 10^{-2} Pa (10^{-4} mbar) as Fig. 3.8 shows. In this test, a variation of the gas pressure was applied by degrading the vacuum with nitrogen and helium gas.

Further details are given in the safety section when discussing tank leaks and failures.

A good MLI performance in a degraded vacuum is promising for overall heat load reduction; however, we need to know what contaminants contribute to the thermal residual conduction which may become a dominant factor. See also Section on residual gas conduction in "maintaining a vacuum quality."

Fig. 3.5 (left) Helium blankets 2NW. (Courtesy of Beyond Gravity)

Fig. 3.6 (right) Hydrogen foil 2NF. (Courtesy of Beyond Gravity)

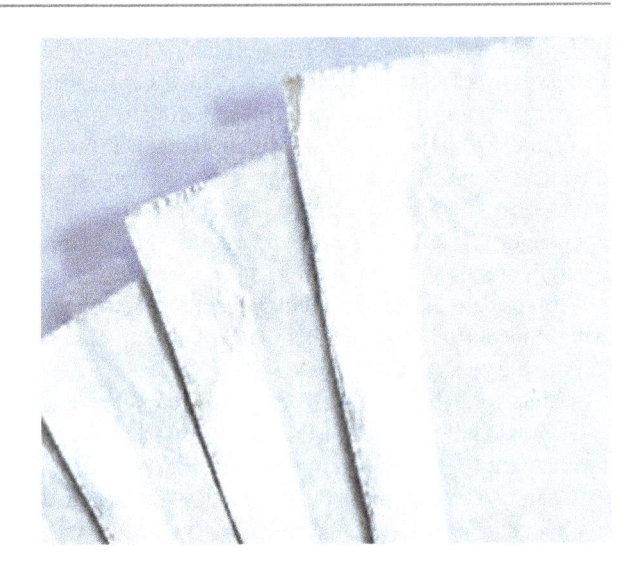

Fig. 3.7 MLI with Velcro fittings. (Courtesy of Beyond Gravity)

3.2.3 Implementing Insulation—Structural Example

Figure 3.9 gives one typical example of the structural build-up of a dual-walled lightweight composite storage tank composed of different tank shells. The main shells are vacuum-separated. For any composite structure that is subjected to

3.2 Cryotankage—Heat Sources

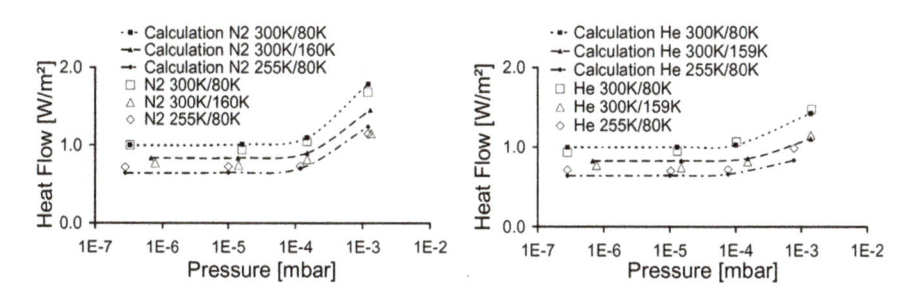

Fig. 3.8 Typical increase of heat flux through insulation with decreasing vacuum (N_2 gas pressure)

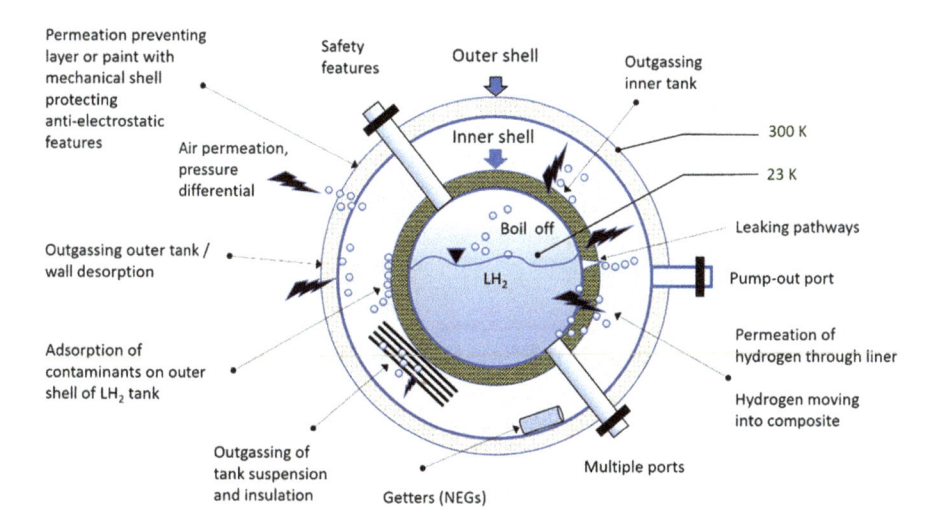

Fig. 3.9 Assembly of storage tank walls

vacuum conditions several phenomena need to be discussed as shown in the figure inset if, for example, vacuum integrity has to be maintained over the lifetime of an aircraft. Since the inner tank carries the liquid cryogen, one needs to know the permeation characteristics of the composite walls, the hydrogen diffusion through the metal liner or through the composite and the desorption of molecules from a composite wall as well as the adsorption of vacuum contaminants at the cryogenic wall. The latter has an impact from the safety perspective (see Chap. 8 on Leak Detection and Appendix 3 on Safety).

The left part of Fig. 3.10 shows a possible design for a liquid hydrogen tank with a metallic liner, a composite wall, a possible thin diffusion barrier to the vacuum side to maintain vacuum integrity, and an outer metallic shell with a composite overwrap. In addition, a protective foil faring can be applied at the outer shell (OVC).

A cut-through section in Fig. 3.10 explains the barrier functions in further detail.

34 3 Cryogenic Liquid Hydrogen Tank Design Aspects: General Overview

System barrier description

| | Varying ambient cruising pressure | Al diffusion/protection barrier II |

300 to 400 K   OVC

Al permeation barrier I   Vacuum   MLI   Al permeation barrier II

20 K Tank

Al diffusion barrier I   Liquid H₂ inventory

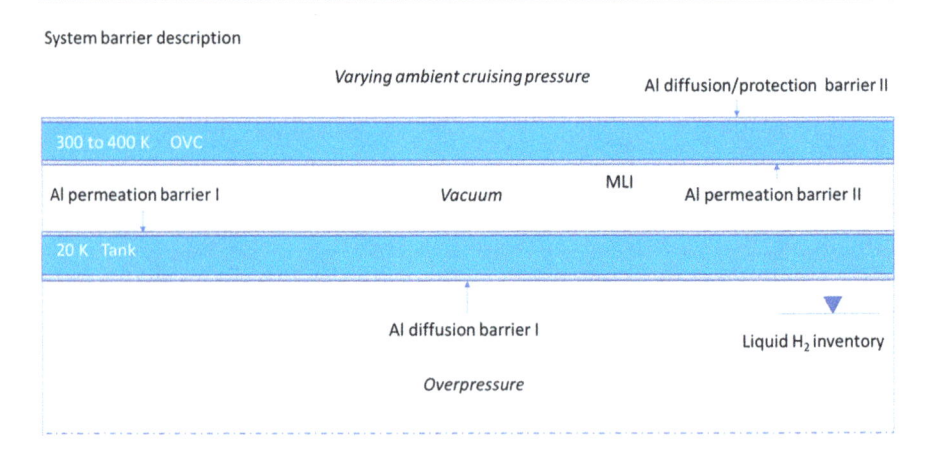

Overpressure

Fig. 3.10 Example tank wall design with aluminum permeation and diffusion barriers.
Barriers:
Al diffusion barrier I: Prevents hydrogen vapor to move into fiber wrap
Al permeation barrier I: Prevents fiber wrap outgassing
Al diffusion barrier II: Prevents air from moving into fiber wrap, also a fiber wrap protection
Al permeation barrier II: Air permeation fiber wrap barrier
Special MLI: Thermal barrier, tailored for aircraft applications

3.3   Cryotankage—Simplified Design Example

2. Maintaining a Vacuum Environment.

If a dual shell tank design is desired as shown in Figs. 3.9 and 3.10 (storage tank walls) one needs to ensure appropriate vacuum conditions are being maintained for the chosen insulation as per Fig. 3.4 (effective thermal conductivity) and whether any residual gas conduction is predictable, well controlled and can be maintained over the lifetime of the storage tank (Fig. 3.11).

Figure 3.4 gives the heat flux to a cryotank depending on the chosen insulation scheme but also gives an estimate of the increasing heat load as vacuum quality worsens. As for using superinsulation the lower vacuum and high vacuum end show only little dependency on the vacuum quality. As we leave the high vacuum pressure range, we enter the Smoluchowski range [16] (see Fig. 3.12) and see a jump in effective thermal conductivity.

An S-shaped curve is typical for powder insulation and MLI as shown in Fig. 3.13. At low pressures, thermal conduction due to residual gas in the vacuum space has subsided [17–19].

As a ballpark figure, in the following example we assume a surface area of 31 m² for a 10,000-liter liquid hydrogen tank. The thermal radiation burden is estimated as follows:

We assume to wrap 20 layers of MLI insulation on the tank surface. For that we refer to Fig. 3.4 and the 2 red dots at a vacuum pressure of 0.0075 mTorr. With a total conductive length through the insulation of 7 mm and an effective thermal

3.3 Cryotankage—Simplified Design Example

conductivity through the insulation at 0.0075 mTorr of 0.026 W/mK for the NW type foil (curve A152) and a given ΔT of 280 K (from RT to 20 K), the expected heat load by radiation through the insulation blankets is approx. 1 W/m², which for the fictive tank surface above would result in 31 W corresponding to a boil-off of - > 3.5 l /h (1 W boils off 113 ml/h hydrogen) see also Table 2.1 in Chap. 2. This gives a very reasonable daily boil-off rate of 0.84% for the tank during aircraft dormancy. As for the inflammable foil (NF type foil), the heat flux density would be higher at 2.36 W/m².

If for some reason the vacuum deteriorates, we move from the medium vacuum range (1 to 10^{-03} mTorr) to the rough vacuum range passing the Smoluchowski regime halfway, for example at 10 mTorr (10 micron), the heat load jumps to 5 mW/mK which now results in a total heat load for the NW type blanket of 200 W/m² which gives an hourly boil-off of 700 l/h that would empty the tank within 14 h.

It is therefore advisable to ensure vacuum conditions do not deteriorate and all material outgassing is well under control and has been taken care of by component bakeout and employing a number of getters and dedicated hydrogen NEGs (non-evaporable getters).

An example of a missing hydrogen getter is given in the following to show the effect of the residual gas conduction as a heat leak, assuming the vacuum around the liquid hydrogen tank as discussed above is around 10^{-3} mbar (7.5 mTorr) region. Given that our tank is maintained at 20 K and all residual gas is frozen out on the outer surface of the inner tank (see Figs. 3.9 and 3.10) we can now estimate the heat load in case of any presence of hydrogen molecules in the vacuum, for the above tank configuration (Table 3.2):

Fig. 3.11 Thermal conductivity through MLI vs pressure

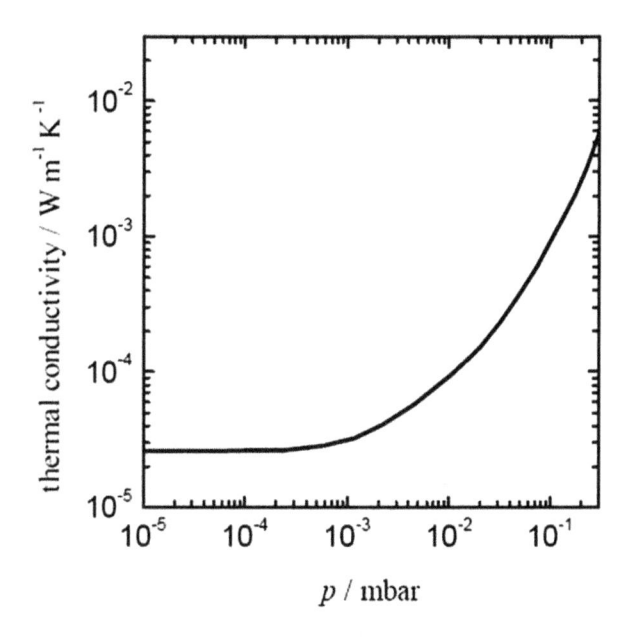

Fig. 3.12 Residual gas conduction in vacuum

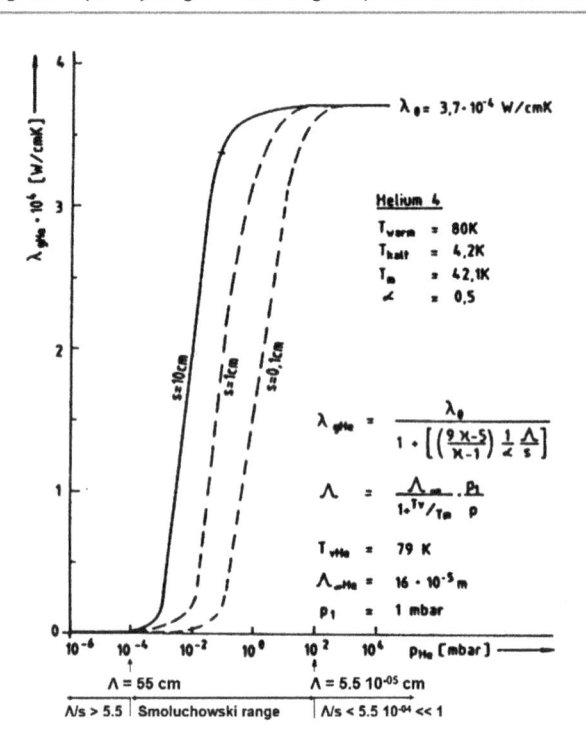

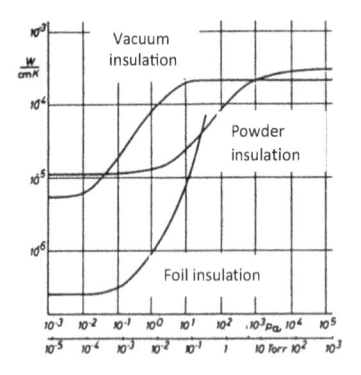

Fig. 3.13 Insulation type and vacuum quality

H (W/cm^2) = constant] α * P [micron/mTorr] (T$_{warm}$—T$_{cold}$) with the constant α being after Corruccini [20]:

$$\alpha = \frac{\alpha_1 \cdot \alpha_2}{\alpha_2 + \alpha_1 (1-\alpha_2) \cdot \dfrac{A_1}{A_2}}$$

We may find the heat load negligible, depending on our insulation specification which drives the need for a suitable getter (e.g., Pd type).

3.3 Cryotankage—Simplified Design Example

Table 3.2 Heat load to tank with worsening vacuum quality

Residual gas (Coruccini)			
H$_2$ constant	0.039	0.039	0.039
Accommodation Coeff. α1 (20 K)	1	1	1
Accommodation Coeff. α2 (300 K)	0.3	0.3	0.3
Accommodation Coeff. α	0.34	0.34	0.34
Pressure (micron)	0.001	0.01	0.1
Pressure (mm Hg)	1.E^{-06}	1.E^{-05}	1.E^{-04}
T$_{cold}$ (K)	20	20	20
T$_{warm}$ (K)	300	300	300
H (W/cm^2)	3.72E^{-06}	3.72E^{-05}	3.72E^{-04}
Heat load to tank (W)	1.15	11.53	115.3

3. Optimize Any Tank Suspensions and Openings, Feedthroughs, or Sleeves for Minimum Thermal Load to the Cryogenic Environment.

For suspending a cold mass or a liquid hydrogen tank a good choice of suspension materials in different shapes is already available. For small neonate systems to big 7 T MRI magnet systems, typically carbon fiber and/or glass fiber rods with pressed metal bushings are used. It is also common to use TiAl6V4 struts with ball-bearing-like ends and Belleville washers. For low-weight tanks, dog bone structures can be cut out of a thin G10 plate. Kevlar slings, depending on type and tank build, can also be modified to be used as a suspension member (not shown).

Suspension elements need to bear dynamic weights and take into account low-cycle fatigue concerns (Fig. 3.14).

Rods (1) are also used as axial suspension for mobile MRI systems to compensate and accommodate high g-loads. A typical suspension system is shown in Fig. 3.15. The figure shows a combination of radial and axial suspension means for a tank with internal slosh baffle configuration.

Tank Supports A more assembly-friendly design approach requires rethinking the cryogenic tank suspension for aircraft. For this, the Heim column designs [21] are attractive as shown in Fig. 3.16. This support member typically consists of 3 different materials, depending on the application. For room temperature down to a first heat intercept a glass fiber tube of G10 or G11 or S-glass is used. The intermediate tube is then made either of aluminum, a high-strength aluminum alloy, or stainless steel or TiAl6V4 (titanium alloy). The third part of a 3-column arrangement conveniently consists of a carbon fiber tube since the thermal conductivity of the carbon fiber is lower than glass fiber for this temperature domain. For the low temperature range below 40 K, G10 or G11 is preferred to minimize the heat burden. Although some plastics expand at low temperatures this is the only mechanical support member known in cryogenics that can expand rather than shrink upon cooldown. Tubes are also capable of carrying high compression loads as in use by CERN. Although 9-column designs have been proposed the 3-column design is the most popular.

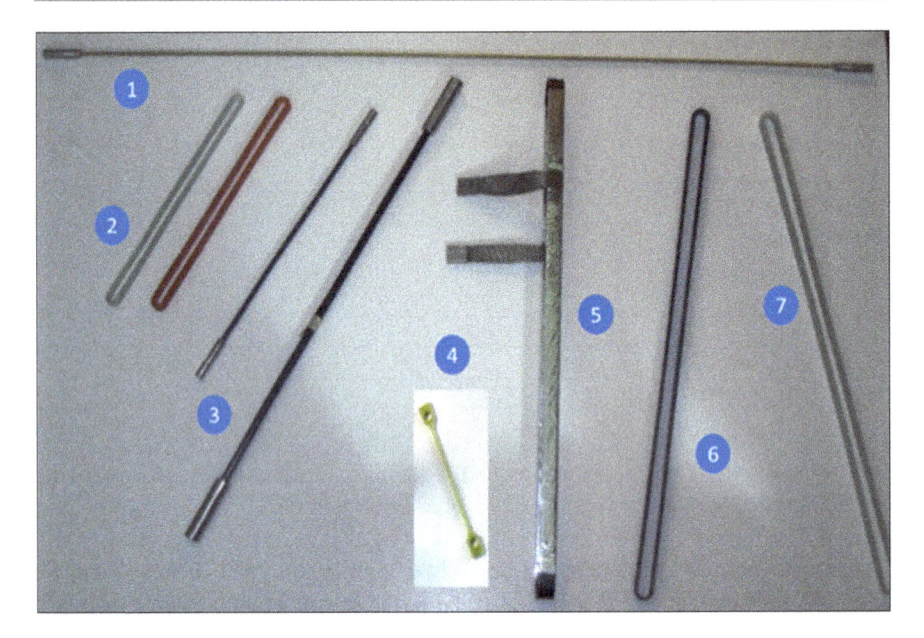

Fig. 3.14 Straps/rods used for cryogenic tanks 1 = G10 rod, 1 m long, 4.5 mm diameter, with ends pressed into bushing, 2 = racetrack-shaped straps for lightweight systems (unidirectionally wound, modified glass fiber composite), 3 = carbon rods with pressed end bushing with bolt holes for radial and axial applications, 4 = tiny strut for suspending cryogenic components, water jet cut G10, 5 = carbon straps for heavyweight MRI, 6 = long, thin carbon strut with small end radius, 7 = long unidirectional glass fiber (GRP) strap

Heim columns do not necessarily need a thermal intercept for liquid hydrogen temperatures.

One further way of supporting a large 10,000-liter tank would be employing a rail guide scheme [22]. The working principle is shown in Fig. 3.17 and is based on a less-known technique to minimize thermal conduction heat loads using spherical contact members.

The outer shell of the inner liquid hydrogen tank is fitted with rail guides. Cryogenic rail guides are advantageous for supporting shell-like structures at low temperatures. The contact members are favorable of spherical shape which can be of any material suitable for cryogenic use. Kaganer [10] was the first to point out the use of ball-bearing-shaped structures at cryogenic temperatures that lines up very well with theory. See also section on Transfer lines and ball bearings.

Heat exchange between spheres has been evaluated by Kaganer [10]. See Fig. 3.18.

Example A 7000-liter tank with a total assumed full tank weight of 515 kg for the tank gives a total weight of 760 kg (including shell) or 7000 N. For a ceramic sphere of 1000 N with a diameter of 10 mm we need to bear a weight of 150 N per sphere we therefore need 46.6 spheres which gives a heat load of 46.6 * 0.8 approx. 37.2 W resulting in a boil-off of 4.3 l/h.

3.3 Cryotankage—Simplified Design Example

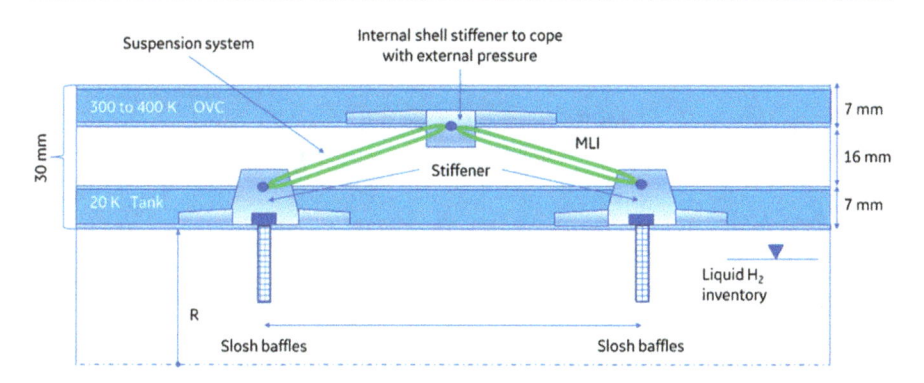

Fig. 3.15 Conventional suspension system

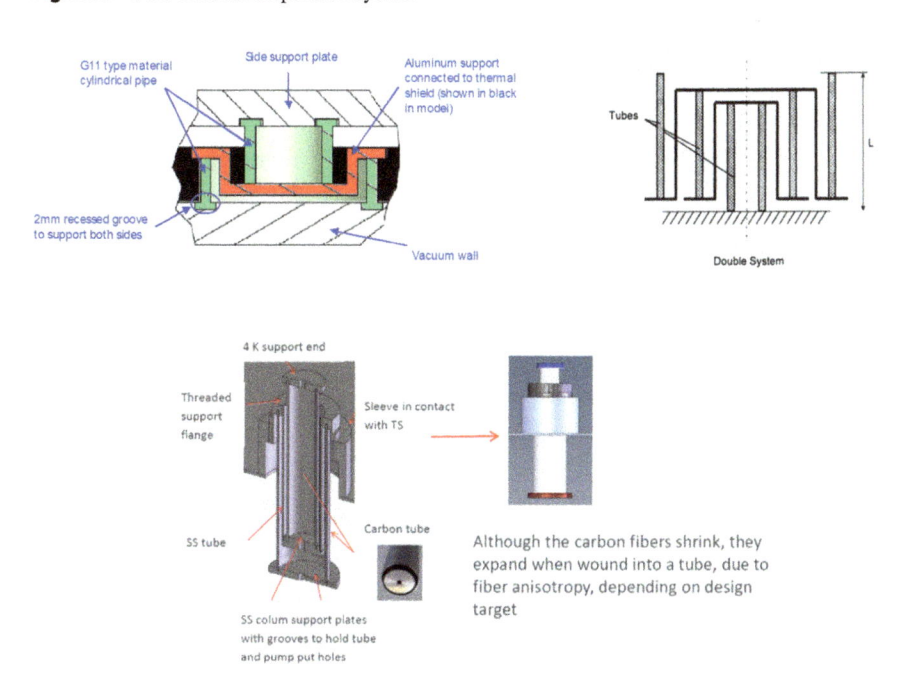

Fig. 3.16 Cryogenic support members of Heim type

4. Ensure Any Outgassing from Composite or Metal Surfaces and Other Components Is Well Controlled.

Point 2 in this section mentioned the vacuum quality that should be maintained during the lifetime of a liquid tank. For that we need to briefly mention outgassing behavior of metals and composites. This is very well-researched by NASA [23].

Liquid Tank Outgassing To protect the liquid tank composite from outgassing, a few measures can be taken to eliminate this. First of all, the inner liquid tank can be

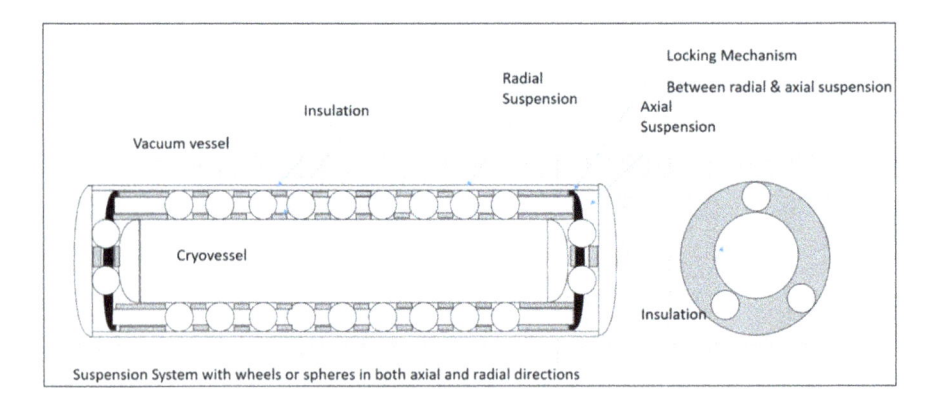

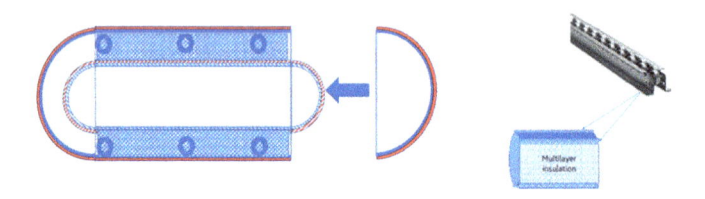

Fig. 3.17 Guide rail support system for liquid hydrogen tanks

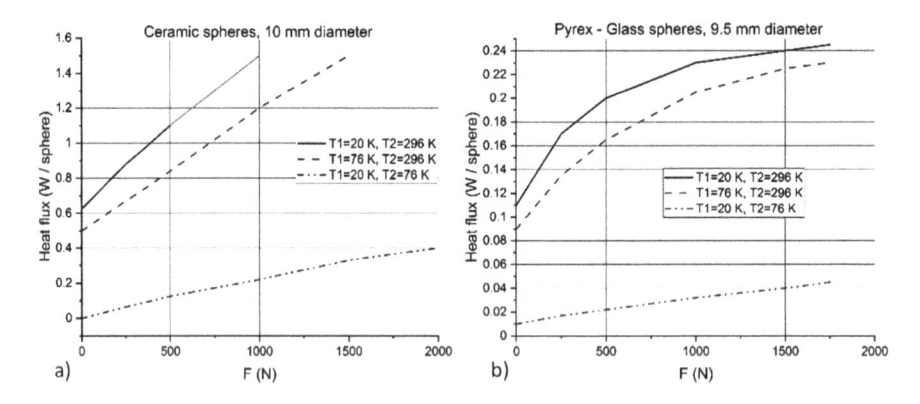

Fig. 3.18 (**a**) Ceramic spheres, (**b**) glass spheres

initially baked out for several hours prior to commissioning in a vacuum chamber with the surface then being sealed off, with either known, effective surface sealing means, e.g., cryogenic paints, or strong metal foils. Second, the vacuum space should be pumped out multiple times following an N_2 pump and flush strategy. N_2 is a so-called carrier gas that allows quick removal of contaminants in a vacuum space. Third, several, well-known, room temperature getters or specific getters for hydrogen (outgassing of weld seams on liners and metallic penetrations) are readily available and can be used for capture. Also, outgassing, high carbon chains, or other

3.3 Cryotankage—Simplified Design Example

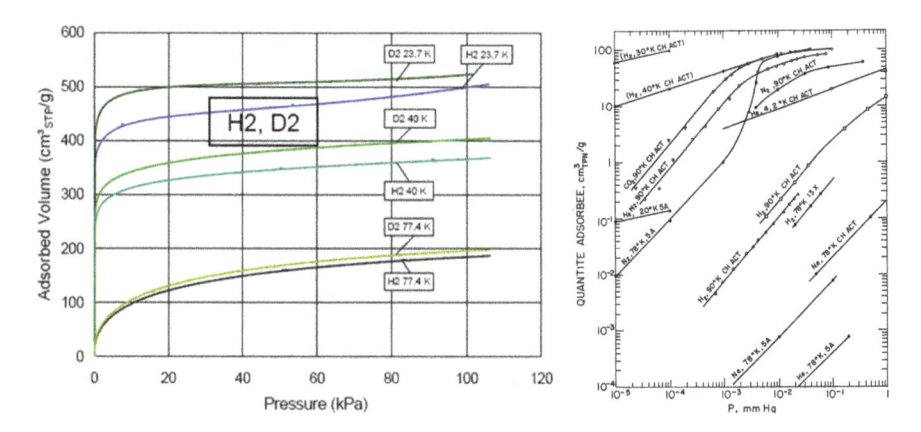

Fig. 3.19 Adsorption/desorption of hydrogen (H_2) using activated charcoal

molecules will freeze out on the outer tank surface. For a 5500-liter CHEETA Tank 3, type 2 design, the surface area is approx. 20 m^2, resulting in a highly efficient cryopump, pumping at 23 K. Furthermore, active carbon pellets can be placed at some areas in the vacuum space, whereas those carbon getters normally can be regenerated without breaking the vacuum. If regenerative sorbs are more likely to be used in a tank, Pd-based non-evaporable getters are a good solution.

Vacuum Shell Outgassing Continuous outgassing from the outer shell into the vacuum space is likely to occur in linerless vacuum shells (room temperature) being subject to a continuous pressure differential. Outgassing can be minimized, if a liner is used at the inner side of the vacuum shell.

Penetrations Penetrations, that is openings, or ports, or turrets connecting the liquid tank with the tank shell can outgas longer. Fortunately, those penetrations are smaller components that can be baked out long enough to mitigate against high outgassing rates.

If all procedures are followed vacuum quality can be maintained for decades, assuming no leaking components. The surface area of the tank, for example, is comparable to most recently built MRI systems operating at 4 K. Those operate without a liquid helium bath but are equipped with a fairly large composite mass. As long as hydrogen is captured very well, no degradation may occur.

A typical measurement of hydrogen adsorbed volume vs pressure (at regeneration conditions) and various regeneration temperatures is given in Fig. 3.19 [24, 25].

5. Calculate the Sloshing Effect Thermal Loads: Compare With/Without Baffles, Optimize Baffles.

Sloshing of Cryogenic Liquids

Introduction

Sloshing is a particular motion of liquid bulk in a tank with respect to the interaction with the tank walls. Depending on the aircraft motion in all its directions, we differentiate between lateral or translational sloshing and vertical sloshing where the fluid motion happens to be normal to the liquid-vapor interface as well as impact motion related fluid motion during sudden changes of aircraft speed and climb and descent. The energy absorbed by the liquid in the sloshing process is dissipated as thermal energy (loss) through the molecular friction with the tank wall surface.

The early cryogenic sloshing research was based on liquid helium semitrailers. A.D. Little [26] finally could prove that rectangular-shaped vessels do not produce any sloshing effect since the liquid can only move in vertical orientation. Based on this, vertical baffles were first introduced on those trailers with good success. It could be shown that the boil-off in those non-venting liquid helium trailers can be reduced by a factor of 80 if slosh baffles are used whenever longitudinal sloshing is to be expected. This was based on a 30-h trip the trailer was exposed to 1600 accelerations and decelerations with an average of 0.04 g. Since it is not possible to maintain only a liquid phase in an aircraft tank (near 100% fill level) as with road trailers, baffles will be necessary, as van Meerbeke has shown [27]. An attempt has also been made for non-venting trailers with given road transport acceleration. In here the energy input is a function of the tank acceleration and length, divided by the fill level height of the liquid and the number of transverse baffles [27]. An attempt has been made for non-venting trailers with given road transport acceleration. Here the energy input is a function of the tank acceleration and length, divided by the fill level height of the liquid and the number of transverse baffles [27].

Researchers currently have no engineering estimate (formula) yet as with trailers, that do not vent to environment, that would allow any prediction to be made on sloshing losses in an aircraft.

Compared to a trailer however, the mission profile and correspondingly the g-spectrum is at least twice as high (a standard aircraft horizontal constant acceleration is around 0.854 m/s^2 (0.087 g) of a duration of 90 s [28], it therefore seems almost unavoidable of a need for dedicated, lightweight baffles in a large cylindrical tank.

Cryogenic sloshing and its modeling [29-32] has also recently gained more importance for large-scale liquid hydrogen transport with ships. Again, the specs for maritime applications are different from those we need to use for aircraft.

Ibrahim [45] gives an extensive overview on theory and applications on liquid sloshing dynamics, covering linear, nonlinear, and parametric sloshing dynamics for different tanks and containers with lateral, pitching, and roll excitation. Some comprehensive work on cryogenic sloshing has been done mainly by NASA in a microgravity environment. Further references are given by Ryali [33].

Fig. 3.20 Compartmentalization of a Jet-A fuel wing tank

Stratification

One advantage of using baffles, however, is as mentioned, the mixing effect that sloshing provides to reduce liquid hydrogen stratification, the latter being of a concern for temperature and pressure build up, whereas controllable sloshing efficiently reduces the pressure in the tank due to gas and liquid phase mixing. A similar experience has been seen in a vibrating liquid hydrogen tank for the automotive industry, initially reducing stratification of the cryogenic liquid. Further information is given in Section Cryotankage—Heat sources paragraph 6 and following.

Aircraft Sloshing

A thorough aircraft flight behavior study may be required to understand how the sloshing force in a huge tank, carrying, e.g., liquid cryogenic bulk, would also affect the aircraft motion. To get to the bottom of that we need to further understand the sloshing physics involved. Sloshing in aircraft has been previously addressed by installing compartmentalization, for example, in much smaller Jet-A fuel wing tanks [34] (see Fig. 3.20).

Some of those tanks are big enough to allow for manual inspection at regular service intervals. Due to the large size of the liquid hydrogen composite tanks with approx. 1 ton of liquid moving around, typical aircraft acceleration loads therefore can cause significant fluid motion to be of concern for a number of reasons:

- Imposes stress concentrations on the liquid tank shell as well as on the suspension on the outer vacuum chamber, especially at penetrations.
- Tank pressure rises due to that movement, caused by corresponding temperature rise due to liquid/vapor mix (composite tank temperature profile circumferentially varies largely when not in touch with liquid surface).
- Increases liquid hydrogen boil-off.
- Increased bubble formation due to liquid vapor mix increases cavitation risk for attached liquid hydrogen pump and puts further strain on downstream phase separator when withdrawing liquid from the tank.

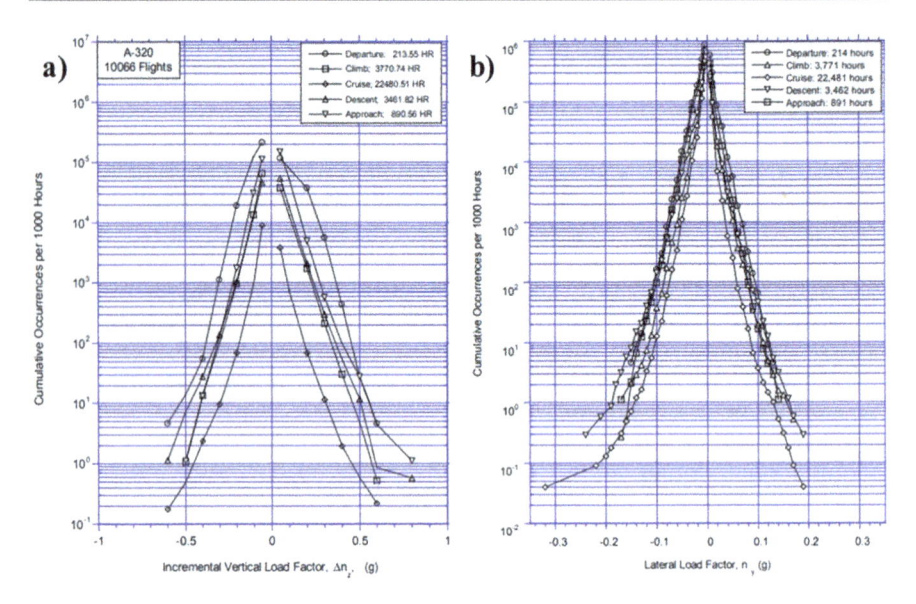

Fig. 3.21 Cumulative occurrences of load factor per 1000 h by different flight phases of Airbus A-320 aircraft [35], (**a**) vertical load factor, (**b**) lateral load factor

- Sloshing creates a yet unknown induced dynamic imbalance force/moment on the aircraft (affects flight stability, lateral tanks as well as left and right tanks as shown in Fig. 3.24).

One major factor that influences the intensity of fuel sloshing is the flight profile of the aircraft. Figure 3.21 shows statistical data for typical accelerations experienced by an Airbus A-320. These plots are based on operational usage data collected from 10,066 flights, representing 30,817 flight hours, as recorded by a single U.S. airline operator [35]. It can be observed from Fig. 3.21 that significant acceleration (both in the vertical and lateral directions) is observed in the aircraft during different phases of flight. Considering a large liquid hydrogen tank as shown in Chap. 5 Cryotankage—Tank Shapes and Airframe Integration, Fig. 5.23, these accelerations, can cause significant sloshing.

Modeling Effort Example
Sloshing of cryogenic bulk liquid in a tank is a fluid-structure interaction (FSI) problem. The Coupled Eulerian-Lagrangian (CEL) technique and dynamic explicit numerical solver in ABAQUS are used to analyze this phenomenon. Several other software packages were found to be unusable. The CEL technique defines the structure (tank) using a Lagrangian mesh and the fluid (liquid hydrogen) using an Eulerian mesh (Fig. 3.22). The equilibrium equations in this formulation are solved in the Eulerian framework and the transport equations of deformation gradient and Jacobian determinant are solved using the Lagrangian framework as depicted in Fig. 3.23.

3.3 Cryotankage—Simplified Design Example

Fig. 3.22 Liquid hydrogen slosh model using Coupled Eulerian-Lagrangian (CEL) technique in ABAQUS

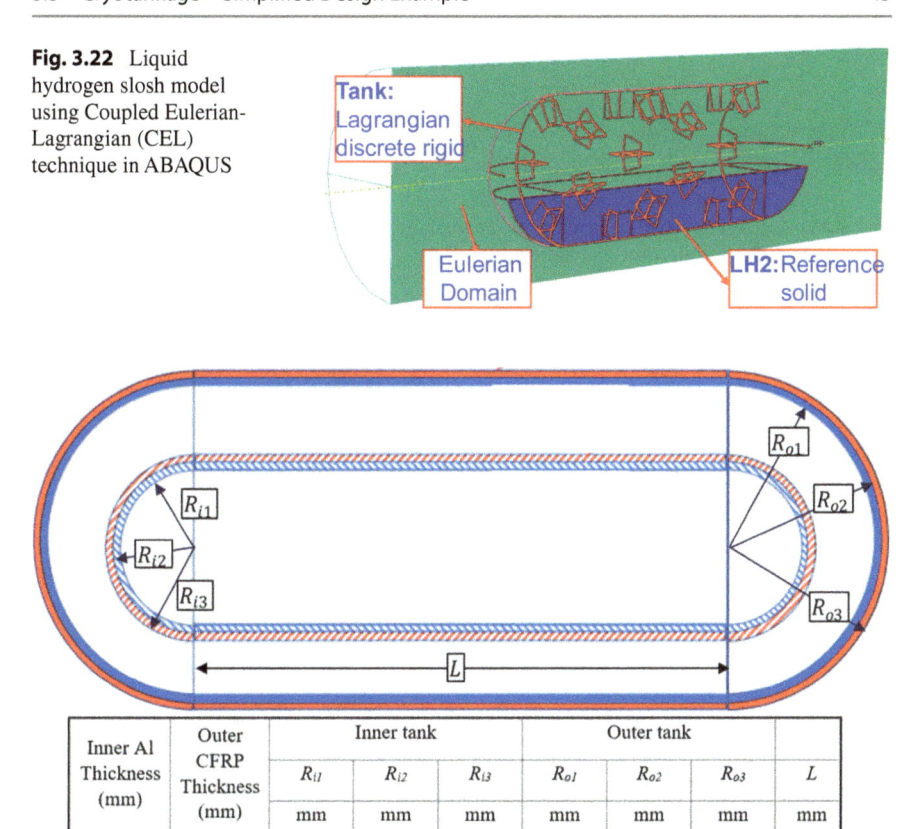

Fig. 3.23 Thin-walled composite cylindrical liquid hydrogen tank design analyzed

Inner Al Thickness (mm)	Outer CFRP Thickness (mm)	Inner tank			Outer tank			
		R_{i1}	R_{i2}	R_{i3}	R_{o1}	R_{o2}	R_{o3}	L
		mm	mm	mm	mm	mm	mm	mm
2	2	837.3	839.3	841.3	872.3	874.3	876.3	3327.4

The tank structure shown in Fig. 3.23 is modeled as a 3D discrete rigid body and fixed in all 6 degrees of freedom. Based on the cryogenic tank conditions (pressure of about 0.6 to 0.8 MPa at a temperature or 22 K), a density of 70 kg/m³ with a dynamic viscosity of 1.24×10^{-5} Pas is used for modeling liquid hydrogen. Gravitational load and body forces (caused due to aircraft accelerations as shown in Fig. 3.24) are applied on liquid hydrogen which is modeled as a 3D deformable body. The loading experienced by the fuel will be complex and changes with time, based on how the aircraft is being maneuvered. To simplify the task two scenarios of longitudinal (during take-off & climb phase, where the aircraft's angle of attack is 15°) and lateral loading are considered, independent of each other. As shown in Fig. 3.24a, b, a nominal load of 0.25 g is chosen in both loading scenarios, this is a reasonable value based on the operational data of Airbus A-320 aircraft shown in Fig. 3.24. Fig. 3.24 describes the acceleration profile used in the model, where the body forces on the fuel are applied only for 1 s, and rest of the simulation time (19 s) is meant to observe the sloshing of liquid hydrogen inside the tank.

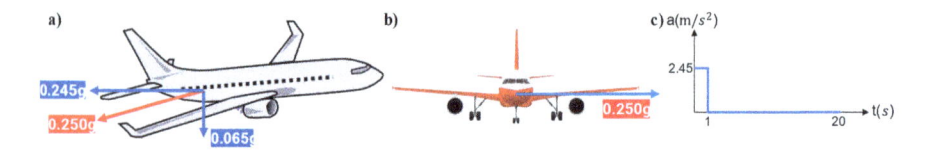

Fig. 3.24 Acceleration loads considered in the sloshing model (**a**) longitudinal (during take-off & climb phase), (**b**) lateral, (**c**) acceleration profile used

Fig. 3.25 Liquid hydrogen aircraft tank with hole plate type baffles

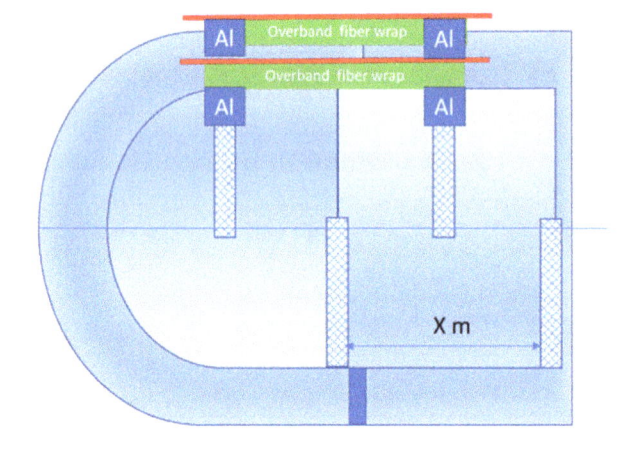

Previous Baffle Designs

Figure 3.25 indicates, in which way known, perforated slosh baffles, adopted from truck designs, could be integrated into a liquid hydrogen tank to limit the wave heights. For further suppressing the wave height, the location of the perforated plates was found more important than the number of plates. These sectioned half plates can be made from lightweight reinforced aluminum foam fitted as shown in Fig. 3.1. The ring structure at the internal tank section can help to reinforce the structure but at an increase in weight. Furthermore, vertical accelerations may not produce sloshing in all conditions but would still add to the total of vibration caused losses.

The following section investigates the predicted sloshing phenomenon in a typical CHEETA fuel tank by considering 70% and 30% fill levels. Initially, the results corresponding to baseline tank design (without any internal baffles) will be discussed to develop a basic understanding of the sloshing phenomenon. Later the impact and effectiveness of various internal baffle designs in suppressing the fuel slosh will be discussed.

Baseline Tank Design—Without Internal Baffles

We first look at the sloshing observed in a baseline CHEETA tank without any baffles as shown in Chap. 5 Cryotankage—Tank shapes and Airframe integration, Fig. 5.23. Figures 3.26 and 3.27 provide visual snapshots of fuel sloshing (at different time instances) under longitudinal and lateral loading respectively. Considering the weight of the fuel that is in motion, (70% fuel fill (~480 kg) and 30% fuel fill (~206 kg)), this is a major cause for concern as it can cause stress concentrations on

3.3 Cryotankage—Simplified Design Example

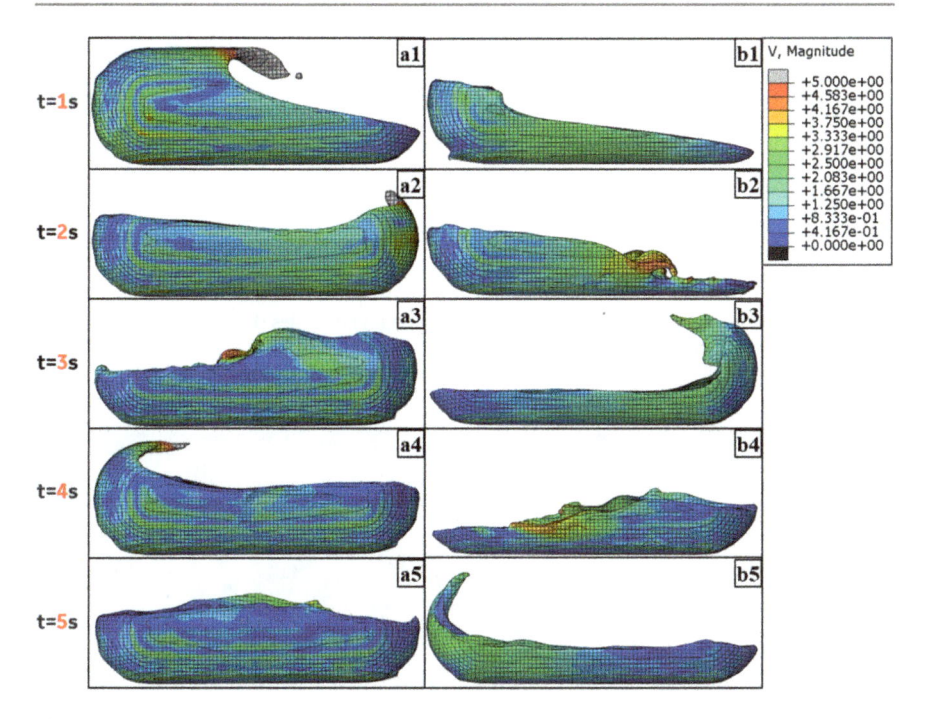

Fig. 3.26 LH2 fuel sloshing in a tank with no baffles (baseline) under the influence of longitudinal acceleration (Fig. 3.24a), (**a**) 70% fuel fill level, (**b**) 30% fuel fill level

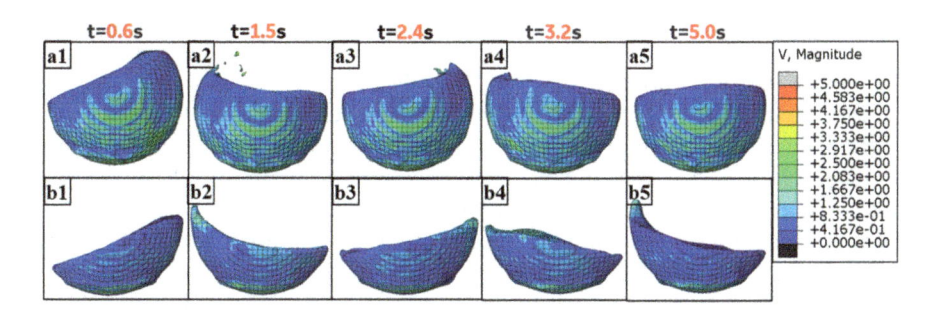

Fig. 3.27 LH2 fuel sloshing in a tank with no baffles (baseline) under the influence of lateral acceleration (Fig. 3.24b), (**a**) 70% fuel fill level, (**b**) 30% fuel fill level

the tank shell, dynamic imbalance force/moment on the aircraft (affects flight stability), bubble formation, and potential boil-off in LH2. Figure 3.28 plots the kinetic energy (K.E.) of the fuel as a function of time, to provide a quantitative representation of the visual phenomenon observed in Figs. 3.26 and 3.27. It can be observed from Fig. 3.28 that the system behaves like a damped system under free vibration. There have been studies in the past [36–46] that have treated the fluid slosh problem as an equivalent spring-mass-damper system. It can also be observed that the lateral sloshing motion occurs at higher frequency and lower kinetic energy, than the longitudinal slosh.

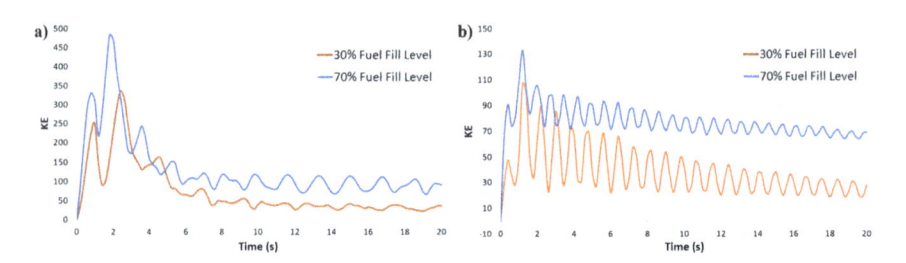

Fig. 3.28 Kinetic energy of the sloshing fuel in a tank with no baffles (baseline), (**a**) longitudinal acceleration, (**b**) lateral acceleration

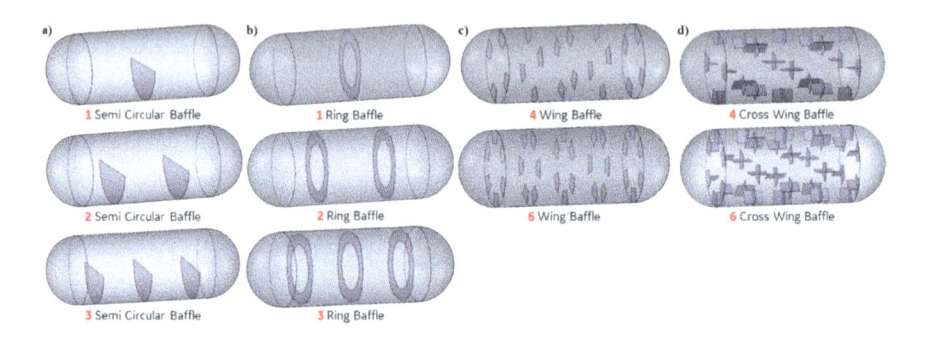

Fig. 3.29 Baffle designs considered in this study, (**a**) Semi-circular baffle (1, 2, 3), (**b**) Ring baffle (1, 2, 3), (**c**) Wing baffle (4, 6), (**d**) Cross-wing baffle (4, 6)

Impact of Internal Baffles

Internal tank baffles exist in different sizes and shapes depending on the application; their main purpose is to suppress and limit the sloshing motion of the fluid. Figure 3.29 shows the different baffle designs considered for the CHEETA LH_2 fuel tank. There is a tradeoff between added mass, reduced total fuel capacity, ease of pumping, and effectiveness in minimization of slosh, that needs to be considered while designing tank baffles. Considering that baffles are made of the same material as the inner layer of the tank (Al), Fig. 3.30 shows the increase in tank mass in percent with respect to the baseline design (tank without internal baffles). This section will discuss the results corresponding to the designs enclosed in green dashed lines, because they are comparable in terms of added mass.

Considering longitudinal sloshing, Fig. 3.31 shows that 3 semi-circular and 4 cross-wing baffle designs are effective in damping the slosh when the tank is 70% filled, but when the tank is 30% filled only the 4 cross-wing baffle design is effective (Fig. 3.32). The semi-circular baffle design is effective only when the fuel level is above the baffle height (fill level > 50%). While considering lateral loading (Fig. 3.33), semi-circular, ring, and wing baffles will be ineffective in suppressing slosh in the lateral direction.

3.3 Cryotankage—Simplified Design Example

Fig. 3.30 Percentage increase in inner tank mass due to different baffle designs described in the previous figure

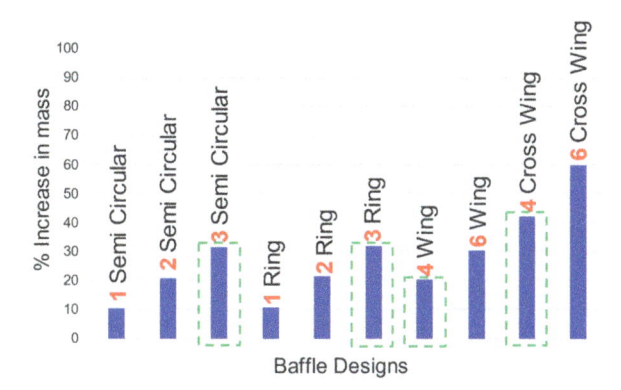

As observed in Fig. 3.33c, d, the cross-wing baffle design is effective in damping sloshing motion in the lateral direction. The cross-wing baffle design (Fig. 3.34) has the following advantages:

- It is effective in damping sloshing motion both in longitudinal and lateral directions.
- As the baffles do not compartmentalize the tank it is more efficient and easier in terms of fueling (no need for additional pumping).
- As the baffles are located along the circumference of the tank, they are effective in suppressing fuel sloshing independent of fill level.

Certainly, further research is required to ensure weight, performance, and tank stability can be maintained in all cruising conditions. Cost, assembly, lifetime, and efficiency paired with low weight are some of the important parameters to look at.

6. Understand the Thermal Load on the Tank Caused by Vibration Being Transmitted to the Tank by Any Airframe Structure.

Introduction What we know: As with many other stored cryogens, any disturbance of the liquid/vapor boundary layer surface will result in increased vaporization of the liquid. When the liquid is disturbed, for example, by continuous vibration, or by accidentally knocking at a liquid cryo tank, the evaporation rate rises rapidly to a high value, and then drops back again quickly. This evaporation spike could also be reproduced repeatedly by tapping the liquid container at fairly lengthy intervals. These boil-off events are called vapor explosions and demonstrate how sensitive the evaporation impedance of the surface is to disturbance [48] as shown in Fig. 3.35.

The liquid/vapor surface interaction itself is only poorly understood but may be of interest when the sloshing effect on the aircraft hydrogen tank is reduced and vibrational characteristics become more dominant during a flight or taxiing at the tarmac. Vibration effects are visible and need to be accounted for whenever a

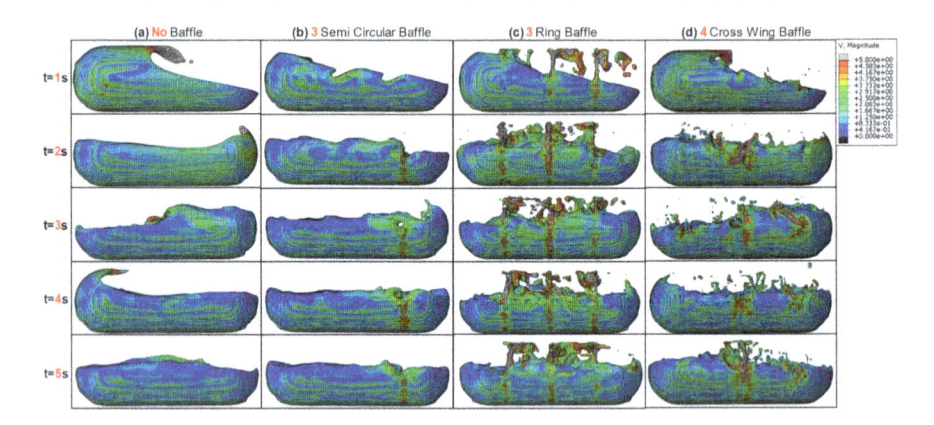

Fig. 3.31 Impact of baffle designs in a 70% fuel fill level tank under longitudinal loading, (**a**) No baffle, (**b**) 3 Semi-circular baffle, (**c**) 3 Ring baffle, (**d**) 4 Cross-wing baffle

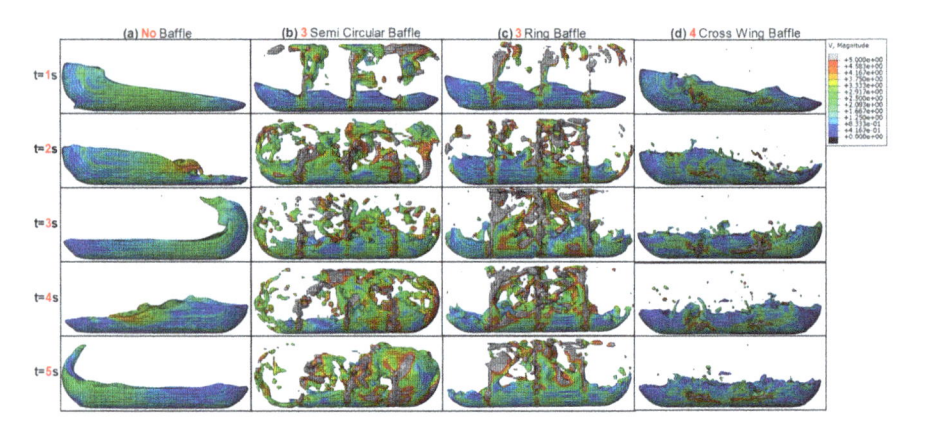

Fig. 3.32 Impact of baffle designs in a 30% fuel fill level tank under longitudinal loading, (**a**) No baffle, (**b**) 3 Semi-circular baffle, (**c**) 3 Ring baffle, (**d**) 4 Cross-wing baffle

cryogenic liquid is transported by road, ship, in aircraft, and space. Likewise, when running adiabatic experiments—any bulk liquid vibration falsifies the measurement result due to a boil-off increase not considered in the overall heat balance: thermal radiation/conduction, gaseous conduction losses, etc. (Fig. 3.36).

Scurlock in detail looked at the evaporation mechanisms and instabilities in cryogenic liquids. "Any disturbance or agitation of this thin conduction/convection region, whereby motion of bulk superheated liquid penetrates through it or replaces it, will result in an immediate, rapid and large, increase in evaporation rate" [48].

The intermittent convection layer thickness for LIN, LA, LO_x, LCH and LNG is about 0.4 mm for the liquids mentioned.

3.3 Cryotankage—Simplified Design Example 51

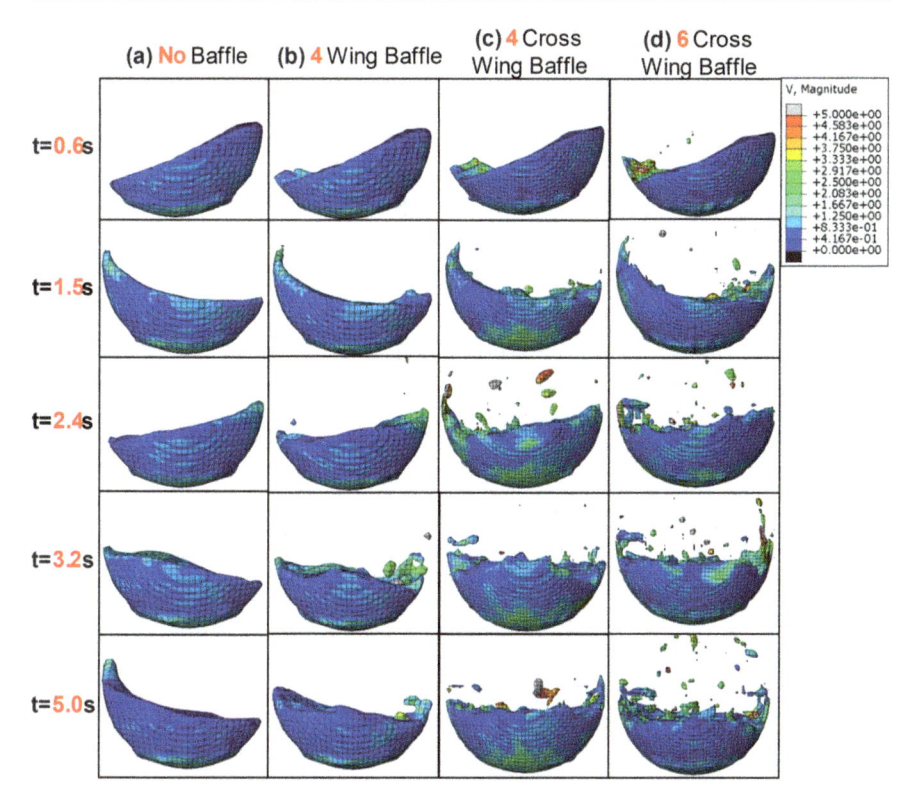

Fig. 3.33 Impact of baffle designs in a 30% fuel fill level tank under lateral loading, (**a**) No baffle, (**b**) 4 Wing baffle, (**c**) 4 Cross-wing baffle, (**d**) 6 Cross-wing baffle

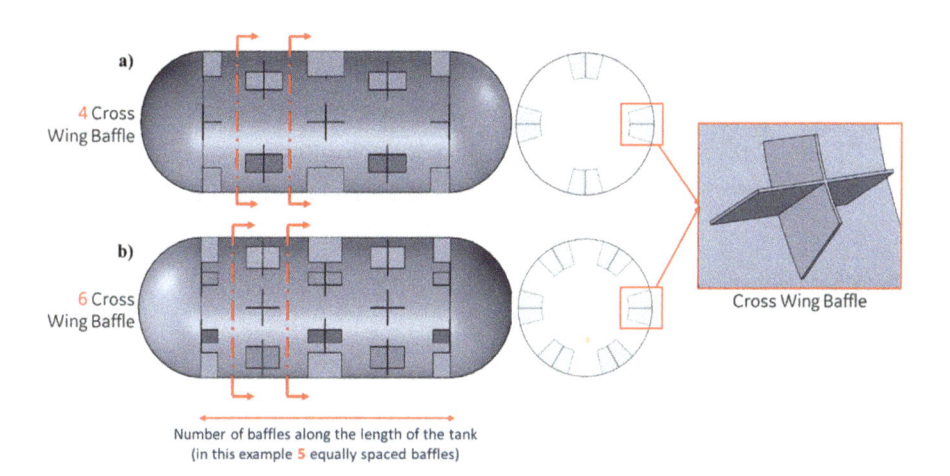

Fig. 3.34 Cross-wing baffle design, (**a**) 4 Cross-wing baffle, (**b**) 6 Cross-wing baffle [47]

Fig. 3.35 Vapor explosion example

Typical time variation of evaporation rate

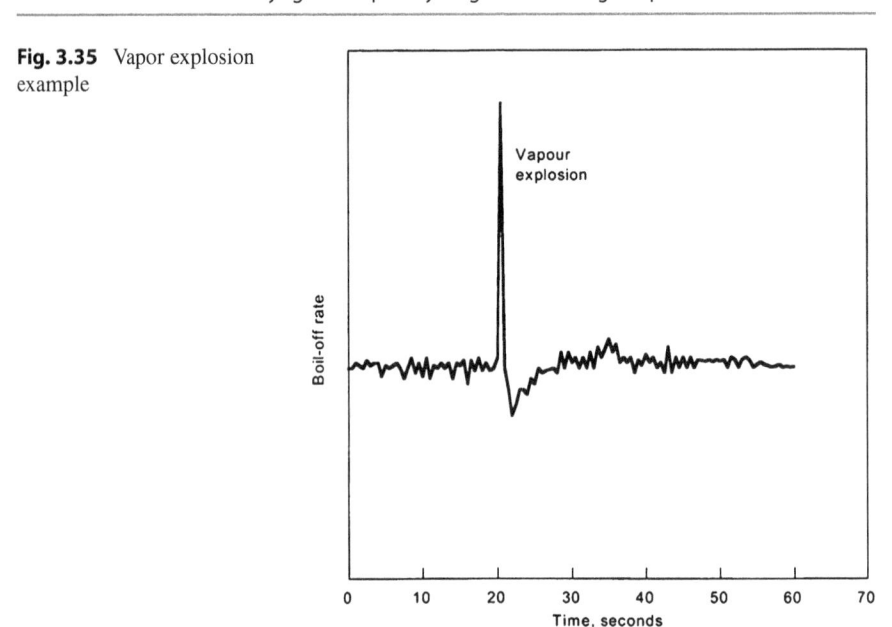

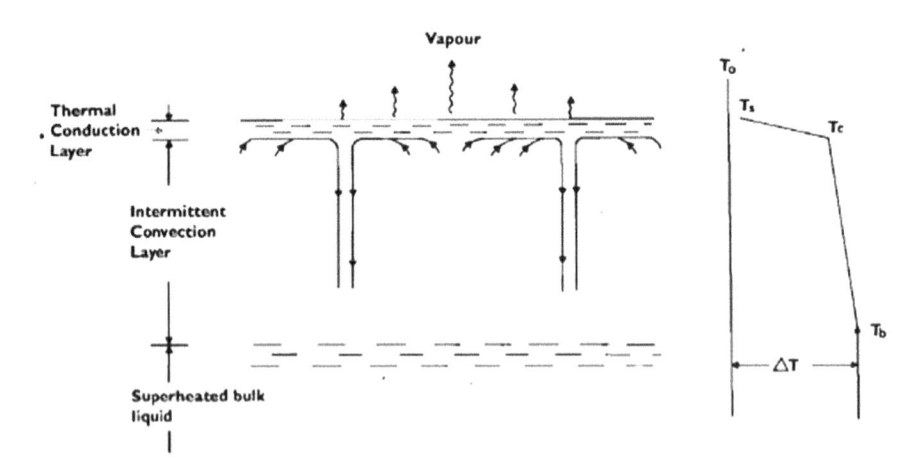

Fig. 3.36 Morphology of surface layers during evaporation as per Scurlock [48]

Hydrogen Tank Boil-off Increase The source of heat is generated by a cryotank low-frequency vibration that is also depending on any critical mechanical system resonance on the tank. It is widely accepted that vibrations even at a low level will increase cryogen boil-off. From road vehicles operation it is known that up to 6 to 8 g acceleration frequencies of up to 30 Hz or higher may occur. Rotenberg et al. [49], for example, conducted measurements on a small 110-liter hydrogen tank

3.3 Cryotankage—Simplified Design Example

Fig. 3.37 Vibration effect on liquid hydrogen tank [49]

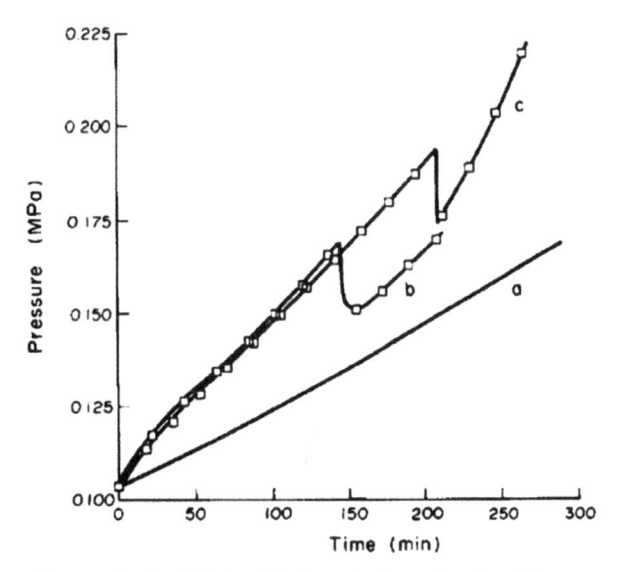

Pressure rise in a full, closed hydrogen tank, as a function of time: (a) numerical simulation, (b) measurements before and during vibration at 2 Hz frequency and 0.02 m displacement amplitude, (c) measurements before and during vibration at 4 Hz frequency and 0.02 m displacement amplitude.

Table 3.3 Vibration effect on LH2 tank pressure rise [49]

Displacement amplitude Y (m)	Acceleration amplitude (for 2 Hz) W^2 (m/s^2)	Acceleration a (m/s^2)	g-load G (m/s^2)
0.02	157.9	3.158	0.32
0.04	157.9	6.316	0.643
0.05	157.9	7.898	0.805

(BMW Hydrogen 7) that is subject to vibrations. At a frequency of 4 Hz corresponding to an amplitude of 0.02 m a boil-off factor of 12 was measured as compared to zero vibration modes. There is a small relevance with respect to the fill level but boil-off behavior in tanks is also substantially different for open vs closed tanks. Clearly, the pressure rise in the tank is much faster than for a liquid tank that is allowed to vent freely. The main reason for that is believed to be closed tank liquid bulk stratification.

The following shows the vibration-induced pressure increase for a closed 110-liter tank with an ullage space of 40 liters of liquid hydrogen (Fig. 3.37 and Table 3.3).

Vibration-induced heat loads to a liquid tank may occur during aircraft dormancy for a limited time. It is assumed vibration-generated heat loads will only very little distribute to boil-off and most likely be overpowered by any tank sloshing movement. In general, not much is known from prior studies yet on the motion effect of hydrogen subject to vibration [50].

Case Study (Historical) Consider a typical bath-cooled, generic, NMR magnet with a liquid helium volume of 40 liters with a consistently measured helium boil-off rate of 35 ml/h in the absence of a cryocooler. You decide the NMR magnet is well in its specification limits and is delivered to the customer site. Customer reports, the NMR magnet is out of spec and close to a boil-off of 100 ml/h. You ask the customer to send the unit back for further investigation. However, the measured boil-off rate is found to be in spec again and you are sure that your measurement technique and setup are correct. What is going on?

Vibrational heating effect of a bath-cooled system:

- Boil-off: 35 ml/h before moving cryostat from test bay to customer.
- Boil-off: > 100 ml/h at customer site (out of spec)—averaged over day and night measurements.
- Boil-off: 35 ml/h (in spec) after cryo was sent back to test bay.
- Boil-off: > 100 ml/h when shipped back to customer site.

Design Helium vessel with superconducting high field magnet suspended on 2 thin-walled steel tubes with 30 mm diameter and 0.5 mm wall thickness, suspended weight approx. 1200 kg on outer shell (outer vacuum case (OVC)). No other vertical suspension, some radial suspension elements.

Installed cryostat is subject to floor vibrations. Measured, merely noticeable floor vibrations transmitted to a liquid surface did result in a high boil-off—an atypical experience since every NMR magnet system already has its own floor vibration-canceling elements to mitigate NMR signal distortion.

Reason Large paper mill drums located ½ mile away from customer site transmitted floor vibrations to customer at daytime. Rotating drums changed the frequency at nighttime shifts lowering the boil-off.

Solution Change operating frequency of pneumatic damper legs on cryogenic tank or change the suspension design. The eigenfrequency of that suspended system however is very difficult to predict and if hit once difficult to control, resulting in excessive high boil-off either for hydrogen due to the large tank surface area, or for helium with a smaller surface area but higher rate of vaporization. Don't hit a resonance spot.

Vibration effects that translate to higher boil-off are also found in bath-cooled MRI magnets. For increased boil-off, both the inner suspension as well as structural components in an MRI suite need to be considered. In some suites, heavy-weight, high-field MRI magnets cannot be sited easily due to additional vibration image distortion effects.

Figure 3.38 gives a good example of a new generation of low cryogen loss NMR magnet.

3.3 Cryotankage—Simplified Design Example

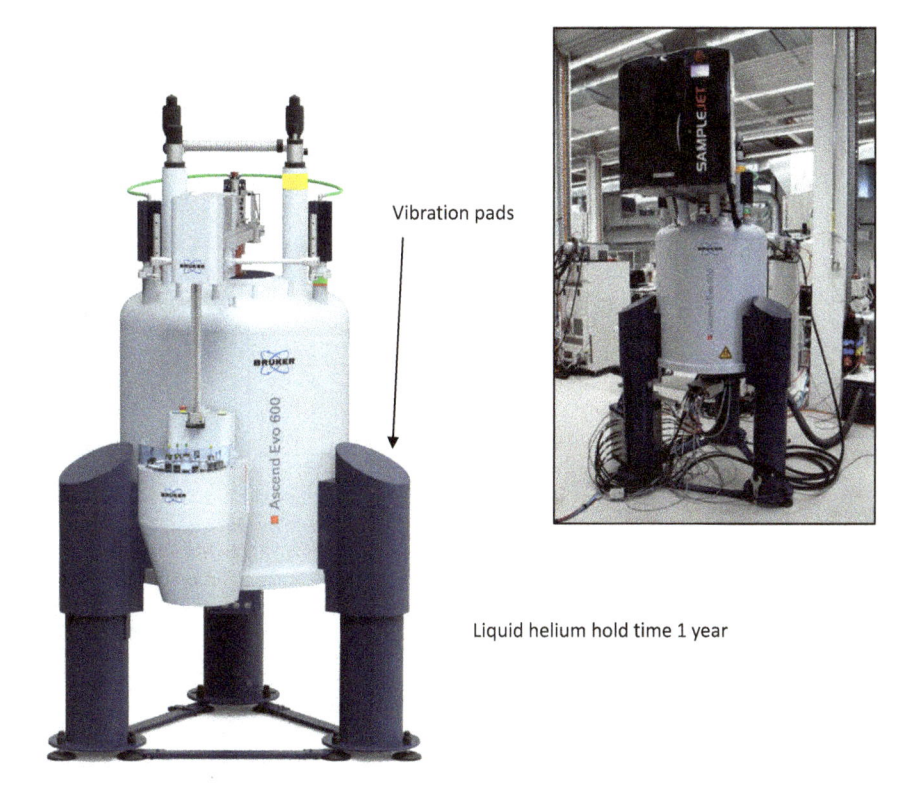

Fig. 3.38 *New* Ascend Evo 600 MHz bath-cooled NMR magnet (2024), liquid helium hold time 1 year, top right, system with automatic probe sampling installed, courtesy of Bruker BioSpin

The new Ascend Evo 600 MHz magnet does not need any liquid helium top up for a year and benefits of a continuous history of 40 years of cryogenic, extreme low loss technology cryostat development, covering all aspects from floor vibration to environmental pressure changes, including many other transient heat sources.

Other attempts to generate vibration-based boil-off data can be found by Kaganer [51]: "When cryogenic vessels and cisterns are transported the rate of cryogenic liquid evaporation is several times greater than under steady-state conditions, for example, 3-8 times greater for liquid helium in a 10-liter vessel."

Experiments were performed involving vibration of a *5-liter liquid helium vessel*, and it was found that the increase in loss depended quadratically on vibration amplitude. Based on shaker results he concludes:

"The containers used for liquid nitrogen on the vibrostand were the 16-liter spherical ASD-16 and the 20-liter vertical cylindrical SDS-20. A VEDS-200 electrodynamic vibrostand was used to create harmonic oscillations with amplitude of 0.1–10 mm and frequency of 5–80 Hz. An SIT vibrostand was used to emulate vibration conditions found in road transport, at frequencies of 4 and 9 Hz with mean amplitude of 6 mm. The amount of liquid nitrogen used was varied from 7 to 16 kg." Results are given in Fig. 3.39.

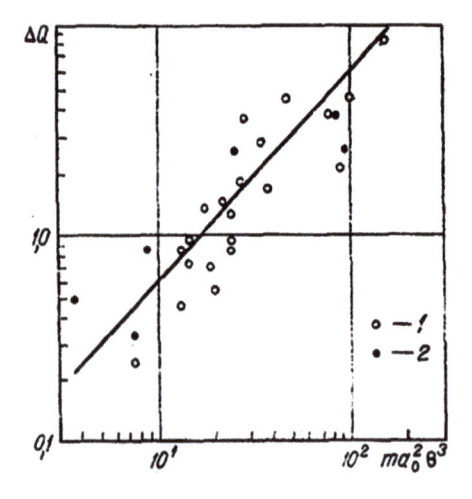

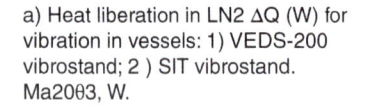

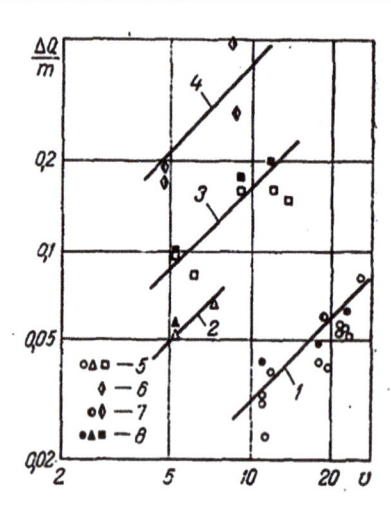

a) Heat liberation in LN2 ΔQ (W) for vibration in vessels: 1) VEDS-200 vibrostand; 2) SIT vibrostand. Ma2θ03, W.

b) Specific heat liberation ΔQ/m (W/kg) in LN2 and helium vs transport rate v (m/s) of vessels in ZIL-130 automobile: 1) high speed road, 2) dirt road, 3) flat cobblestone, 4) profiled cobblestone, 5) TsTK-1.6/0.25 cistern, 6) ASD-16 vessel, 7) SDS-20 vessel, 8) STG-10 vessel.vtt

Fig. 3.39 (**a**) Thermal load increase due to vibration on vibrostands, (**b**) specific heat release in W/kg for different road conditions

$$\Delta Q = \frac{\Psi_0}{4\pi} m a_0^2 \theta^3. \quad \Psi_0 = 2\delta \text{ (fudge factor)}$$

a_0 = amplitude of forced vibration in m,
θ = frequency (1/s),
Ψ_0 = oscillation energy absorption coefficient in liquid; approx. 0.2.
dQ = oscillation energy scattered in liquid (heat liberation in W),
m = liquid mass in kg.

Williamson et al. [52] shared insight into their experiments with the shaker tables on a 4-liter cryogenic liquid helium tank and concluded (Fig. 3.40):

For displacements of up to approximately 0.014 cm, the loss rate increase is less than 30%. By way of example a displacement of 0.014 cm would occur for a 3 g load at 250 Hz or a 250 g load at 1000 Hz. For larger displacements the loss rate rises to as much as ten times the zero displacement value.

Such enhanced rates are certainly of importance in the design of dewars for use in large vibration environments such as in rocket payloads. Furthermore, the structural resonance in the inner vessel makes it imperative to consider displacements of the inner vessel and not simply the driving field (Fig. 3.41).

3.3 Cryotankage—Simplified Design Example

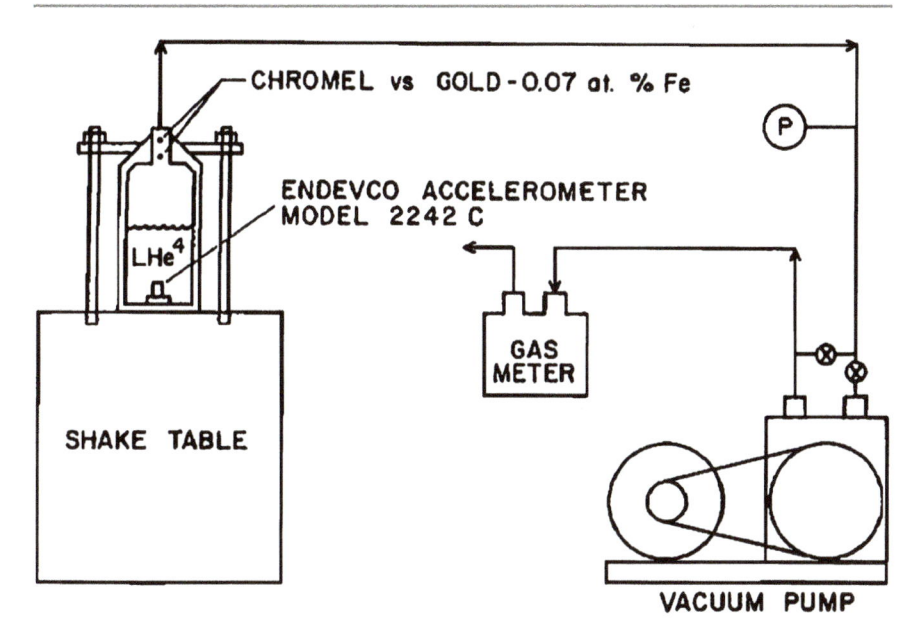

Fig. 3.40 4 liter LHe cryotank on shaker table [52]

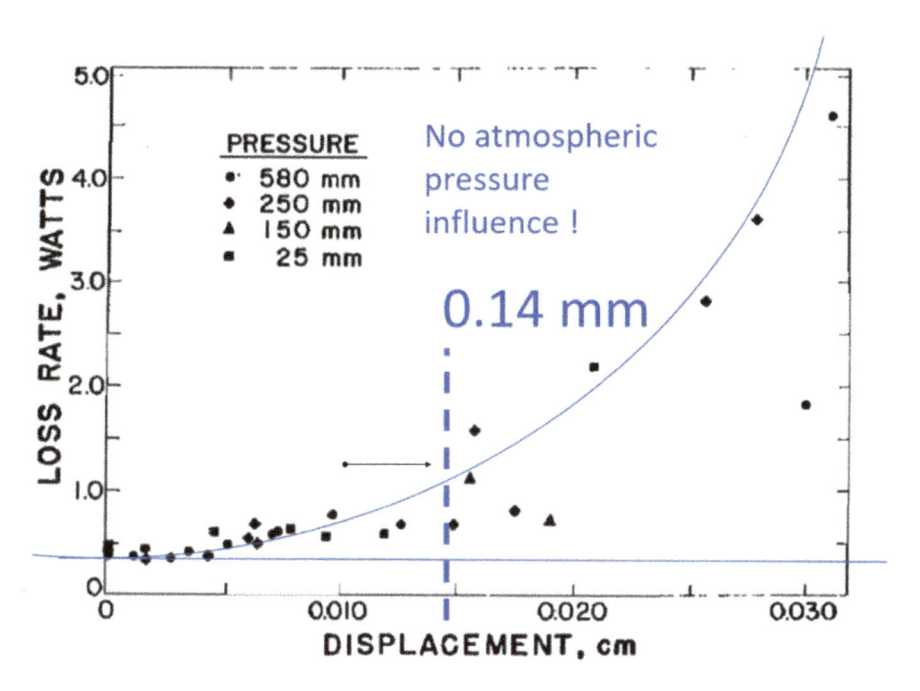

Fig. 3.41 Boil-off increase of the Terrier Sandhawk rocket [52]

Table 3.4 Pressure altitude changes LH2 properties

H (ft)	H (km)	P (Pa)	P (mbar)	T_{sat} (LH$_2$) (K)	H (J/kg)	Lv (J/kg)
40,000	12.19141	18672.53316	186.7253	15.686	2.318E+05	4.5185E+05
37,000	11.27705	21640.66331	216.4066	16.012	2.342E+05	4.5223E+05
30,000	9.143554	30068.86375	300.6886	16.782	2.400E+05	4.5268E+05
20,000	6.095703	46539.40079	465.394	17.911	2.491E+05	4.5210E+05
10,000	3.047851	69656.05737	696.5606	19.075	2.593E+05	4.4985E+05
5000	1.523926	84281.6145	842.8161	19.671	2.648E+05	4.4799E+05

Collected data indicate that for inner vessel displacements of less than about 0.0127 cm, boil-off increases by less than a factor of 2 in this case.

From the sparse previous literature data available and experience, any cryogenic aircraft liquid tank may need to be supported by dedicated damping elements that ideally compensate for any vibration effect. The design of damping elements is based on the tank design, tank suspension, e.g., fatigue design and external vibration modes acting on the tank assembly.

So far, gravity of increased vibration on boil-off under flight conditions is not well understood yet and sloshing effects instead may become the dominant factor.

However, there could be certain cruising conditions or when tarmac taxiing for prolonged times, where vibration effects may become more dominant and need to be counted it.

7. Avoid stratification in single-component liquids [26].

    In case of density stratification in the storage tank a less dense fluid layer resides over a higher density layer suppressing local convective mixing and thus temporarily reduces boil-off. This however will cause a boil-off instability in the liquid that may also lead to sudden, large vapor ejection out of the tank or even so-called rollover phenomena. The risk for a large hydrogen tank that is pumped and emptied may be low but so far has not been investigated yet in detail. The effect of stratification has been noticed by Rotenberg [49] even for small liquid hydrogen tank sizes. Given the flight profile of an aircraft, it is assumed stratification is a rather rare, uncommon event.

8. Non-conformal tank designs need to consider liquid tank internal convection heat losses depending on liner and build design. Conformal tanks in addition require good knowledge of external convective flows around the shell of the tank within the fuselage as well as a good understanding of any electrostatic loads (see also Chap. 8 on Safety).

9. Tank pressure changes based on a mission profile. For some cryogens, boil-off increases as the aircraft undergoes altitude changes. For hydrogen, the latent heat of vaporization however does not change much with pressure as shown in Table 3.4 and Fig. 3.42, in most cases this increased boil-off can be neglected for large tanks.

With respect to the tank fill, the mission starts with an underground liquid storage fill, the time it takes to fill the tank, staying on ground, taking off and climbing as partially shown in Fig. 3.42, and ends with diverting the flight to the taxiing in the procedure.

3.3 Cryotankage—Simplified Design Example

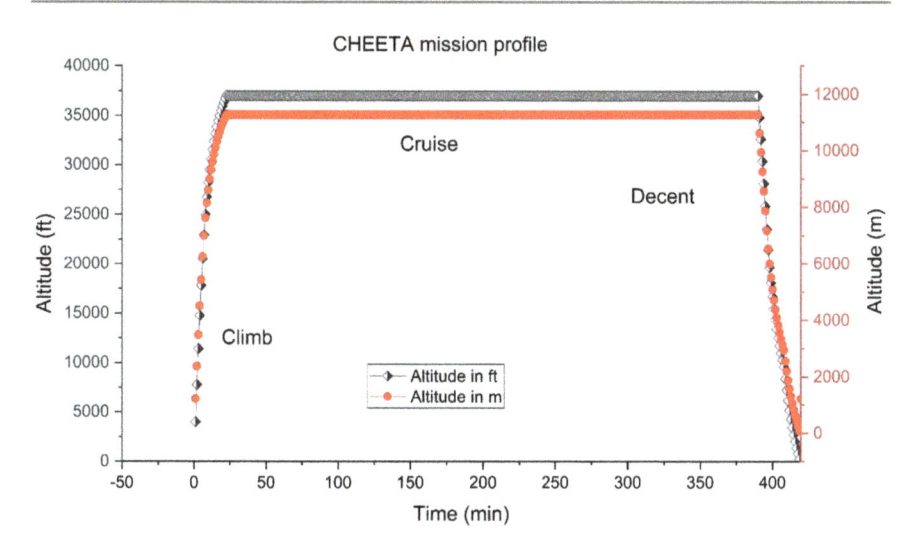

Fig. 3.42 Mission profile

For air pressure calculations the standard barometric altitude formula can be used, however the temperature reduction with increasing cruising height has to be considered and the equation can be rewritten as: p = 1.013 10^5 (Pa) [(1−6.5 / 288 km * H altitude (km)] $^{5.255}$ (valid up to 11 km) [53].

Figures 3.43 and 3.44 depict enthalpy and latent heat change within the 2-phase tank region subject to aircraft altitude. Data has been compiled with GasPak 3.35 [54].

It does seem possible that a pre-slush hydrogen phase may be created at high altitudes if the initial tank pressure design is too low (at approx. 1000 mbar). This could be avoided by using a wide-range absolute atmospheric pressure regulator that maintains a more desirable pressure within the tank. Otherwise, the operating state of the mission profile needs to be communicated to the tank interfaces, e.g., pumps or others (see also Chap. 7 on liquid hydrogen pumps).

The complete mission profile starts with an initial tank purge if required, a chill down, filling of tanks, consuming the fuel, and defueling. Fueling of liquid hydrogen into the aircraft can either be executed with underground fill storage or using liquid hydrogen trucks. An economical assessment of hydrogen short-range aircraft with the focus on the turnaround procedure has been studied by Mangold [55].

Taxiing on the ground, take off, climb, cruising, descent, and touch down accompanied by further taxiing then completes the mission profile.

For further details, please see also Chap. 11. In summary, any kind of heat transfer mode to the tank leads to an increase in tank pressure. For determining the pressurizing period see Hofmann [56].

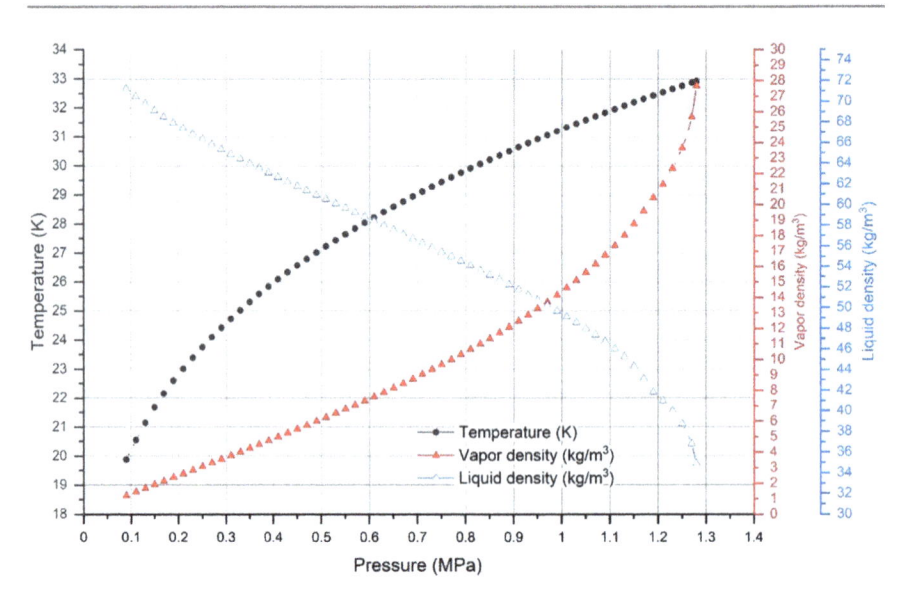

60   3   Cryogenic Liquid Hydrogen Tank Design Aspects: General Overview

Fig. 3.43 Saturated pressure and density of hydrogen liquid and gas phase. *Last value before leaving saturated region. (Source: GasPak v 3.35 / Horizon)

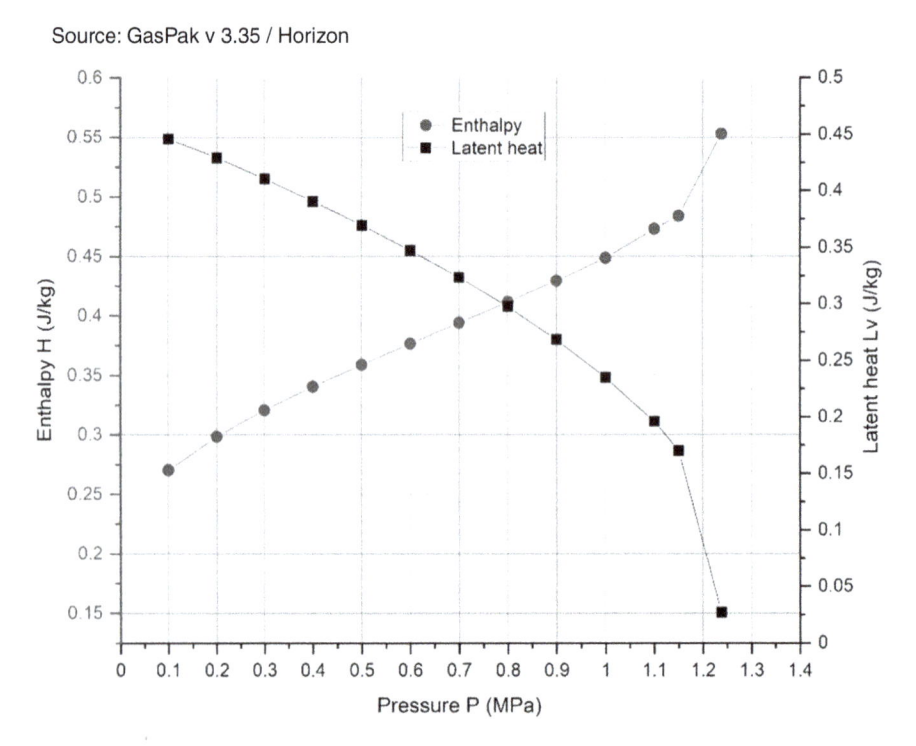

Fig. 3.44 Liquid hydrogen tank pressure and physical parameters Generated with GasPak v 3.35 [54]

References

1. M. Boll, "Hybrid-Electric Re-equipment of an A321 XLR aircraft", Rolls-Royce Electrical Seminar at University of Strathclyde, 2020
2. P. Wheeler, K. Haran, "Electric/Hybrid-Electric Aircraft Propulsion System", Proceedings of the IEEE, vol. 109, No. 6, 2021
3. T. Balachandran, K. Haran, "A fully superconducting air-core machine for aircraft propulsion", ICMC 2019, IOP Conf. Series: Materials Science and Engineering 756 (2020) 012030, IOP Publishing, doi:https://doi.org/10.1088/1757-899X/756/1/012030
4. K. S. Haran, S. Kalsi, T. Arndt, H. Karmaker, R. Badcock, B. Buckley, T. Haugan, M. Izumi, D. Loder, J. Bray, P. Masson, E. W. Stautner, Topical Review "High power density superconducting rotating machines—development status and technology roadmap", *Supercond. Sci. Technol.*, vol. 30, no 12, pp. 1–41, 2017. Doi: https://doi.org/10.1088/1361-6668/aa833e.
5. S. Smith, "Superconducting technologies for hybrid-electric aerospace propulsion", University of Manchester, 2022, priv. communication
6. T. Wallnera, H. Lohse-Buscha, S. Gurskia, M. Duobaa, W. Thielb, D. Martin, T. Korn, "Fuel economy and emissions evaluation of BMW Hydrogen 7 Mono-Fuel demonstration vehicles", International Journal of Hydrogen Energy, 33, 2008 7607–7618
7. L. Klebanoff, Hydrogen Storage Technology, CRC Press, 2012
8. M. Mckee "Fallen foam stalls launch of Rosetta", NewScientist Newsletters, 2003 https://www.newscientist.com/article/dn4726-fallen-foam-stalls-launch-of-rosetta/
9. G.D. Brewer, Hydrogen Aircraft Technology, CRC Press, 1991
10. M.G. Kaganer, Thermal insulation in cryogenic engineering, Israel Program for Scientific Translations, Jerusalem, 1969, IPST cat. No. 2200, IPST Press
11. J. Fesmire, A. Swanger, Advanced cryogenic insulation systems, 25th IIR International Congress of Refrigeration, Montreal, 2019
12. Beyond Gravity, priv. communication
13. Beyond Gravity, Thermal insulation products
14. J.Y. Faudou et al, Cryogenic reservoir with helically wound thermal insulation US patent 4741456 1988
15. C. Laa, C. Hirschl, J. Stipsitz, "Heat flow measurement and analysis of thermal vacuum insulation", in: Weisend II, JG eds. Adv. Cryog. Eng. Vol. 53 B pp 1351–1358 2007
16. M. Smoluchowksi, Ber. II. Internationaler Kältekongress, Wien 1910
17. W. Lehmann. Thermische Isolation in der Kryotechnik—VDI Seminar Karlsruhe, 1997
18. W. Lehmann, Thermische Isolation in der Kryotechnik, KFK report 030304P11B, 1985
19. W. Lehmann, priv. communication, 1982
20. R. J. Corruccini, Advances in Cryogenic Engineering, vol 3, 1960, pp 353–366
21. J.R. Heim, National Accelerator Laboratory, Report TM-334A, 1971
22. W. Stautner et al. Suspension system for a cryogenic tank US patent application 2024/0052976 A1
23. NASA Outgassing Database, source: https://outgassing.nasa.gov., data last updated: 08/19/2015 https://outgassing.nasa.gov/
24. C. Day, "Vacuum pumping", Fusion Summer School, ITER, KIT 2007
25. R.R. Conte, "Elements de Cryogenie", Masson & Cie, 1970, p 68
26. Kropschot R H Technology of Liquid Helium NBS Monograph 111 1968 p 192
27. van Meerbeke R *Thermal stratification and sloshing in liquid helium trailers* Advances in Cryogenic Engineering Vol. 13 1968 pp. 199–206
28. Minas C priv. communication 2023
29. van Foreest A *Modeling of cryogenic sloshing including heat and mass transfer* Cuvillier Verlag 2021
30. Roberts J R Basurto E R Chen P *SLOSH DESIGN HANDBOOK I* Northrop Space Laboratories for George C. Marshall Space Flight Center, NASA-PDF file NAS 8-11111 19660014177.pdf
31. https://www.cambridge.org/us/academic/subjects/engineering/thermal-fluids-engineering/liquid-sloshing-dynamics-theory-and-applications?format=HB&isbn=9780521838856

32. Lui Z Feng Y Lei G et al *Fluid sloshing dynamic performance in a liquid hydrogen tank* International Journal of Hydrogen Energy 44 2019 13885–13894
33. Ryali L Stautner W et al 2024 *IOP Conf. Ser.: Mater. Sci. Eng.*1301 012068 *Impact of Internal Baffle Designs on Liquid Hydrogen Sloshing in Cryogenic Aircraft Fuel Tanks*. https://doi.org/10.1088/1757-899X/1301/1/012068
34. A. Nizamudeen *A Study on Sloshing using Smoothed Particle Hydrodynamics (SPH)* B.A. degree paper, Heriot-Watt University UK 2019
35. Rustenburg J Skinn D Tipps D *Statistical Loads Data for the Airbus A-320 Aircraft in Commercial Operations* 116 2002
36. Li Yu-C, Gou Hong-Liang *Modeling Problem of Equivalent Mechanical Models of a Sloshing Fluid* Shock and Vibration Vol. 2018 Article ID 2350716 2018
37. Li Y Wang J *A supplementary, exact solution of an equivalent mechanical model for a sloshing fluid in a rectangular tank*, Journal of Fluids and Structures Vol. 31 pp. 147–151 2012
38. Roberts J R Basurto E R Chen P *SLOSH DESIGN HANDBOOK I* Prepared under Contract No. NAS 8-11111 by NORTHROP SPACE LABORATORIES Huntsville Ala. for George C. Marshall Space Flight Center NASA 1966
39. *Liquid hydrogen as a potential low-carbon fuel for aviation* IATA conference August 2019
40. Demirel E Aral M *Liquid Sloshing Damping in an Accelerated Tank Using a Novel Slot-Baffle Design* Water 10 (11):1565 2018
41. Cho I H Choi J S Kim M H *Sloshing reduction in a swaying rectangular tank by an horizontal porous baffle* Ocean Engineering Vol. 138 pp 23–34 2017
42. Liu Z Feng Y Lei G Li Y *Fluid sloshing dynamic performance in a liquid hydrogen tank* International Journal of Hydrogen Energy Vol. 44 Issue 26 pp. 13885–13894 2019
43. Kim S West J *Sloshing In The Liquid Hydrogen And Liquid Oxygen Propellant Tanks After Main Engine Cut Off* JANNAF 5th Spacecraft Propulsion Subcommittee Meeting 2011
44. Liu Z Feng Y Lei G et al *Sloshing Behavior Under Different Initial Liquid Temperatures in a Cryogenic Fuel Tank* J Low Temp Phys Vol. 196 pp. 347–363 2019
45. Raouf A I *Liquid Sloshing Dynamics: Theory and Applications* Cambridge University Press 2005
46. Smith J R Gkantonas S Mastorakos E *Modelling of Boil-Off and Sloshing Relevant to Future Liquid Hydrogen Carriers* Energies 15 (6) 2046 2022
47. Stautner W Ryali L Mariappan D *Multi-Directional Baffles for Aircraft Fuel Tanks* US patent application 2024/0343407A1
48. Scurlock R G *Low-loss storage and handling of cryogenic liquids: The application of cryogenic fluid dynamics* Monographs on Cryogenics Kryos Publications 2006
49. Rotenberg Y Burrows M McNeil R *Vibration effects on boil-off rate from a small liquid hydrogen tank* Int. J. Hydrogen Energy Vol. 11 no 11 pp 729–735 1986
50. Peschka W *Liquid hydrogen as a vehicular fueling challenge for cryogenic engineering* Int. J. Hydrogen Energy Vol. 9 No. 6 pp 515–523
51. Kaganer M G Zhukova R I *Evaporative loss of cryogenic liquids during vibration and transport in vessels and cisterns* Journal of Engineering physics and Thermophysics 1986 50 (4) pp 416–418
52. Williamson K D Liebenberg D H Edeskuty F S *Vibration enhanced boil-off rate from a liquid helium dewar* Los Alamos Publication Proceedings XIII International congress of refrigeration 1st commission 1971 pp 565–568
53. E. Hering, R. Martin, M. Stohrer, Taschenbuch der Mathematik und Physik, Springer, 4th edition, 2004 page 198
54. Gaspak® Software v. 3.35, Horizon Technologies, 2007
55. J. Mangold, "Economical assessment of hydrogen short-range aircraft with the focus on the turnaround procedure, Master Thesis, 2021
56. A. Hofmann "Determination of the pressurization period for stored cryogenic fluids". Cryogenics, Vol 46, Issue 11, pp 825–830, 2006. https://doi.org/10.1016/j.cryogenics.2006.06.003

Cryotankage: Structural Thoughts

4.1 Cryotankage—Structural Thoughts

4.1.1 Early Liquid Hydrogen Tank Designs

Since the beginning of liquid hydrogen storage in the 1960s, many different design concepts were pursued for different modalities. In this book, we refer to groundbreaking design concepts that can be extrapolated for use with aircraft, if modified and tested extensively. It is this available data source of liquid and gaseous hydrogen applications that gives us confidence in the safe handling of this cryogen [(see also Appendix 3 on Safety)].

Simply storing liquid hydrogen was one of the first challenging tasks, well described by Elrod [1], a guidebook even for today's applications. Nowadays, impressive big storage tanks of the size of 4.7 million liters have been realized, as recently shown by Fesmire [2]. See also Appendix 4.

A further tank design evolved out of the need of transporting liquid hydrogen to the end user, as for example described by Kropshot [3] and Peschka [4]. The automotive industry then introduced a novel storage design with the BMW Hydrogen 7 car. For further details, see also Appendix 4.

Table 4.1 shows some of the first efforts with the NASA Composite Cryotank Technology Demonstration (CCTD) project to design and build a composite liquid hydrogen tank and of the European CHATT project:

1. Boeing Phantom Eye—HALE tank designs [5].
2. DC-XA composite LH2 cryotank [6].
3. X-33 composite LH2 cryotank [7].
4. Space Launch Initiative (SLI) composite cryotank [8].
5. CCTD composite cryotank [9].
6. CHATT project—investigate CFC material for cryogenic fuel tank applications [10].

Table 4.1 Some basic design parameters for space applications

System	Size (m) (diameter/length)	Volume (liter)	Pressure (MPa)	Structural details	Comment
HALE ball/ Boeing			0.655	Aluminum sphere with SOFI (spray on foam insulation)	4 vertical slosh baffles 90° apart/ heat leak around 600 W
DC-XA McDonnell Douglas	2.43/4.88			IM7/8552 toughened epoxy	Passed ground and flight tests
X-33 Lockheed Martin	6.09 width, 4.26 height, 8.68 long			Carbon fiber composite skin sandwiched between nonmetal honeycomb core, aluminum foil liner	Sandwich construction, see also section on safety, failed validation test
SLI Northrop Grumman	1.8/4.5		0.82	Thin plies to microcracks OM 5320–1 prepreg	Resistant to microcracks
CCTD	8.8	634,297	0.32	IM7/977–2 lamina Cytec CYCOM 5320–1 prepreg	Short cylindrical design with domes
CHATT	1/2.4	1900		Cylindrical tank	GFRP and glass fiber combined, huntsman araldite LY 564 with Aradur 22,962
	0.29/0.57	33	1.2	Cyl. Tank with liner	T-700 carbon fiber, PE liner
				Cyl. Tank without liner	Thin-ply laminates, large void content, leaked

The mentioned liquid hydrogen tanks above for next-gen launch vehicles to a certain extent provide some synergies for aircraft designs. Some of the above design solutions rely heavily on the application of composites. The fundamental issues of all liquid hydrogen tanks tested at NASA were hydrogen permeability and understanding the mechanical composite properties at cryogenic temperatures.

We appreciate the work at NASA on liquid hydrogen storage which is of outstanding value for all applications.

4.1.2 Tank Wall—Composites

Designing a liquid hydrogen tank is not as obvious as it is for other cryogens, e.g., with liquid helium. Here are a few thoughts on structural issues. There may be more than one good solution depending on the operating conditions and the following

4.1 Cryotankage—Structural Thoughts

paragraphs by no means cover the subject matter completely but do point out a few key aspects.

Table 4.2 summarizes the favorable weight advantage when using composite tank structures, with tanks of an aluminum shell, set as nominal 1 as shown by Canfer and Evans [11]:

Still, any cryotankage using a stored cryogen needs to keep the comment of Canfer/Evans [11] in mind that holds true since it was first written down in 1997: "*A large knowledge gap exists in material science as to the effect of cryofuels on the long-term performance of polymeric composites.*" Not much has changed since then.

Although most of the mechanical properties of metals and composites at cryogenic temperatures will work in favor for the cryotankage, they need to be considered very carefully, especially for multiload conditions, and in particular with respect to microcracking and fatigue.

Figure 4.1 shows composite strands (not for pressure vessels) hours to failure under tensile stress [12] at room temperature and should be a reminder to carefully choose any design with composites and take time to failure into consideration. The plots show the median 0.5 probability, the shaded area indicates the region used for the Weibull fit.

One can therefore conclude that for GRP glass composite strand wraps, the probability of failure for 25 years under tensile stress is about 100% as compared to CFC fibers (after Klebanoff [12]).

Table 4.3 gives some typical values for epoxy-fiberglass laminates at cryogenic temperatures.

Table 4.2 Properties and dimensions of aluminum and composite structures at room temperature (RT)

Material	Modulus (GPa)	Density (kg/m^3)	Thickness for equal bending stiffness	Mass for equal bending stiffness
Aluminum alloy	70	2700	1.00	1.00
CFRP, HS 75 vol% fiber	74	1720	0.98	0.63
CFRP, HS 70 vol% fiber	68	1690	1.01	0.63
CFRP, HS 60 vol% fiber	58	1620	1.07	0.64
CFRP, HM 75 vol% fiber	114	1720	0.85	0.54
CFRP, HM 70 vol% fiber	106	1690	0.87	0.55
CFRP, HM 60 vol% fiber	90	1620	0.92	0.55
S-glass, 60 vol% fiber	25	1970	1.42	1.03

HS High Strength, *HM* High Modulus

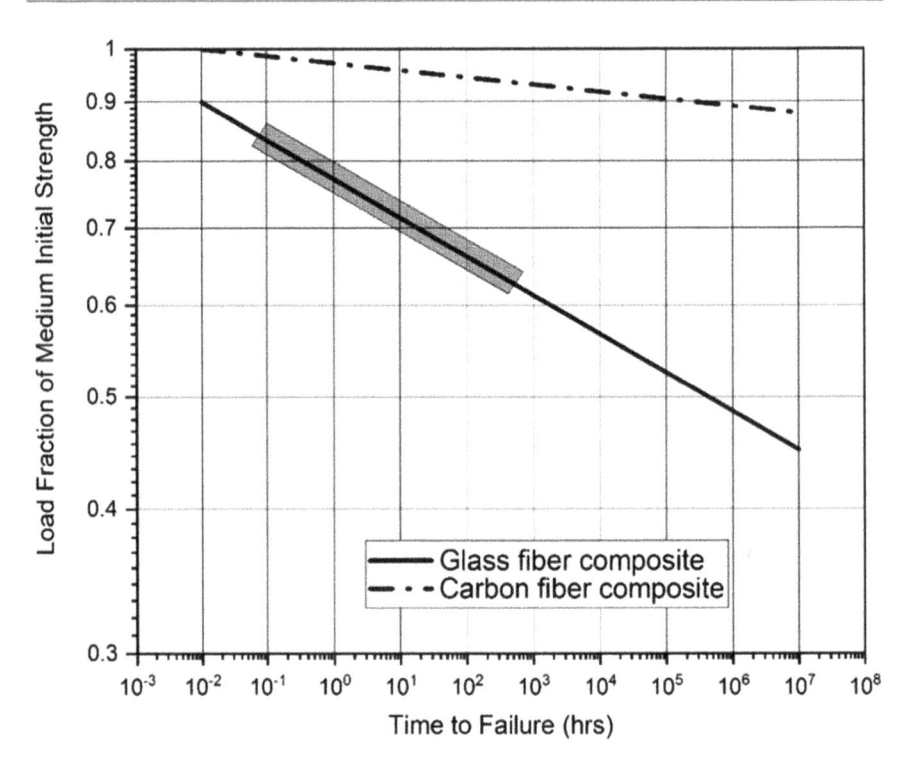

Fig. 4.1 Composite fatigue

Table 4.3 Mechanical properties of epoxy-fiberglass laminate (FMNA with 35% glass-content by weight)

	Temperature (K)		
Property (N/mm^2)	**300**	**77.3**	**4.2**
Tensile strength	294.2	632.5	701.2
Compressive strength	294.2	755.2	735.5
Flexural strength	549.2	1029.7	980.7
Fatigue strength			
10^3 cycles	294.2	559.0	696.3
10^4 cycles	231.4	484.5	554.1
10^5 cycles	189.3	407.0	386.4
10^6 cycles	160.8	328.5	245.2

Figure 4.2 on the right compares epoxy-fiberglass laminates with typical cryogenic materials and their respective stress increase that leads to a higher fatigue strength at cryogenic temperatures [13].

4.1 Cryotankage—Structural Thoughts

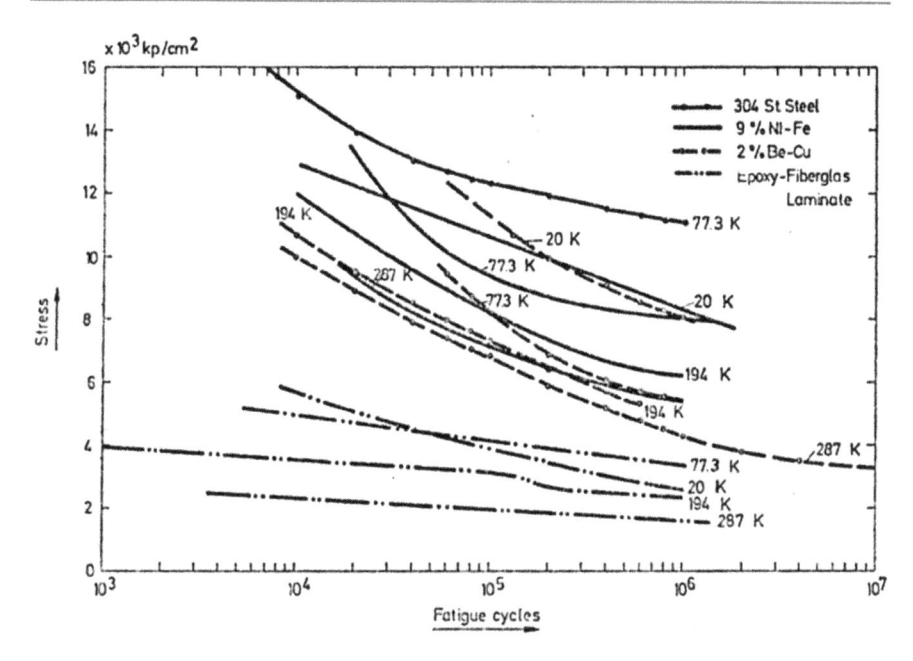

Fig. 4.2 Fatigue strength of some cryogenic materials as compared to epoxy-fiberglass laminate

Testing of Low-Pressure Liquid Cryogen Tanks

Any tank design for use in a cryogenic environment needs thorough testing. As of today, no series production of those new tanks has been made, hence no Weibull statistics can be applied to gauge the lifetime of a liquid tank.

Chiao reported on fatigue lifetime tests of pressure vessels with liner and composite structure [14]. One of the very few test results that have been made public. In Fig. 4.3, a modified test setup is shown for multiple testing of pressure vessels at room temperature with and without liners using high pressure helium of 1 MPa. The testing procedure allows sustained biaxial tension of this composite tank with different liner designs. A similar configuration can be applied for cryogenic testing of low-pressure tanks.

As mentioned in Appendix 4, there are gas storage tanks available for a wide range of high-pressure gas cylinders and tanks (tank V structure) or GEN V type, but so far only little has been published for lightweight and low-pressure range, or for ultra-long liquid storage conditions, e.g., as those required for aerospace applications, which are distinctively different, e.g., from short-term, space microgravity applications. Besides, any aircraft tank will need to work for thousands of fill cycles over an aircraft lifespan.

In cryogenics, there are therefore no shortcuts to end up with a successful design, if material property data is lacking. The same applies to the cryotankage.

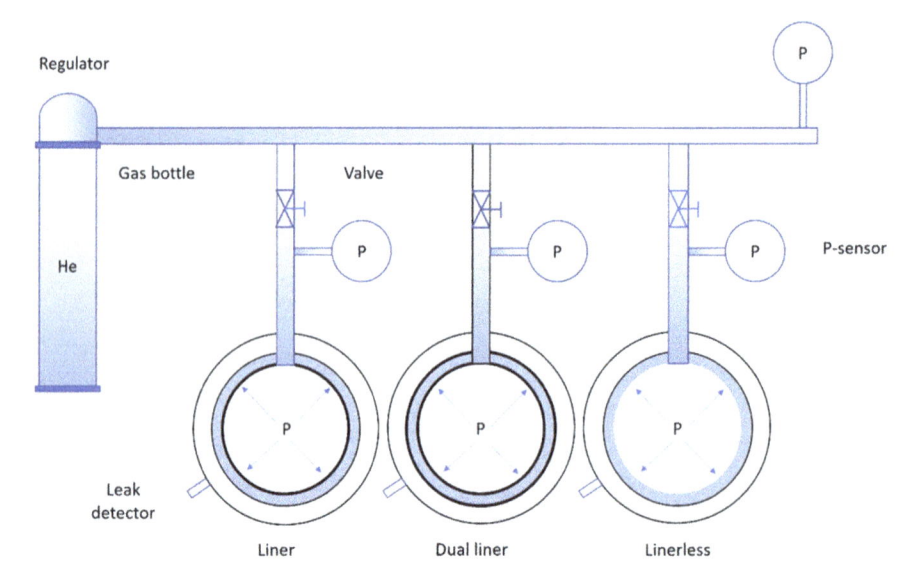

Fig. 4.3 Composite pressure vessel testing -exemplary test setup

Fig. 4.4 Copper foil on composite tank

A literature survey lists the comprehensive efforts on weight-reduced composite shell designs for liquid hydrogen tanks, e.g.:

- All CFRP liquid tank designs.
- Titanium alloy foil in-between laminates in an all CFRP design [15].
- Complete liner-less CFRP tanks [16] with a defined leak path pressurization test on CFRP liner-less tanks at liquefied N_2 temperatures.
- Liner-less liquid cryogen tanks composed of PEEK do not need a liner [11, 17].
- With liner (foil) on the inside of the CFRP tank.
- With liner or foil on the outside of a CFRP tank, e.g., copper, Fig. 4.4 [18].
- With special epoxide liners fitted to a GRP or CFRP structure.
- With special aluminum foil, 0.08 mm thick inserted into cross plies [19].

4.1 Cryotankage—Structural Thoughts

- With polymer-layered silicate nanocomposites [20].
- With liner of a certain Al thickness, and PEEK [21, 22].
- With high strength liner at inner radius and CFRP over wrap—see HEE index in Appendix 1.
- With high strength aluminum alloy liner at outer radius on CFRP over wrap.
- Same as above and with diffusion barrier (e.g., modified epoxide-based paints, aluminum, or other foil) at external radius.
- Same as above and cryogenic high strength CFRP (carbon nanotube/graphene reinforced).
- Composites with carbon nanotubes and graphene—lower tank weight fraction.
- Inner liquid tank with liner and outer vacuum shell without reinforcement structure or liner.
- Tank insulation used as a structural tank wall element [23].

Hydrogen Permeability through Liners

As discussed in Chap. 3 on Cryogenic liquid hydrogen tank design aspects, if a liner is mandatory for a certain liquid hydrogen tank configuration, the liner then needs to be thick enough for preventing hydrogen from diffusing through it and into the composite shell or from even permeating further into the vacuum space surrounding the inner tank, as shown in Chap. 3, Fig. 3.9, and Fig. 3.10). Any chosen liner further needs to comply with any hydrogen material embrittlement (see Appendix 1) for materials at cryogenic temperatures, and has to bear mechanical stresses as well as temperature transients, especially considering lifetime chill down (fueling/defueling) cycles.

A liner must be made such that it can be assembled and welded together, etc. or using other means, especially to connect to penetrations. Welding a liner seam is not always simple. Figure 4.5 shows a seam failure of an aluminum liner on a composite pressure vessel [24]. Attempts using Mylar foil as a liner material have failed due to the differential contraction/expansion of Mylar and GRP.

Fig. 4.5 Seam liner fail

Further research on how hydrogen affects composites has been analyzed by Geiss [25].

Metallic Liners

As compared to CcH2 (cryocompressed liquid hydrogen tanks operating at, e.g., 700 bar, see Chap. 2 Fig. 2.2) the liner thickness of low-pressure liquid hydrogen tanks is primarily defined by the permeability and intrinsic fatigue properties over the lifetime of the tank. For those low-pressure tanks however, metallic liners can also work as a primary shell of the liquid tank if the CTE mismatch between liner and reinforcing composite shell is acceptable or optimized with respect to repeated chill-down and warm-up cycles.

Still, for aircraft weight reduction this liner should be made as thin as possible, even when using light-weight metals, e.g., aluminum. In this case the minimum liner thickness is primarily defined by the permeability of hydrogen through the liner but keeping the use of a secondary liner requirement in mind.

More recently, metallic liners with respect to hydrogen permeation have been researched well by Schultheiß [18]. Table 4.4 compares various, thick, different metals, and metallic combinations with some of the key findings. Electroplating a thin layer of nickel however was successful due to the ductility of nickel during temperature cycles.

The table shows metal liners on the external surface of a cylindrical tank (substrate) to ensure any shell-permeating hydrogen does not diffuse into the vacuum space.

Schultheiß [18] narrowed down several possible liner solutions and gave their permeation properties. CFRP substrate is given with thickness l_s = 1.8 mm with P_s = 10^{-16} mol/ms Pa. Q_p is the permeation gas flow through the CFRP inner tank.

Remarkably, a 25 μm Cu foil adhered on CFRP and three welded, 1 mm thick Al sheets did not result in any measurable permeation. At 20 K, no reasonable permeation through the metal liner is expected. The thinner the foil, the higher the chance that those liners will show surface material defects.

Composite Liners

For further weight reduction and ease of manufacture, a growing interest in liner-less tank designs or additional, modified composite liners has evolved. There is only little data available on hydrogen permeability through those composite liners. It seems some prepregs are promising, however any solution needs to be thoroughly

Table 4.4 Liner results

Liner	l_1 (μm)	P_1 (RT) mol/ms Pa	Q_p (RT) mbar l/s
Copper foil	25	< 1.87 10^{-22}	< 1.75 10^{-07}
Aluminum foil	30	< 1.83 10^{-22}	< 1.43 10^{-07}
Copper coating	50	< 8.00 10^{-22}	< 3.74 10^{-07}
Copper coating	100	< 8.00 10^{-22}	< 1.87 10^{-07}
Aluminum sheet (Al5083)	1000	< 1.71 10^{-22}	< 4.00 10^{-08}

4.1 Cryotankage—Structural Thoughts

Table 4.5 Hydrogen permeability of epoxy resin samples

Sample	P_{H2} Permeability Coefficient cm³ STP cm cm⁻²· s⁻¹ cm Hg⁻¹	Barrer unit
Epoxy resin	1.820 10⁻¹¹	0.182
Epoxy resin + graphite (5% vol.)	2.350 10⁻¹¹	0.235
Epoxy resin + oxidizer (5% vol.)	3.220 10⁻¹¹	0.322
Epoxy resin + fly ash (5% vol.)	1.770 10⁻¹¹	0.177
Epoxy resin + fly ash (30% vol.)	1.774 10⁻¹¹	0.177

Table 4.6 Permeability through GRP tubes

Type	Air	Helium	Hydrogen
Pre-impregnated fabric tube	1.5 x 10⁻⁹	4.1 x 10⁻⁸	3.1 x 10⁻⁸
After 10 cycles in liquid nitrogen	1.5 x 10⁻⁹	4.0 x 10⁻⁸	3.3 x 10⁻⁸
Vacuum-impregnated fabric tube	9.9 x 10⁻¹⁰	1.0 x 10⁻⁷	6.0 x 10⁻⁸
After 10 cycles in liquid nitrogen	8.1 x 10⁻¹⁰	8.5 x 10⁻⁸	5.2 x 10⁻⁸

tested and verified for its cryogenic suitability [26]. Table 4.5 shows some modified epoxy resin samples. As with all composites, any change in the resin composition can change one or more of its cryogenic physical and mechanical properties that may then render the composite liner incompatible. Table 4.5 highlights the effect of additives to the resin on hydrogen permeability (for RT).

As mentioned, a liner might be a suitable candidate at room temperature, but may develop hairline or microcracks during cooldown, opening leakage paths into a vacuum space with non-integral tanks, or even worse, leaks into the fuselage in the case of integral tanks (see also Fig. 3.9 in Chap. 3).

Besides any hydrogen moving into hairline cracks within the epoxy composite layer, there is still another effect to consider, that is mainly related to liquid hydrogen temperature, namely the fatigue of epoxies (see also Section "Tank wall-Composites") that may particularly suffer at 20 K, as compared to an operating temperature of 77 K. Evans repeatedly stresses the need for further research on long-term effects on the mechanical properties of carbon-fiber reinforced polymers at 20 K [11].

Recently, Yokozeki et al. [27] evaluated the gas permeability through damaged CFRP laminates for use in a cryogenic tank, with the conclusion that damage simulation at crack intersections and the quantitative evaluation of gas leakage must be investigated further for the prediction of gas leakage and the permeability-based design of composite tank structures.

The amount of measurement and control technology required for sniffing hydrogen and detecting leaks for the latter tank types should not be underestimated (latest Hindenburg results [28]).

Hydrogen Permeability through Composite Tubes

Table 4.6 shows some typical measured, permeability values (mol / m s bar) for liner-less GRP tubes at room temperature and after thermal cycling down to liquid nitrogen temperatures that may mislead to the conclusion there is no cryogenic effect on permeability.

Evans and Reed [29] thoroughly analyzed the hydrogen permeability of composites from pre-impregnated fabrics of a 2 mm thickness at room temperature, concluding that indeed voids in the composite play a very important role in the gas passage through a resin-based composite panel, suggesting further work needs to be done in this field.

A further, even more detailed study by Bechel [30] reveals some key design parameters for composite tanks and for 3 carbon composites, namely IM7/977-2, IM7/977-3. and IM7/5250-4 with a higher curing temperature and correspondingly higher maximum service temperature.

Figure 4.6a shows the permeability test results as a function of the thermal cycle profile, (b) as a function of the lay-up, (c) as a function of chosen material, whereas (d) shows the permeability as a function of minimum crack-density.

Tests have also been executed for those composites from LN_2 temperatures up to room temperature (RT) and higher, up to 177 °C.

One conclusion is that the choice of lay-up influenced the permeability to a great extent. Samples with $[0/90]_{2S}$ and $[0/90/45/-45]_S$ lay-ups resisted leakage at least 100 cycles more than the $[0/45/-45/90]_S$ samples.

Most likely we may have to assume an increasing permeability at liquid hydrogen temperatures of around 23 K (see Fig. 4.6) not only because of cycling and

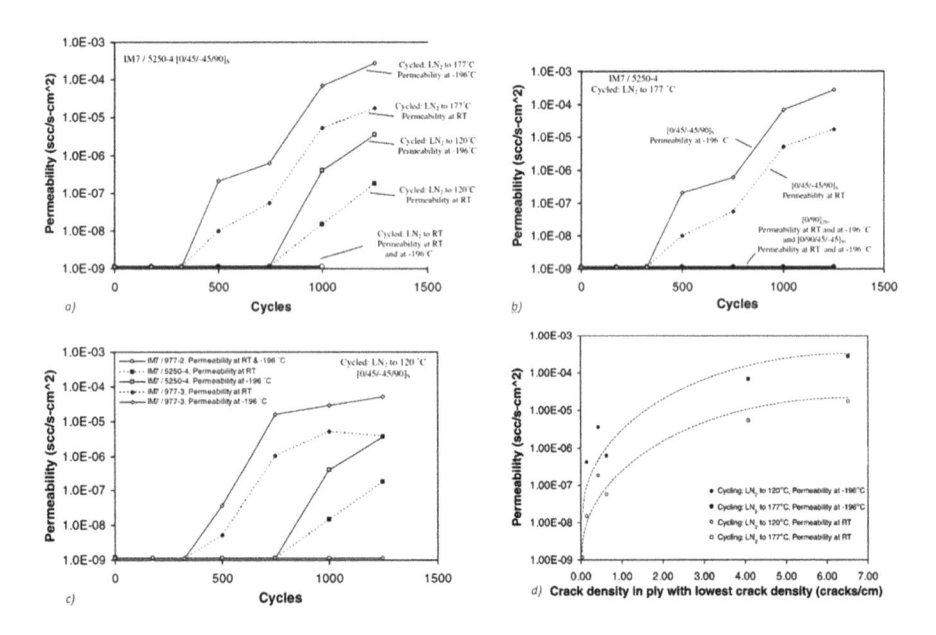

Fig. 4.6 Limiting the permeability of composites for cryogenic application (**a**–**d**)

internal pressure stresses but also due to an LH_2 tank being subject to cycling g-loads and vibrational effects on its shell.

Recently, Grogan [31] presented and analyzed an optimized tank lay-up using a numerical method to ensure satisfying resistance to microcrack formation and fuel leakage through the tank walls under operating loads. Further information is needed to understand how microcracking at cryogenic temperatures can be avoided.

A good review of the current research status on composites at cryogenic temperatures is given by Hohe et al. [32].

Resins for Composite Tanks—Physical Properties

Case Study

Design a liquid hydrogen tank with an aluminum or copper liner that matches the shrinkage of the structural composite during cooldown. Avoid stress concentrations in the tank shell. Select a filled resin with a reasonable Roving material and start with the winding process of a scaled-down tubular container.

Introduction

The design of a composite tank requires detailed knowledge of mechanical, thermo-mechanical, and electrical properties of the material used. Composites are designed for a specific cryogenic use. Based on those specifications we may have to decide to use either unfilled or filled resins together with a reinforced material, for example, glass or carbon fiber. The goal is to develop a composite matrix that falls into the desired, specific design goal limits knowing that the change of one parameter may have a detrimental effect on either thermal or mechanical performance. In the following, some guidelines are given on how to choose the composite matrix parameters.

Once a glass or carbon fiber reinforced epoxy resin has been chosen, thorough testing down to cryogenic temperature is mandatory. Data taken from literature at best give you only an estimate on what to expect. In the following, some guidelines are given, inherited from an extensive use of metals together with composites as, for example, used in superconducting magnet technology. For a recent review of superconducting coil impregnation methods and materials see also [33].

The following section allows for preselecting resin and fiber composites for a composite liquid tank, preferably for use with liquid hydrogen down to 20 K or lower. As an example, data is given for early Ciba Geigy resins. Although those resins are no longer available, adequate resin qualities can be found.

Depending on the resin composition, material properties of the resin composite can vary as shown in Table 4.7:

With respect to flexibility mainly semi-flexible resin compositions are in use. The flexible one mentioned in the table is almost rubber like and shows a larger thermal contraction down to 4 K. Semi-flexible resins generally do not differ much in their thermal contraction.

Table 4.7 Resin compositions after [34, 35]

Composition	Parts per weight	Chemical classification	Flexibility at RT	Young's modulus MN/m^2	Fracture stress MN/m^2	Fracture strain %	Poisson's ratio
X 183 / 2476 (LMB 234) HY 905	100 130	Glycidyl ester Anhydride	Rigid	7650	142	1.9	0.37
CY 221 HY 979	100 30	Aliphatic Aromatic Amine	Semi-flexible	7300	179	2.1	0.37
CY 221 HY 956	100 20	Aliphatic Aromatic Amine	Flexible	7350	–	–	0.36
MY 740 Jeffamin D 230	100 44	–	Semi-flexible	8170	200	2.4	0.37

4.1 Cryotankage—Structural Thoughts

Unfilled and Filled Common Resins

Thermal Contraction

Figure 4.7a shows the Integral thermal contraction of typical unfilled resins [34], and (b) the normalized integral thermal shrinkage as a function of filling factors from 293 to 4.2 K [35].

Reinforced Resins

To achieve high mechanical strength, resins need to be reinforced with either glass, carbon, or other fibers. The thermal shrinkage of an Owens Corning Fiberglass, embedded in an early Ciba Geigy resin with two layers at different fiber angles, is given in Fig. 4.8a and three layers in Fig. 4.8b [34].

Practical Result

A tube of 92 mm inner diameter with a 5 mm wall thickness was wound using multilayer winding. Tubes were produced with Araldite X 183/2476 (LMB 234), hardener HY 905 with DX 062 and 100:130:2 parts per weight. Use a Roving Type OCF 859–885 tex of Owens Corning Fiberglass.

Due to winding-related techniques pipes with winding angles of ±45° up to ±90° were produced. The glass fiber content was held constant at 70 vol. % [37, 38].

Case Solution Select resin with a reinforced cloth winding angle that matches the liner to ensure shear forces acting at the liner/composite interface are negligible.

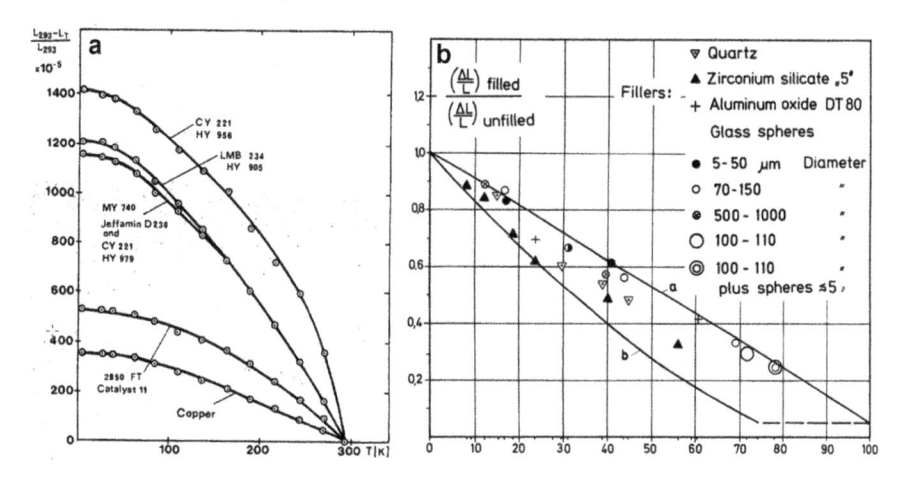

Fig. 4.7 Integral thermal contraction of unfilled and filled (2850 FT) resins relative to copper [34] (**a**) filled/unfilled resins vs fill factor [35] (**b**)

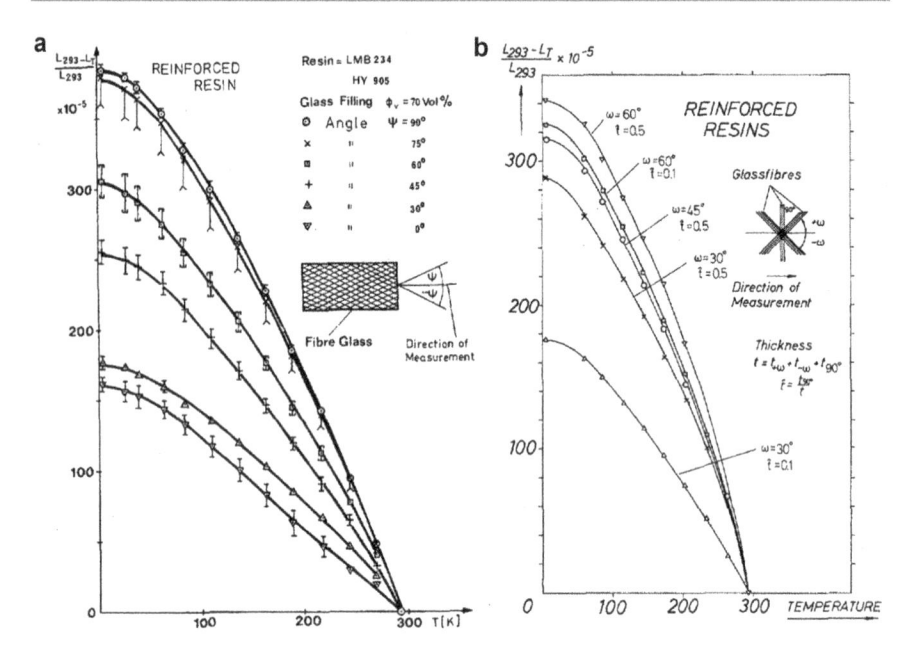

Fig. 4.8 Integral thermal contraction versus temperature of reinforced resins **a**) two-layered and **b**) three-layered

The theory as described by Hartwig et al. [37] agrees very well with experimental results so that predictions can be made for a preliminary design of a composite tank. Also, one notices that any change in the epoxy resin hardly changed the total shrinkage [38].

When using carbon fibers however, note that shrinkage can be positive depending on the winding angle and reaches a positive maximum around 20 K for 25 and 35°, as shown in Fig. 4.9b [38].

Other parameters to observe when designing a composite liquid cryogen tank are the Young's modulus given below that shows the impact of fillers (for example, quartz) as compared to unfilled resins and base material fatigue.

Young's Modulus

A further important parameter to look at is the tank elasticity. Figure 4.10 shows data of a typical epoxy used in the past, namely Cy 221 with hardener Hy979 (Ciba Geigy) [39].

In case of high stress on the resin especially at penetrations, cracks and microcracks will occur. As known within the magnet superconductivity community the work of fracture for resins is an important parameter to keep in mind. Typical values for that are in the range of 300 J/m^2 at 4 K to 420 J/m^2 at 77 K [34].

When running modeling tasks for carbon and Kevlar fiber be aware of the unusual negative Poisson's ratio as compared to glass fibers.

4.1 Cryotankage—Structural Thoughts

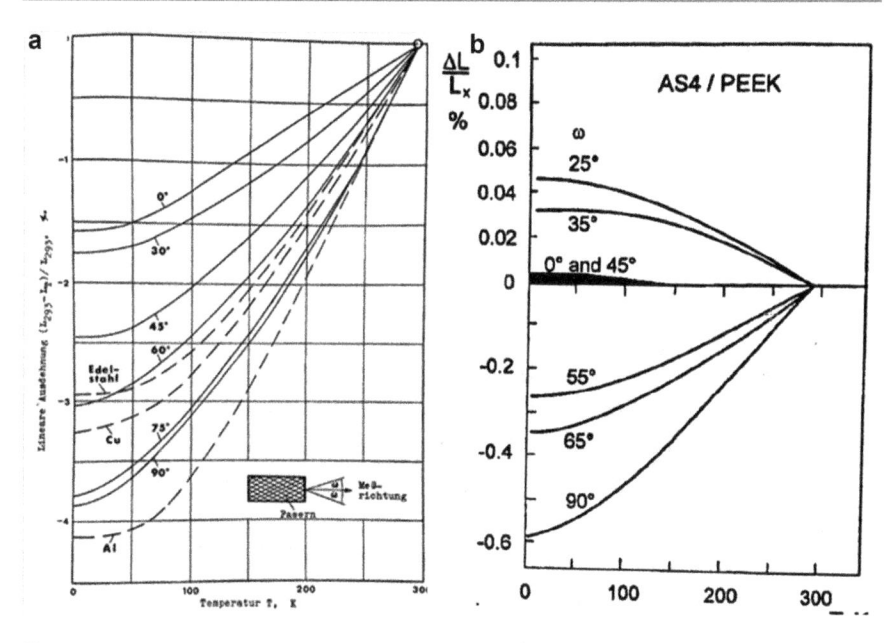

Fig. 4.9 (a) Relative length change in ‰ compared with common metals [34], (b) relative length change in % of an AS4 Carbon (Herkules)/PEEK matrix with 60% vol fiber content at different winding angles [38]

Fatigue

Epoxy resins are generally brittle at cryogenic temperatures. It is therefore of utmost importance on how long a particular material can be used dynamically, for example, in fuel tanks that experience thousands of thermal cycles during liquid fill and withdrawals. Figure 4.11 gives some insight into what to expect. For that the Wöhler line was measured at 77 K for a common epoxy on round samples. The ultimate number of cycles is indicated at the point where there is an asymptotic change in the curve arriving at the fatigue limit for this material.

Also given in Fig. 4.11 are the fatigue crack growth and propagation in typical CT specimens. This is the type of measurement that needs to be done for mechanical analysis of the resin matrix as this is the weakest element. Data is available thanks to prior "pulsed superconducting magnet" applications.

For more recent data on fatigue of epoxies, see also Lüders [36].

Besides mechanical aspects, thermal properties also matter when selecting the shell matrix.

The circumferential thermal conductivity defining the tank temperature homogeneity is an important factor that potentially impacts boil-off and can be a cause of stress concentrations reducing the mean time between failure of the tank.

As with earlier thermal expansion findings, thermal conductivity as well does not change much with temperature for unfilled resins as depicted in Figs. 4.12 and 4.13.

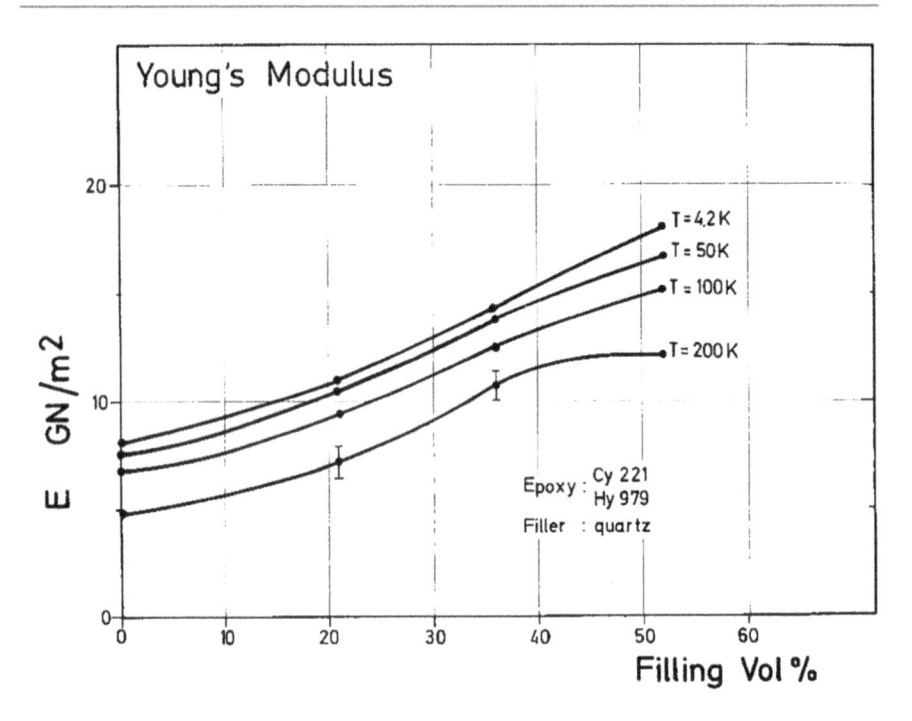

Fig. 4.10 Modulus of elasticity of filled and unfilled resins [39]

Electric Breakdown
For typical resin plate of CY221 HY979, the typical breakdown voltage for unfilled resins down to 77 K is approx. 25 kV as compared to 18 kV at room temperature [34].

Figure 4.12 shows the influence of copper powder on the thermal conductivity of an epoxy resin CY221 with HY979 at a 10:3 weight mix ratio with 27 vol. % of copper, with varying copper particle size. Note that below 20 K one can clearly see the Kapitza effect starting to reduce the thermal conductivity [40].

Figure 4.13 gives further details on general thermal conductivities [41] for unfilled and reinforced epoxies at different fiber orientations. Figure 4.14 defines the matrix fiber orientation.

Radcliffe and Rosenberg measured the thermal conductivity of glass and carbon fibers with a 20 µm diameter confirming the predictability of this property for carbon fiber/epoxy composites above 10 K [42].

Specific Heat of G10 for Sample Tank
The specific heat is important to arrive at a first estimate on how much energy initially has to be extracted from a composite tank shell. The following example gives the enthalpy based on Fig. 4.15 with data taken from Cryocomp Version 5.2 [43]. Maximum internal energy removal would be the initial fill with 140 kJ/kg. For refills with an assumed max temperature rise to 80 K would be around 10 kJ/kg.

4.1 Cryotankage—Structural Thoughts

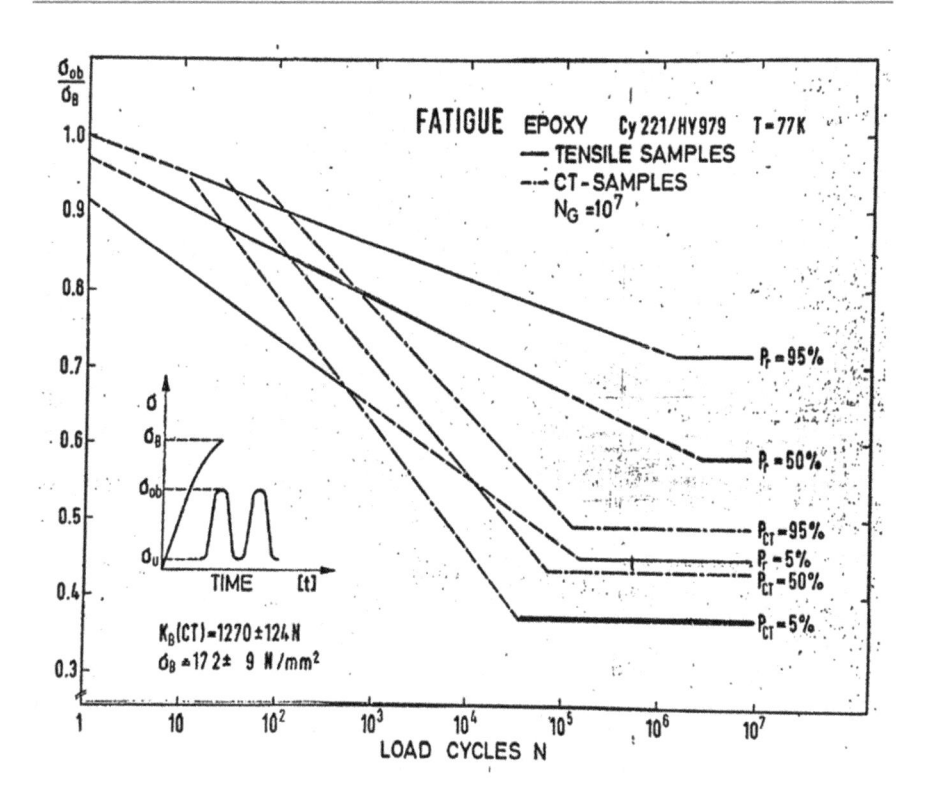

Fig. 4.11 Fatigue of epoxy resins at 77 K [39]

Figure 4.16 explains the volumetric definition for this layered tank with some volumes for given tank designs in Table 4.8.

Example When choosing inner tank 1 design type 3 as shown in Fig. 4.16, but without a liner and a composite wall thickness of 4 mm, this would result in a composite solid volume of 105.43 liters for a given fill volume of 9730 liters, assuming a material density of 1600 kg/m^3 and correspondingly we arrive at 24 MJ when cooling down to 20 K from room temperature. This corresponds to an equivalent copper mass for cooldown of approx. 300 kg. For refills, assuming the temperature drop from 80 K to 20 K this value reduced to 1.68 MJ, corresponding to 2 kg of copper only for this lower temperature range.

Chill Down Speed with Composite Tanks
For quick fills and liquid withdrawals some first estimates can be made. A detailed FEM analysis however needs to be conducted to include all heat transfer mechanisms to get the complete picture.

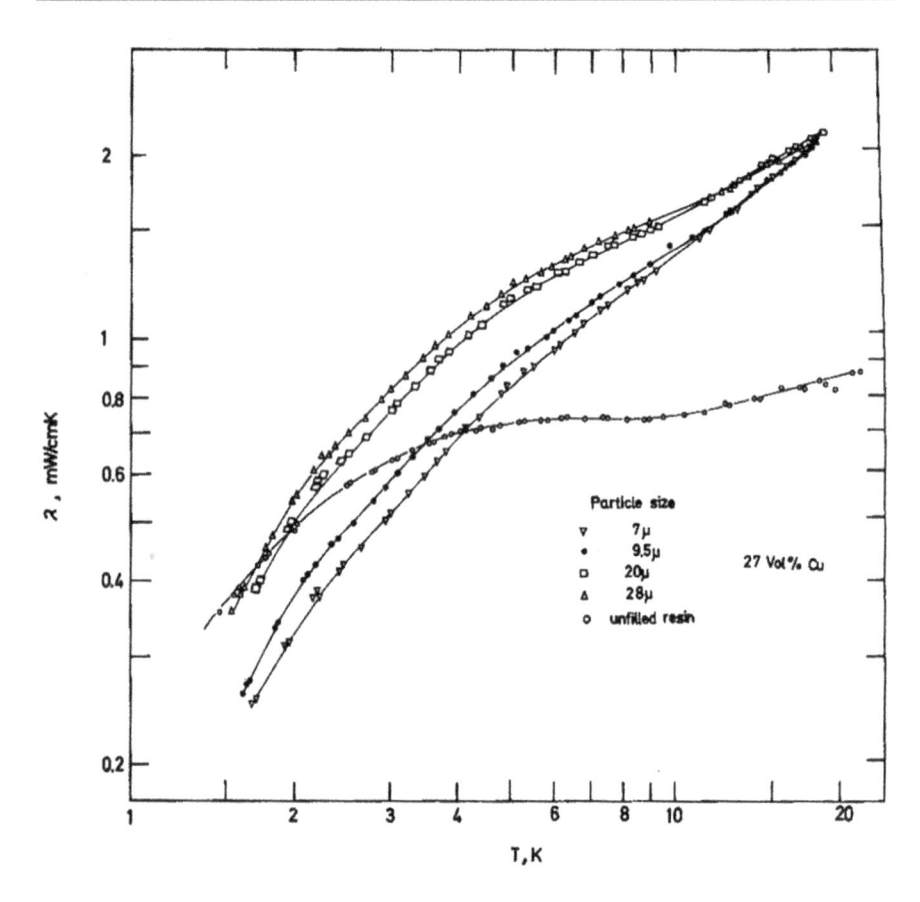

Fig. 4.12 Thermal conductivity of a copper filled vs unfilled resin relating to particle size [40]

For the tank design knowledge on the thermal response time is helpful. A simplified estimate takes the thermal diffusion time constant $\tau_{diff} = L^2/(4 * \alpha)$ with L as the wall thickness of the tank, and α as the thermal diffusivity. For G10, a typical fiber glass/epoxy bond, the thermal diffusivity is in the range of approx. $3 \cdot 10^{-7}$ to $1 \cdot 10^{-6}$ m^2/s at 20 K [38].

For a shell thickness of 4 mm, the typical thermal response time based on the heat penetration formula is maximal approx. 4 to maximum approx. 13 seconds, that in most cases is acceptable. Note that from room temperature down to 20 K the heat capacity plunges orders of magnitude thus generally drastically reducing thermal response times.

Permeation Through a Composite Tank Wall
Reinforced epoxy resin naturally acts as a permeation barrier. Interestingly, permeations of helium and hydrogen are so similar that we can tap into previous research on helium permeation through composites as shown in Fig. 4.17. From the atomic

4.1 Cryotankage—Structural Thoughts

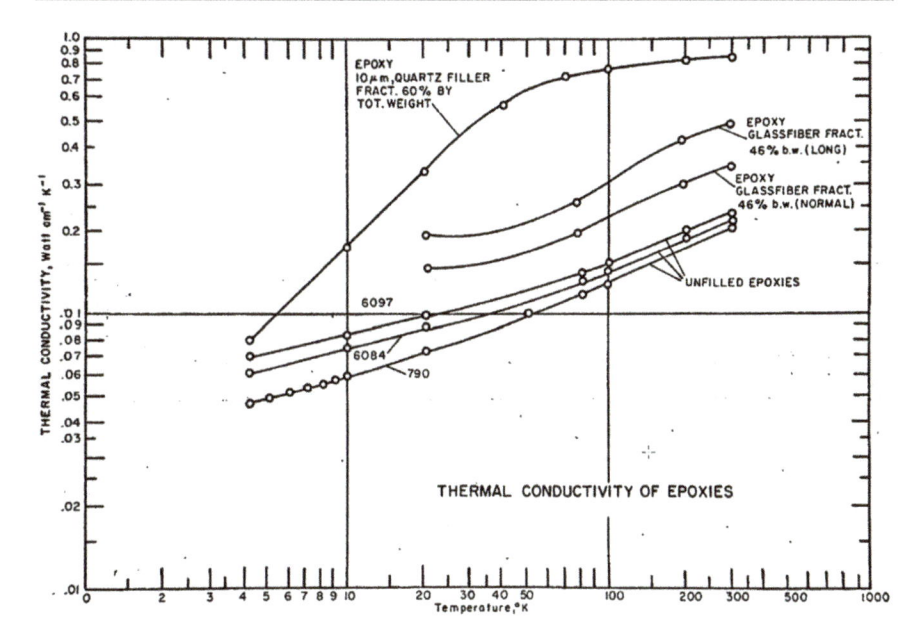

Fig. 4.13 Thermal conductivity vs temperature of filled, unfilled, and reinforced epoxies [41]

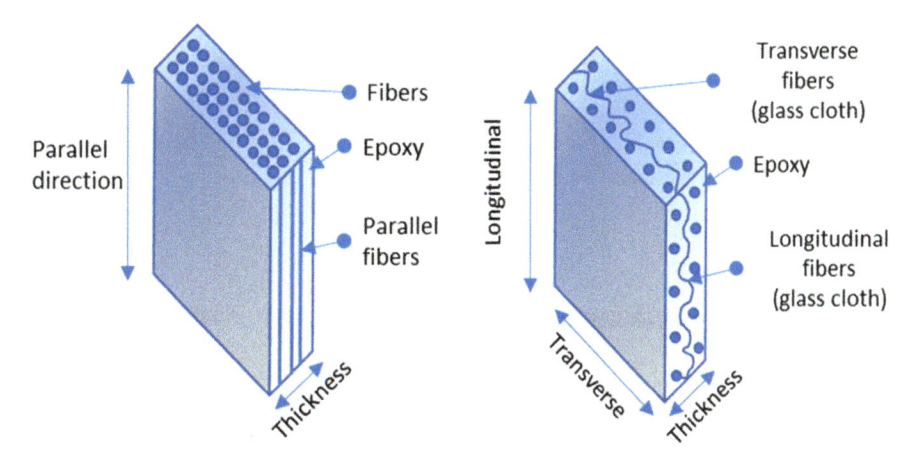

Fig. 4.14 Definition of uniaxial and glass cloth epoxy matrix fiber orientation

radius point of view, hydrogen is specified as 53 pm (picometer) and helium as 31 pm. Hydrogen should therefore be less inclined to go through a barrier than helium. The following references seem to confirm that.

For the barrier, Ep of glass/Ep crossplies with a fiber volume content of approx. 56%, the measured value for Ep is 20 kJ/mol with helium, 26 kJ/mol with hydrogen, and 50 kJ/mol with methane [38]. Note that this value strongly depends on the fiber volume.

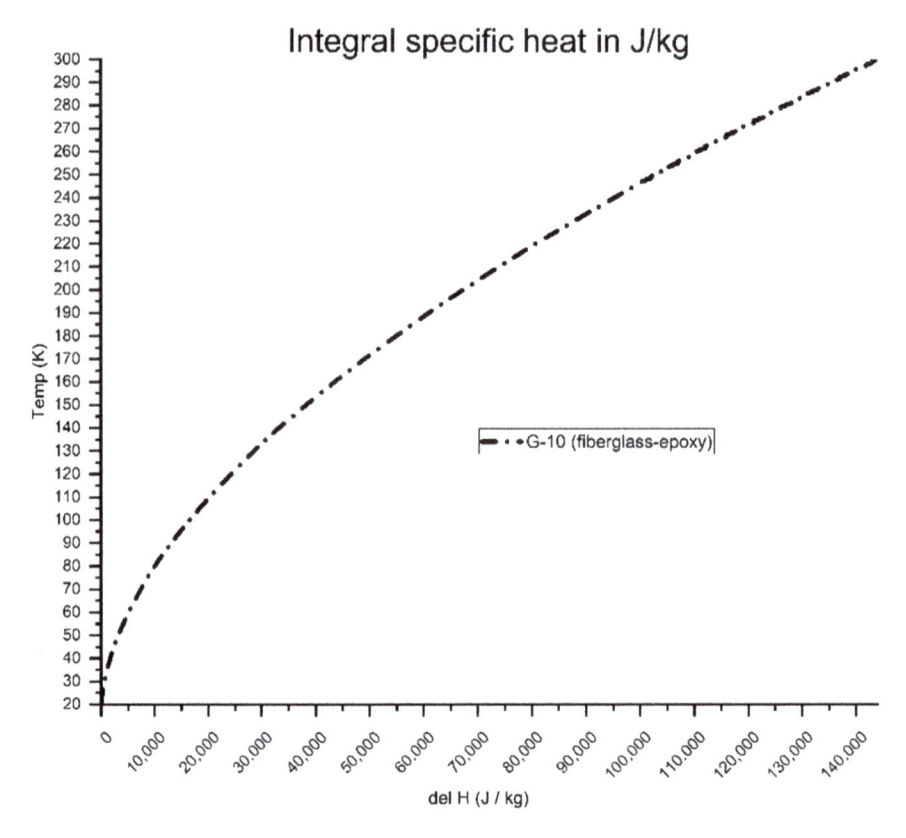

Fig. 4.15 G10 integral enthalpy for initial and subsequent fills

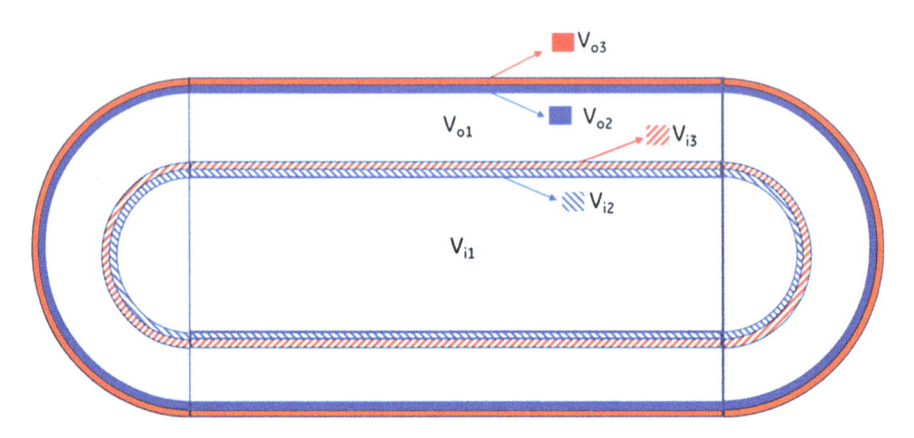

Fig. 4.16 Nomenclature for material tank volumes

4.1 Cryotankage—Structural Thoughts

Table 4.8 Example tank designs

Tank type	Inner tank (IT) Enclosed volume of the inner tank V_{i1} (liters)	Solid volume of the inner Al layer of the inner tank V_{i2} (liters)	Solid volume of the outer CFRP layer of the inner tank V_{i3} (liters)	Outer tank (OT) Enclosed volume of the outer tank (vacuum space between the OT and IT) V_{o1} (liters)	Solid volume of the inner Al layer of the outer tank V_{o2} (liters)	Solid volume of the outer CFRP layer of the outer tank V_{o3} (liters)
Tank 1 Type 1	9840.10	26.42	52.97	842.62	27.86	55.85
Tank 1 Type 2	9787.38	52.71	52.88	841.31	55.68	55.85
Tank 1 Type 3	9734.84	26.25	105.43	839.99	27.78	111.54
Tank 2 Type 1	7379.86	27.18	54.52	873.12	29.05	58.26
Tank 2 Type 2	7325.67	54.19	54.41	871.41	58.04	58.26
Tank 2 Type 3	7271.70	26.96	108.38	869.70	28.94	116.30

OT outer tank, *IT* inner tank

What we know so far is that below 200 K typical fiber composites are impermeable and in the permeability range P of 10^{-14} m²/s, whereas P is the permeation coefficient that depends on the temperature using an Arrhenius-type equation:

$$P = P_0 \exp\left(\frac{Ep}{RT}\right) \tag{4.1}$$

with R as the gas constant, P_0 as a constant factor (known permeability at a temperature T), and Ep for the barrier against permeation. Permeation through the matrix is the dominant factor here, given that fibers are nearly impermeable.

In summary, fatigue loading as discussed earlier seems to increase P by less than 30%, concluding that thermal cycling seems to have less of an effect on permeation.

The Liner Question Revisited

Efforts have been made and are underway to use metallic liners or organic liners to further reduce permeability to ensure residual gas pressure in vacuum is under control and under no circumstances should hydrogen permeate through to a vacuum wall as discussed. It is known for example that aluminum foils delaminate in composites under thermal and fatigue loading, unless some special surface treatment is used. Hartwig and Humpenöder report on good adhesion properties for tin foils, whereas a 100 μm tin foil can be laminated with the fiber composite [44]. The key

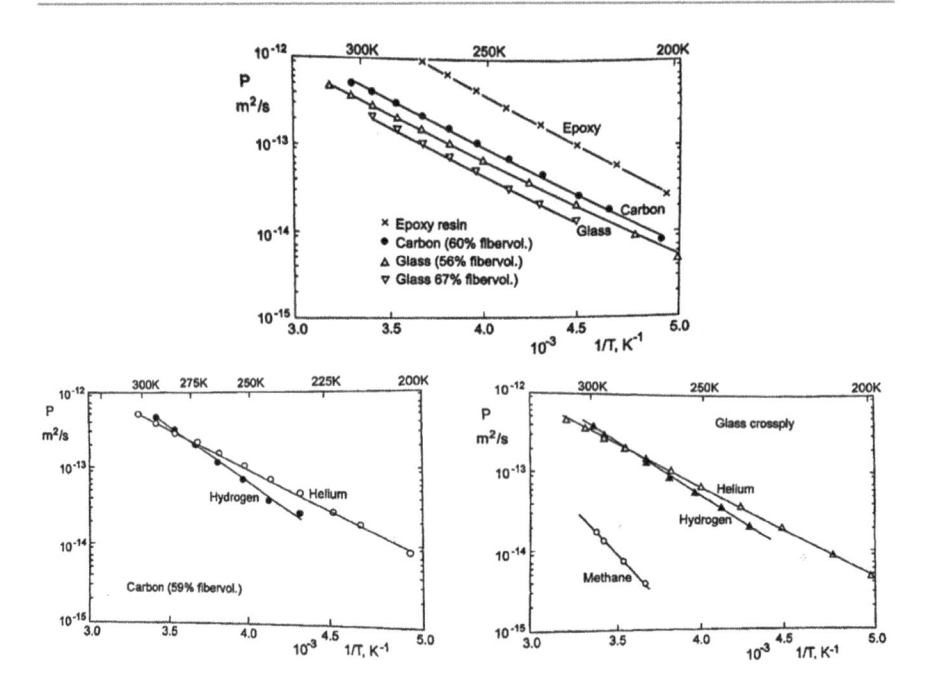

Fig. 4.17 Permeabilities P of glass crossplies with different gases for glass and carbon fibers [38]

point is to define the special surface treatment process and understand the weld seam processing.

As Humpenöder concludes on tanks for propulsion systems using hydrogen and methane, gas permeability of fiber glass and carbon fiber reinforced epoxy resins cannot be neglected at room temperature [44].

Okada and Nishijima report on gas permeation on an FRP cryostat and conclude leaks arising from bubbles enclosed in the composite matrix during the manufacturing process, that means either VPI or other filament winding methods are required to eliminate those [45]. Evans et al. give permeation results for composites and different resins and the need for further established technologies for long-term use of composites in a cryogenic environment [46]. Haberstroh et al. are now preparing to set up a permeation test facility as the need for further data is increasing and deeper insight into the permeation process is required [47].

Epoxy Piping

The General Electric Research and Development Center researched the use of cryogenic piping for composite AC transmission cables testing the flexural strength of E-glass epoxy composites at 77 K on 20 cm diameter pipes. Particular emphasis was placed on joining technologies and in particular glass fiber reinforced epoxies (GRE) pipe-to-metal flanged joints, e.g., GRE to aluminum bonds, while reporting high lap-shear strengths as shown in Fig. 4.18 [48]:

4.1 Cryotankage—Structural Thoughts

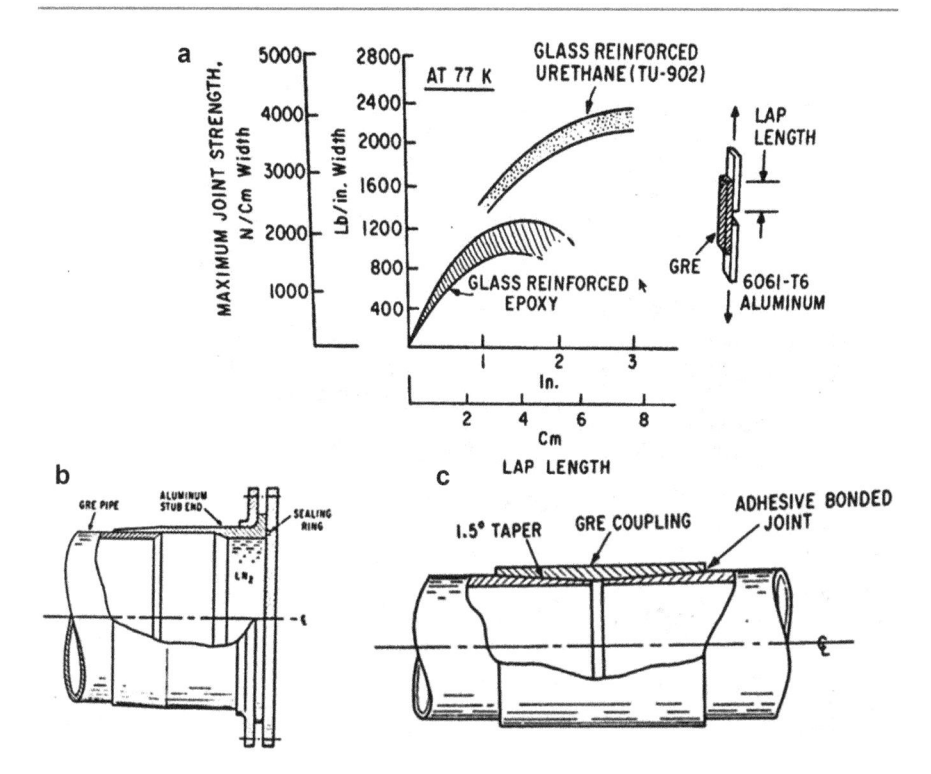

Fig. 4.18 (a) Composite piping designs with scarfed joint, (b) adhesive pipe joint, (c) lap shear test results

Stand-alone glass fibers without resin backing fall into the rare category of cryogenic materials that maintain their elasticity down to 77 K and lower and are therefore suitable for glass fiber-based sensor technologies (see also *"Supercables" technology*, for further references), e.g., for measuring cryogenic deflections or elongations, when attached to a base plate, or directly embedded in the composite wall enabling data readout from room temperature down to cryogenic temperatures [49].

There is a vast pool of data available on many reinforced epoxies, mainly thanks to designs for structural supports. A good topical overview is given by Reed and Golda [50]. However, the findings need to be verified when applying them to pressure vessel technology and other tank designs, since those specifications may differ from conventional cryogenic suspension, struts, and supports.

Summary

Good progress has been made and is underway to create composite liquid hydrogen tanks with low mass fraction. Success can be expected if we include all neighboring modalities to get to a much higher TRL level, including space technology, and the automotive industry on composite liquid hydrogen tanks.

We need two materials, one material that ensures liquid hydrogen does not creep into micro pathways of any composite structure or diffuses through any composite matrix (permeates). Most of the systems looked at were designed for composite matrices only that allows one to wind the pressure vessel into different shapes and create the maximum liquid storage conditions. This is challenging since it does not safeguard one against the aforementioned hydrogen creep into a matrix. Some designs on the other hand assume a stainless-steel foil and other foils applied on the internal surface of the tank. Another route is to combine all the aforementioned and choose a high-strength material that at the same time acts as an efficient hydrogen barrier but also gives structural ability. In the following case therefore, many calculations were run to obtain the minimum weight fraction for the inner and outer tank while safeguarding the tank against pressure-bearing capability and buckling.

As can be seen, high pressure tanks need excessive thickness of aluminum and additional carbon fiber braided structure to bear the internal pressure.

CHEETA Tanks

In the following pages we simulated both methods, carbon fiber on the outside with aluminum on the inside and carbon fiber on the inside and aluminum on the outside for a low/pressure tank.

To our surprise the first approach actually gave even better results with respect to buckling. Now we just had to combine a high-strength material with a high-performance carbon fiber overwrap to minimize the weight fraction. The design details are described in the following sections.

For the pressure-stabilizing internal layer we chose a material that is considered acceptable for liquid hydrogen use and is well-known in cryogenics, meaning, many test data on mechanical and thermal properties have been reported.

It must be said that this may not be the only choice, but other alloys will be suitable likewise in the near future. The same applies for the carbon fiber overwrap. It is perfectly thinkable that the carbon fiber matrix will be tailored to the exact requirements of a hydrogen tank. Several figures of merit may exist that allow one to tailor mechanical properties to the liquid hydrogen tank specifications. One example is the use of epoxy fillers to enhance the elastic modulus or the ultimate tensile strength of enhanced fatigue behavior. Several attempts have been made in the literature to increase the thermal and mechanical properties of the epoxy matrix. Whether those would be applicable to cryogenics remains to be seen but those are almost certainly a good choice for the outer vacuum vessel.

A good overview on the properties of cryogenic and low temperature composite materials is given by Sápi and Butler [51] and at great length in references [52–93].

Conformal tanks, that is tanks that are adapted to the airframe structure can accommodate at least 20% higher fuel storage capacity. However, many of these tanks are preferably made liner less and even manufacturing those tanks is not as simple yet but may well deliver the lowest possible weight fraction.

As has been shown in here and in Appendix 1, the structural material choice is important as summarized in Fig. 3.2, Chap. 3 "Liquid hydrogen tank designs aspects."

4.1.3 Optimized Liquid Tank Structure

A tank design requires consideration of tank faults and their possible operating conditions. These requirements need to be balanced with respect to the goal of achieving the optimal mass tank weight fraction as will be shown in the following sections. Based on the agreement of adhering to the fault conditions we can now present an exemplary design that evolves from a detailed stress analysis of 3 sample tanks.

It is obvious that composite tanks help us for a lightweight design. For a first ballpark figure we can design the composite tank according to AD 2000-Merkblatt N1, "Pressure vessels in glass-fiber reinforced thermosetting plastics," as issued on May first, 2018 [94] and the respective pressure equipment directive guidelines.

For a more detailed look on design and fault conditions, we need to work with an FEM program, e.g., ANSYS, or others.

4.2 Cryotankage—Design for Tank Faults/ Operating Conditions

As mentioned, a design balance needs to be achieved between tank faults and various operating conditions as well as for initial or recurring leakage acceptance tests and routine inspection. Below is a brief summary on what to consider for both of the enclosures as shown in Chap. 3 Figs. 3.9 and 3.10 [Cryotankage—Simplified design example]. This then lays the groundwork for the stress analysis following this section. We can therefore summarize the following:

1. We assume the outer vacuum case (OVC) remains evacuated and is at zero pressure.
2. Initially, the inner liquid storage tank is in a <u>warm</u> state and is being filled with liquid hydrogen from the filling station.
    (a) Vacuum break may be detected during this stage and hence will be prevented since sensors attached to and embedded in the composite structure record vacuum/cryo quality and abort the fueling process in case vacuum quality is being compromised, e.g., fuel leakage into the vacuum space.
3. Inner tank chill-down is complete and being filled/topped up with liquid hydrogen with up to 0.8–1 MPa internal pressure, maintaining an ullage space of 10–20%.
4. If during cruising, for some reason the inner tank goes subatmospheric there is no pressure difference between both pressure envelopes and safe operation is guaranteed.

(a) For the aforementioned integral or said conformal tanks without a vacuum enclosure this requires a control circuit between all tanks that ensures that if one tank goes subatmospheric any neighboring pressurized tank will compensate for this pressure loss and will maintain normal operating pressure between all tanks. This is not required for tanks with a vacuum enclosure.

5. In case there is a breach of vacuum in the outer vacuum chamber (OVC) the inner liquid tank does not experience any buckling pressure as long as there is overpressure in the liquid tank.

(a) The inner tank needs to be designed for buckling in the rare case where vacuum has been lost and the inner tank is running subatmospheric.
(b) This instance however can be mitigated and prevented by following design 4 a.
(c) Vacuum integrity can be compromised, in case of a tank rupture or a smaller tank leakage into the vacuum space.
(d) In this case, excessive boil-off occurs that needs to be vented externally (flared), although the performance of the liquid supply to the engines may not be inhibited.

Some thoughts on ensuring tank integrity with leak checking during assembly:

- Checking for cryotank leaks is possible, if OVC is at ambient pressure and liquid tank is being evacuated, integral helium test, OVC with helium atmosphere.
- Leak check possible, prior to inserting the tank into the OVC, liquid tank overpressure leak check, integral leak test.
- Leak check possible, if vacuum maintained in OVC and tank is being pressurized, sniffer tests.

The above strategy results in a maximal safe tank operation. If a further reduction in tank mass weight fraction is required, the inner liquid tank may be designed to be collapsible in case of a vacuum breach bearing in mind some initial leak checking detriments.

There is no reason to believe any vacuum failure would be a safety concern. However, the shock heat load would empty the tank in several seconds, regardless of whether there is an additional foam placed in between MLI and cold liquid helium tank.

4.2.1 Creating a Composite Pressure Vessel Shell

Before we now look closely at the stress analysis of a shell, we need to understand how composite shells are typically made for cryogenic use. There are some very small containers for cryogenic applications that require a non-magnetic environment [95], but as of today there is no such commercially made composite tank for

4.2 Cryotankage—Design for Tank Faults/Operating Conditions

cryogens in their liquid phase, like helium, hydrogen, or neon. A good first start would be to look how commercial composite tubes are made.

G10 tubes are traditionally wound, which means base materials such as glass, cotton, paper, Aramid type, or carbon fibers are used that are then being impregnated using a resin matrix (e.g., epoxy, phenol, silicone, or polyester) and by applying a laminated coating. In the following, pressure and temperature is applied while those materials are steadily wound around a slightly tapered mandrel. The tube built in this way then undergoes a curing process in a furnace. When removing the tube from the furnace, the tube is pulled down from the mandrel thanks to the taper and is then subject to a following surface treatment process. In this process, the tube is cut and honed and polished, with ends hemmed, and then undergoes a further process, in which the final dimensions of the tube are mechanically processed on a lathe, if necessary. The process is not simple since the temperature in the oven as well as in the environment can drastically influence the tolerance accuracy. It goes without saying that cryogenic temperatures affect the tolerances of the tube to metal contacts.

G10 and G11 type tubes are known for their cryogenic compatibility, in particular their enhanced compressive strength is of exceptional benefit for many applications. Those tubes consist of an epoxy resin reinforced by a glass cloth. One however needs to keep in mind that mechanical properties are highly anisotropic, which means reinforcement is more efficient in parallel than perpendicular fiber direction. For cross plying, the laminations give a good reinforcement but modulus and strength are only one third of a parallel fiber arrangement. It is known that the bond at the fiber/matrix interface is weak and large stress concentrations can occur. Any material used for GRP fuel tanks can therefore become porous.

4.2.2 CHEETA Stress Analysis Example

In this section, we present a structural assessment of liquid hydrogen tanks for aircraft applications [96]. Three different tank designs made of aluminum and composites (CFC and GRP) have been considered. Each of the tank designs consists of an outer tank, a vacuum space, and an inner liquid hydrogen tank. The lightest weight design that provides adequate factor of safety against two failure modes is being considered and evaluated for practical use. The following figure cuts through

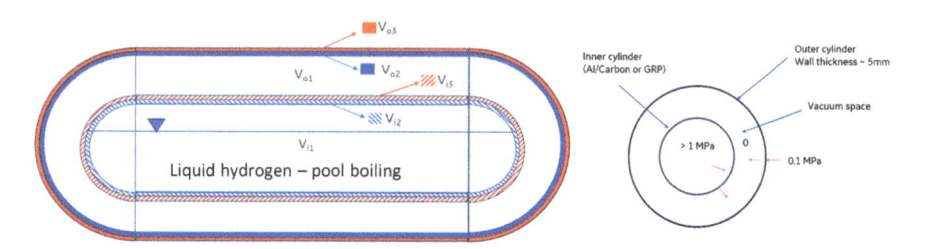

Fig. 4.19 Tank geometry—cut through—V = volume indices

Fig. 4.20 IM7–8552 3D Laminate

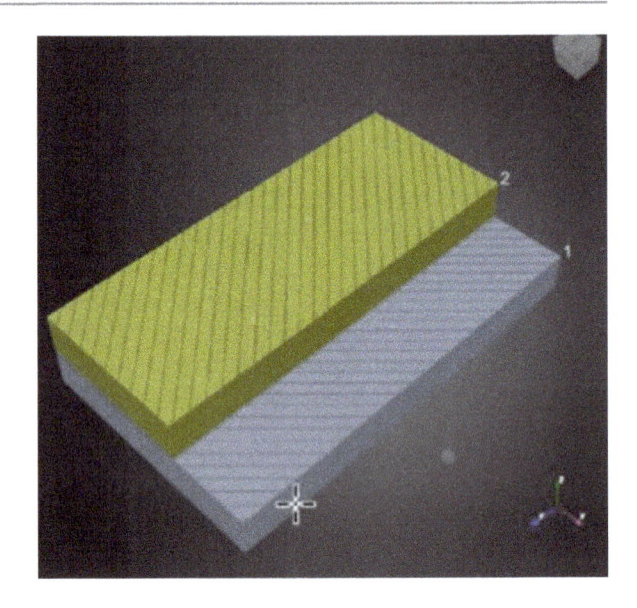

Fig. 4.21 Liner and shell build up

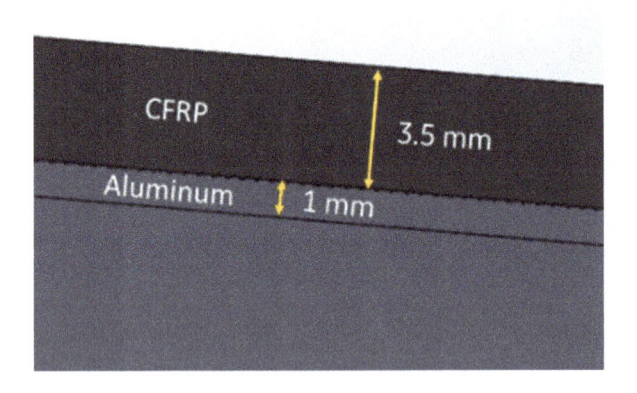

the tank and gives the pressure range applied to the outer and inner shell (see also Fig. 4.19). The inner shell is subject to internal pressure (i.e., where the emphasis is placed on cryogenic tensile strength), whereas the vacuum shell is calculated for buckling (elastic modulus sensitive).

For the structural analysis 2 aluminum liner alloys and 2 composite structures (Figs. 4.20 and 4.21) have been selected with the following properties as in Tables 4.9 and 4.10.

The laminate construction was designed and built at GE Research using the Helius® [Autodesk® Helius® Composite 2017] material and design database [97]. The Carbon Fiber composite IM7/8552 is made out of two layers. The layers consist

4.2 Cryotankage—Design for Tank Faults/Operating Conditions

Table 4.9 Yield strength

Al alloy	T (K)	Yield strength (MPa)
6061-T6	297	279.9
2219-T87	297	388
	20	501

Table 4.10 Tank-type wall composition

Type	Al (mm)	CFC (mm)
1	1	2
2	2	2
3	1	4

of carbon fibers arranged at +45° and −45° directions. After mapping cartesian to cylindrical coordinates for hoop, axial, and radial direction, the lamina properties are given in Table 4.11, with the material properties [98] as shown in Table 4.12.

Inner Tank—Internal Pressure

As an example of the structural analysis, results are given for tank 3 of type 2 (see Table 4.10 for reference).

The stress distribution and specifically the equivalent stress (von Mises stress) on the walls of the inner tank are calculated as a function of the internal pressure. The relationship between the equivalent von Mises stress and the tank internal pressure is linear as shown in Figs. 4.21, 4.22, 4.23, 4.24 and 4.25.

Figure 4.23 refers to the pressure-bearing capacity of a 2 mm aluminum liner of Al 6061 and a 2 mm CFC laminate for tank configuration 3 (tapered tank) at room temperature. Without safety factor the inner tank can bear 10 bar, with an assumed safety factor of 1.6 the pressure-bearing capacity reduces to 5.73 bar.

Figure 4.24 shows the effect of a high-strength aluminum liner with a 2 mm CFC laminate and a safety factor of 1.6 as above, which gives an increase in pressure rating of up to 7.94 bar.

In Fig. 4.25 we finally apply the advantage of the cryogenically increased strength of the alloy at 20 K to the tank with the same safety factor of 1.6. In this case the pressure rating reaches 1 MPa.

The stress in the aluminum layer approaches the yield strength of the aluminum before the stress in the CFC layer approaches its yield strength value. Therefore, the stress in the aluminum layer is used as the design criteria for the inner tank. The choice of the liner material is therefore important if high safety factors are needed.

Table 4.11 Composite properties

Property/Parameter	Symbol (units)	Value
Modulus of elasticity (X direction)	Ex (MPa)	1.37E+04
Modulus of elasticity (Y direction)	Ey (MPa)	1.70E+04
Modulus of elasticity (Z direction)	Ez (MPa)	1.70E+04
Shear modulus (XY)	Gxy (MPa)	4.32E+03
Shear modulus (XZ)	Gxz (MPa)	4.32E+03
Shear modulus (YZ)	Gyz (MPa)	3.63E+04
Poisson's ratio (XY)	Nuxy	8.13E-02
Poisson's ratio (YX)	Nuyx	1.01E-01
Poisson's ratio (XZ)	Nuxz	8.13E-02
Poisson's ratio (ZX)	Nuzx	1.01E-01
Poisson's ratio (YZ)	Nuyz	7.87E-01
Poisson's ratio (ZY)	Nuzy	7.87E-01
Density	ρ (kg/m^3)	1.59E+03
Thickness	t (mm)	2.00E+00

Table 4.12 Material properties of the composite (IM7/8552) laminate

Property	Units	IM7/5230–1	IM7/977–2	IM7/8552
0° tensile strength	MPa	2703	2690	2650
0° tensile modulus	GPa	156	165	168
Poisson's ratio	–	0.34	–	–
90° tensile strength	MPa	81	75	–
90° tensile strength	GPa	9.7	7.6	–
0° compressive strength	MPa	1737	1580	1690
0° compressive modulus	GPa	143	152	150
Short beam shear strength	MPa	119	112	128
Compression after impact (CAI) strength	MPa	176	262	234
Open-hole tensile strength	MPa	498	448	–
Open-hole compressive strength	MPa	386	310	–

Outer Tank—External Buckling Pressure

The outer vacuum chamber (OVC) requires a number of stiffeners to prevent shell buckling. For the tapered tank we assumed equidistant stiffeners. Figure 4.26 gives the result for Al6061 liner with a 2 mm CFC (IM7/8552) composite.

By fixing the dimensions of the outer surface including the diameters and the overall length between the two hemispherical end caps, different designs were evaluated by calculating the lowest critical pressure for buckling. To perform this study, another design parameter that was varied was the number of ring stiffeners (Fig. 4.27).

4.3 Minimum Tank Mass Weight Fraction

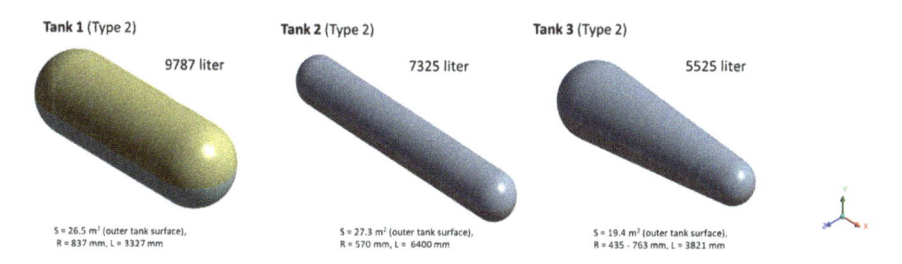

Fig. 4.22 CHEETA tank types

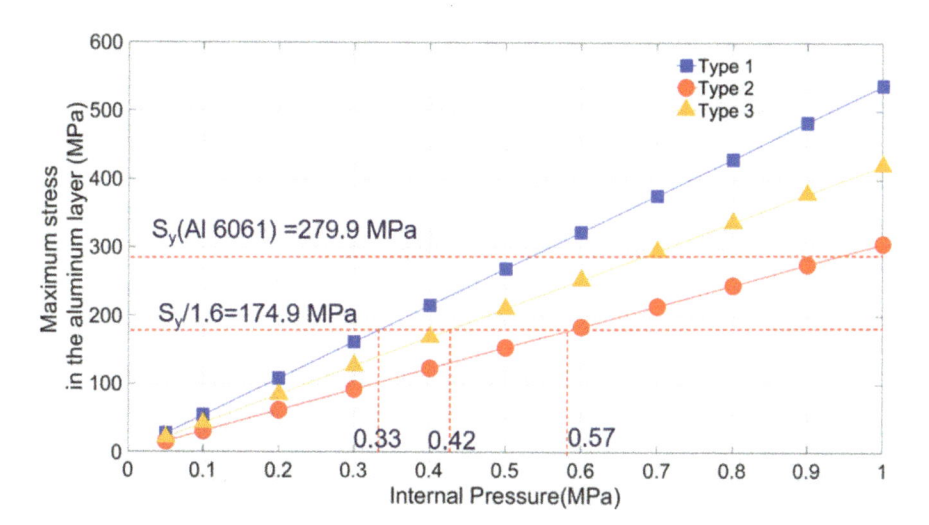

Fig. 4.23 Maximum internal pressure values that can be withstood with safety factor of 1.6 for tanks made with aluminum alloy 6061 (properties at 297 K)

Comparison of Results—Carbon Fiber Vs Glass Fiber

Fig. 4.28 shows the comparison of the variation of maximum stress in the aluminum layer for tank 3 (type 2) made with glass fiber and carbon fiber (IM7/8552) composites with applied internal pressure.

The difference, as shown in Fig. 4.28, is negligible between both materials, however, given the better long-term fatigue properties of CFC, the latter may be more suitable.

The analysis of rapid liquid hydrogen tank chill down did not reveal any stress concentrations in the shell and was found to be within tolerable limits.

The pie chart in Fig. 4.29 shows the weight percentage of each component for the tank 3, type 2 example. Note the small percentage of the tank insulation and liners, indicating a very small proportional contribution to the mass weight fraction.

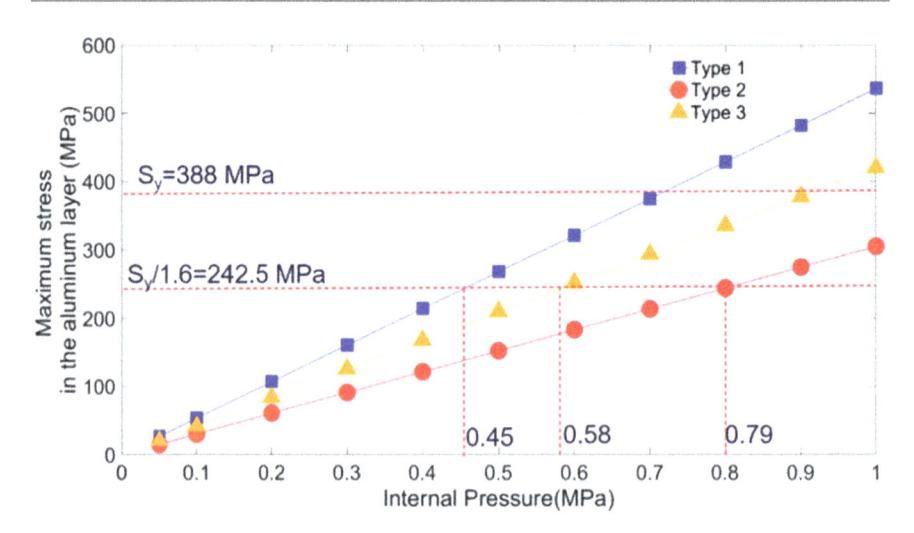

Fig. 4.24 Maximum internal pressure values that can be withstood to have a factor of safety of 1.6 for tanks made with aluminum alloy 2219-T87 (properties at 297 K)

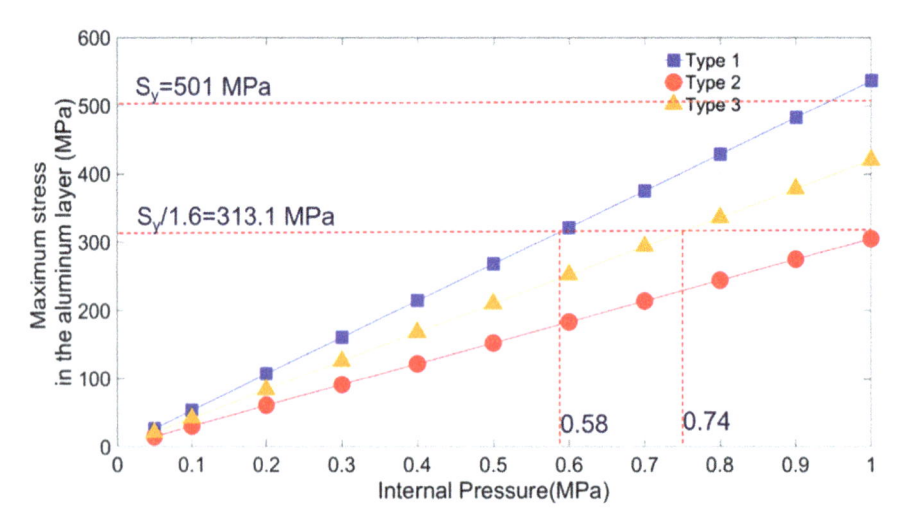

Fig. 4.25 Maximum internal pressure values that can be withstood to have a factor of safety of 1.6 for tanks made with aluminum alloy 2219-T87 (properties at 20 K)

4.3 Minimum Tank Mass Weight Fraction

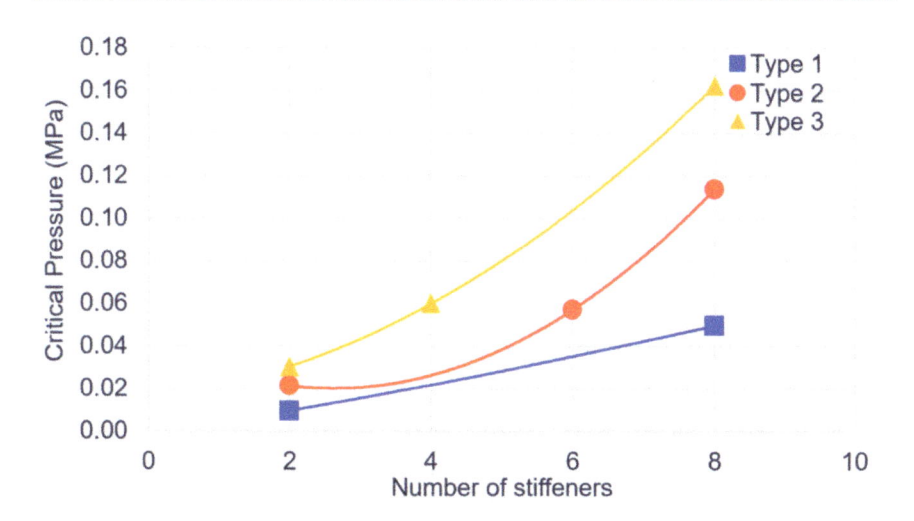

Fig. 4.26 Variation of critical buckling pressure with number of stiffeners for outer tank 3 with the inner aluminum layer and outer carbon fiber (IM7/8552) layers

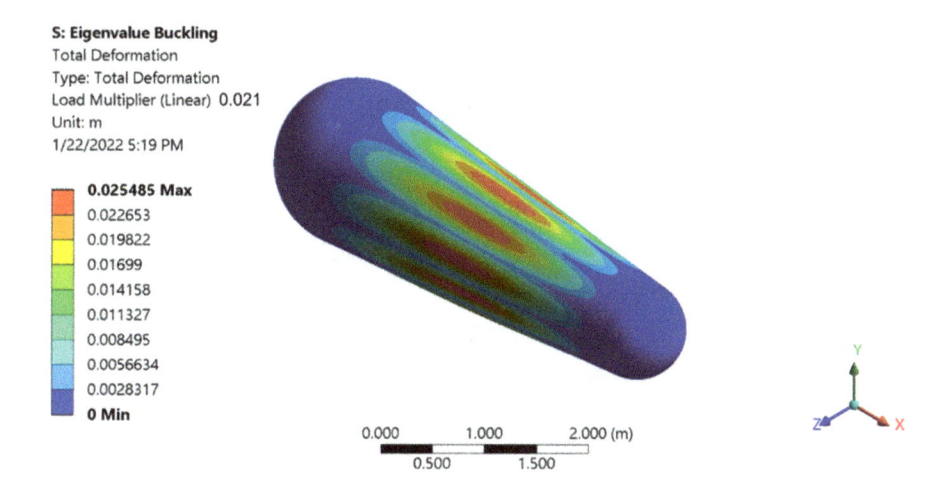

Fig. 4.27 Tank 3 type 2 ANSYS buckling mode

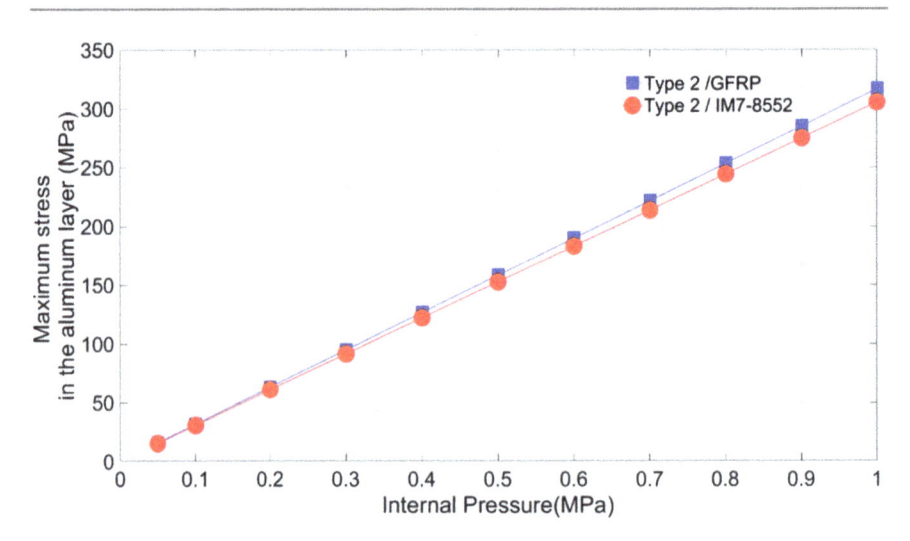

Fig. 4.28 Comparison of the variation of maximum stress in the aluminum layer for tank 3 (type 2) made with glass fiber and carbon fiber (IM7/8552) composites with applied internal pressure

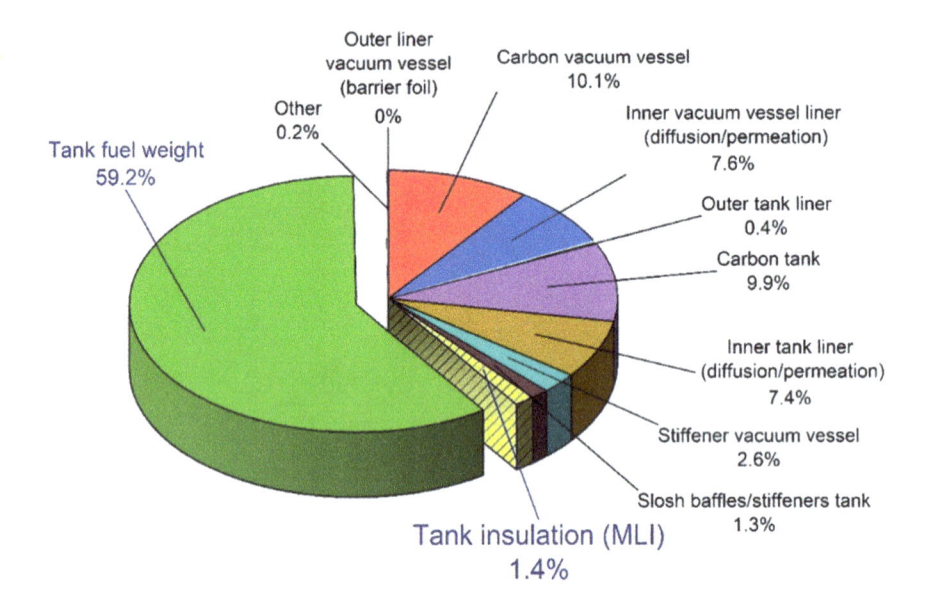

Fig. 4.29 Design example—final pie chart

Results Summary (Fig. 4.29)

4.3 Minimum Tank Mass Weight Fraction

The important design goal is to store liquid hydrogen at maximum density and lowest volumetric density. Figure 4.30 gives an overview of the gravimetric and volumetric density comparing several fuels [99].

4.3 Minimum Tank Mass Weight Fraction

Fig. 4.30 Gravimetric vs volumetric density of fuels

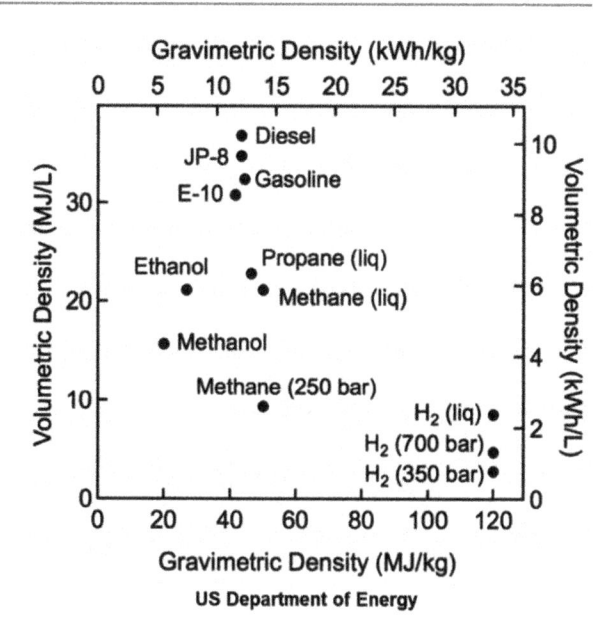

A useful parameter for defining low-weight tanks is the concept of a "Tank Mass Weight Fraction" that compares various liquid hydrogen tank designs. The definitions are as follows:

Minimum tank mass weight fraction (conventional) (MWF)

$$MWF = W_t / W_f$$

W_f = mass of "fuel" in tank.
W_t = total tank mass (this includes outer shell and inner liquid container and components).

Gravimetric storage density (after Verstraete [100]) (GSD)

$$\eta^{grav} = W_f / W_f + W_t$$

W_f = mass of "fuel" in the tank.
W_t = total tank mass (this includes outer shell and inner liquid container, as well as attached components).

Represents the fraction of liquid H_2 stored in the tank for the overall liquid tank system mass. In that case we can call this storage density.

Table 4.13 gives the achieved values for 3 tank types and 3 wall thickness types (T1–T3) for aluminum lined carbon fiber tanks comparing lowest weight fraction with max achievable storage density (Fig. 4.31).

Table 4.13 Minimum MWF vs GSD for CHEETA

<table>
<tr><th colspan="3">Minimum tank mass weight fraction</th><th colspan="3">Gravimetric storage density</th><th colspan="3">Aluminum layer</th><th colspan="2">Carbon fiber layer</th></tr>
<tr><th></th><th>Type</th><th>Weight fraction</th><th></th><th>Type</th><th>Storage density</th><th>Type</th><th>mm</th><th></th><th>mm</th></tr>
<tr><td rowspan="3">Tank 1</td><td>T1</td><td>0.45871881</td><td rowspan="3">Tank 1</td><td>T1</td><td>0.68553308</td><td>Type 1</td><td>1</td><td></td><td>2</td></tr>
<tr><td>T2</td><td>0.67181766</td><td>T2</td><td>0.59815136</td><td>Type 2</td><td>2</td><td></td><td>2</td></tr>
<tr><td>T3</td><td>0.71219429</td><td>T3</td><td>0.58404587</td><td>Type 3</td><td>1</td><td></td><td>4</td></tr>
<tr><td rowspan="3">Tank 2</td><td>T1</td><td>0.63377023</td><td rowspan="3">Tank 2</td><td>T1</td><td>0.61208117</td><td></td><td></td><td></td><td></td></tr>
<tr><td>T2</td><td>0.92964309</td><td>T2</td><td>0.51823055</td><td></td><td></td><td></td><td></td></tr>
<tr><td>T3</td><td>0.98705383</td><td>T3</td><td>0.50325763</td><td></td><td></td><td></td><td></td></tr>
<tr><td rowspan="3">Tank 3</td><td>T1</td><td>0.59843685</td><td rowspan="3">Tank 3</td><td>T1</td><td>0.62561120</td><td></td><td></td><td></td><td></td></tr>
<tr><td>T2</td><td>0.87746614</td><td>T2</td><td>0.53263278</td><td></td><td></td><td></td><td></td></tr>
<tr><td>T3</td><td>0.93124427</td><td>T3</td><td>0.51780089</td><td></td><td></td><td></td><td></td></tr>
</table>

4.3 Minimum Tank Mass Weight Fraction

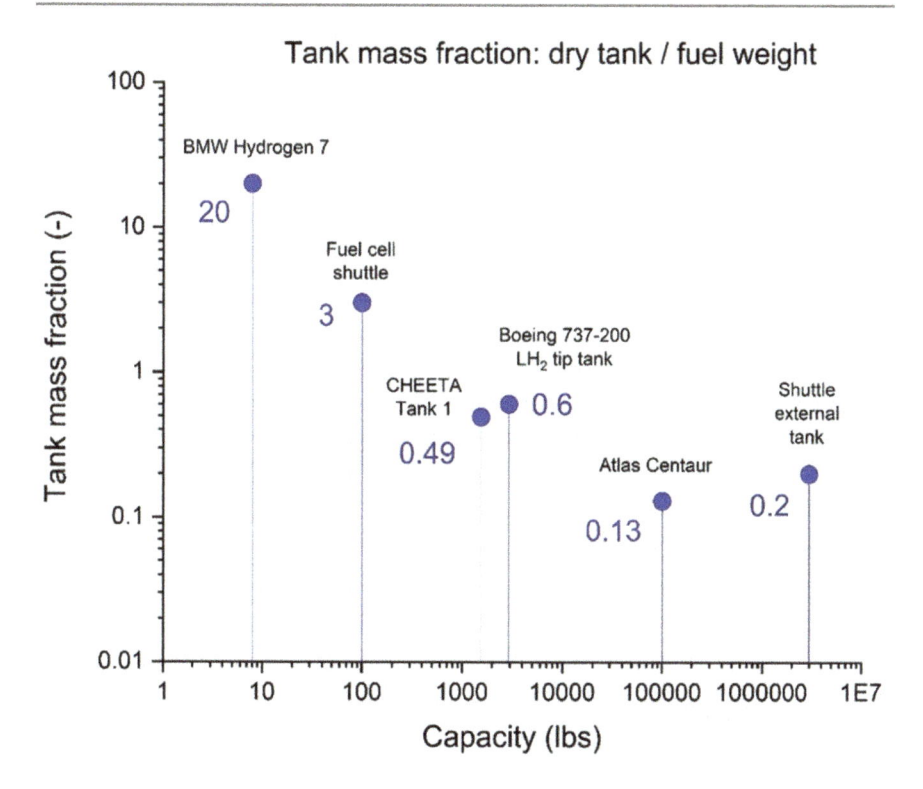

Fig. 4.31 Tank mass fraction examples for different designs

Above, the known tank mass fractions of systems are shown and compared as a function of storage capacity of liquid hydrogen and calculated from Penland [[101]]. Except for CHEETA, all those design approaches use liquid hydrogen as a combustible fuel source of a thermal engine. Besides that, there is a wealth of past neighboring modalities that all use liquid hydrogen, including NASA-designed liquid storage tanks for rockets with 500,000 liters liquid hydrogen storage capacity in 1972, configurations for transport aircraft developed by Gary Brewer of Lockheed 1976, and BMW Hydrogen 7 cars introduced in 2005.

We can summarize as follows:

Key design drivers:

- Tank design—Modular/Non-integral.
- Tank weight fraction.

Material choice:

- High-strength, light-weight cryogenic aluminum alloy 2219-T87—aviation grade—UTS typically 665 MN/m^2, qualified and safe use with LH$_2$, see Appendix 1.

- Carbon fiber woven ply reinforced shell, well-known cryogenic material properties for use with superconducting magnets and many other cryo applications.
- Cryobarriers (liners) and suspension/support structures.

Non-integral solutions

- Tank weight fraction down to 0.5 may seem feasible, shell structures are currently modeled and optimized, current design at TWF 0.65, for some tank types.

Integral solutions

- Tank weight fraction of <0.5 seems feasible, tanks designed as integral part of the airframe structure may prevent vacuum enclosure from buckling.
- Tank weight fraction of 0.2, as cited in the literature, may be unrealistic—generally not suitable for cryogenic aircraft designs.

References

1. C.W. Elrod, "Design handbook for liquid and gaseous helium handling equipment", 1963, ASDTD061-226.1-505
2. J.E. Fesmire, A. M. Swanger, et al "Energy efficient large-scale storage of liquid hydrogen" CEC/ICEC 2021
3. Kropshot R H Bermingham B W Mann D B *Technology of Liquid Helium* National Bureau of Standards Monograph 111 US Government Printing Office 1968
4. W. Peschka, "The status of handling and storage techniques for liquid hydrogen in motor vehicles, International Journal of Hydrogen Energy, 12, 1987, pp. 753–764
5. G. L. Mills, B. W. Buchholtz, "Design, fabrication and testing of a liquid hydrogen fuel tank for a long duration aircraft", AIP Conference Proceedings 1434, 2012, p 773
6. C. Wilkerson, "Acoustic Emission Monitoring of the DC-XA Composite Liquid Hydrogen Tank During Structural Testing", NASA Technical Memorandum 108520
7. R. Goetz, R. Ryan, A.F. Whitaker "Final report of the X-33 liquid hydrogen tank test investigation team", NASA technical report, Huntsville, Marshall Space Flight Center, NASA, 2000
8. G. Warwick, "Leak-proof Composite Tank Could Hold Fuel for Astronauts", Flight International, 2004, 166, 4951; SLI Northrop Grumman
9. T. F. Johnson, D. W. Sleight, R. A. Martin, "Structures and Design Phase I Summary for the NASA composite cryotank technology demonstration project" (CCTD) 54th AIAA/ASME/ASCE/AHS/ASC Structures, Structural Dynamics, and Materials Conference, 2013 https://doi.org/10.2514/6.2013-1825
10. M. Sippel, A. Kopp, "Progress on Advanced Cryo-Tanks Structural Design Achieved in CHATT-Project CHATT, https://www.semanticscholar.org/paper/Progress-on-Advanced-Cryo-Tanks-Structural-in-Sippel-Kopp/a209de6c11d27cf7dd7535773423841ef86708d4
11. S. J. Canfer, D. Evans, "Properties of materials for use in liquid hydrogen containment vessels", Advances in Cryogenic Engineering (Materials), Vol. 44, 1998, pp. 253–259
12. L. Klebanoff, Hydrogen Storage Technology, CRC Press, 2012
13. H. Brechna, Superconducting Magnet Systems, "Technische Physik in Einzeldarstellungen", Band 18, Springer Verlag, page 484, 1973
14. T.T. Chiao, C.C. Chiao, et al, "Lifetimes of fiber composites under sustained tensile loading" UCRL 78367 report, 1976

15. T. Ogasawara, N. Arai N, R. Fukumoto, T. Ogawa, T. Yokozeki, A. Yoshimura, *Titanium alloy foil -inserted carbon fiber/epoxy composites for cryogenic propellant tank application* Advanced Composite Materials 23:2, pp 129–149 2014
16. T. Morimoto, T. Ishikawa, T. Yokozeki, Y. Hayashi, T. Shimoda, Y. Morino Y, *Pressurization test on CFRP liner-less tanks at liquefied nitrogen temperature* Advanced Composite Materials 13:2, pp 81–88 2004
17. D. Evans, S J Robertson, S Walmsley and J Wilson, "Measurement of the Permeability of Carbon Fibre/PEEK Composites" in: "Cryogenic Materials 1988 Volume 2 Structural Material" R P Reed ed. ICMC Boulder Colorado p 755
18. D. Schultheiß, "Permeation barrier for lightweight hydrogen tanks", PhD Thesis, 2007
19. R. Wei, X. Wang. C. Chen, X. Zhang et al "Effect of surface treatment on the interfacial adhesion performance of aluminum foil/CFRP laminates for cryogenic propellant tanks", Materials and Design 116 pp. 188-198 2017
20. S. G. Miller, M.A. Meador, "Polymer- Layered Silicate Nanocomposites for Cryotank Applications", 48th AIAA/ASME/ASCE/AHS/ASC Structures, Structural Dynamics, 2007, pp 1–9
21. K. Ahlborn, "Mechanische Eigenschaften von kohlefaserverstärkten Thermoplasten für die Anwendung in der Tieftemperaturtechnologie", KFK Bericht 4487, 1989
22. M. Flanagan, D.M. Grogan, J. Goggins, "Permeability of carbon fibre PEEK composites for cryogenic storage tanks of future space launchers" Composites: Part A 101, 2017, pp. 173–184
23. S. Dye et al. "Integrated and load responsive multilayer insulation", Advances in cryogenic engineering: Transactions of the Cryogenic Engineering Conference, vol. 55, 2010 pp 946 to 953
24. M.P. Hanson, "Static and dynamic fatigue behavior of glass-fiber filament-wound pressure vessels", Adv. Cryo. Eng. Vol. 17, 1972, pp. 16–175
25. G. Geiss, Einfluss von Tieftemperatur und Wasserstoff auf das Versagensverhalten von Glasfaser-Verbundwerkstoffen unter zyklischer Belastung, FZKA 6581, Dissertation, 2001
26. D. Gajda, M. Lutynski, "Hydrogen Permeability of Epoxy Composites as Liners in Lined Rock Caverns—Experimental Study", Appl. Sci. 11, 2021, 3885, pp. 1–12
27. T. Yokozeki, T. Ogasawara, T. Aoki, T. Ishikawa, "Experimental evaluation of gas permeability through damaged composite laminates for cryogenic tank" Composites Science and Technology 69 2009 pp. 1334–1340
28. Hindenburg results -link https://www.pbs.org/wgbh/nova/video/hindenburg-the-new-evidence/
29. D. Evans, R.P. Reed, "The permeability of resin based composite materials to radiolytic gases", Cryogenics 38, 1998, pp. 149–154
30. V. Bechel, M. Negilski, J. James "Limiting the permeability of composites for cryogenic application", Compos. Sci. Technol. 2006, 66, pp. 2284–2295
31. D.M. Grogan, C.M. Ó Brádaigh, J.P. McGarry, S.B. Leen, "Damage and permeability in tape-laid thermoplastic composite cryogenic tanks", Composites: Part A 78 2015 pp. 390–402
32. J. Hohe, A. Neubrand, S. Fliegener, C. Beckmann, M. Schober, K.-P. Weiss, S. Appel, "Performance of fiber reinforced materials under cryogenic conditions—A review", Composites: Part A 141 2021 106226
33. Feldman J Stautner W Kovacs C Miljkovic N Haran K *Review of materials for HTS magnet impregnation* Supercond. Sci. Technol. 2024 https://doi.org/10.1088/1361-6668/ad1aeb
34. Hartwig G *Low temperature properties of potting and structural materials for superconducting magnets* IEEE Transactions on Magnetics vol Mag-11 No 2 1975
35. Hartwig G *Low-Temperature Properties of Epoxy Resins and Composites*. In: Timmerhaus, K.D., Reed, R.P., Clark, A.F. (eds) Advances in Cryogenic Engineering, vol 24. Springer, Boston, MA. 1978 pp 17–36 https://doi.org/10.1007/978-1-4613-9853-0_2
36. Lüders C, Sinapius M. *Fatigue of fibre-reinforced plastics due to cryogenic thermal cycling*. J Compos Mater 2019 53:2849–61

37. Hartwig G Puck A Weiß W *Thermische Kontraktion von glasfaserverstärkten Epoxydharzen bis zu tiefsten Temperaturen*, Kunststoff Bd. 64, 1974 Band 1
38. Hartwig G *Properties of fibre composites* in Seeber B Section F2 Handbook of Applied Superconductivity p 1043
39. Hartwig G in Maurer W *Entwicklungen auf dem Gebiet der Hochstromsupraleitung im Kernforschungszentrum* Karlsruhe KFK 2290 page II-5-11 1976
40. Schmidt C *Influence of the Kapitza resistance on the thermal conductivity of filled epoxies* Cryogenics 1975
41. Jensen J E Tuttle W A Stewart R B Brechna H Prodell A G Brookhaven National Laboratory Selected Cryogenic Data Notebook *Thermal conductivity of epoxies* BNL 10200-R Vol VII 1980
42. Radcliffe J Rosenberg H M *The thermal conductivity of glass-fibre and carbon-fibre/epoxy composites from 2 to 80 K Cryogenics* vol 22 issue 5 pp 245–249 1982
43. Eckels Engineering Cryocomp Version 5.2 2012
44. Humpenöder J *Gas permeation of fibre reinforced plastics* Cryogenics Vol 38 Issue 1 January 1998, Pages 143–147
45. Okada T Nishijima S *Gas permeation and performance of an FRP cryostat* Advances in Cryogenic Engineering vol 34, pp 17–24 1988
46. Evans D Morgan J T *The permeability of composite materials to hydrogen and helium gas* Advances in Cryogenic Engineering vol 34, pp 11–16 1988
47. Just T Will J Haberstroh C *Hydrogen Permeability Testing of Fibre Reinforced Thermoplastics under Cryogenic Conditions - Validation of a Test Rig Concept* CEC ICMC conference Honolulu C3Or3B03 2023
48. Schoch K F Bergh D D *Glass-reinforced epoxy piping for liquid nitrogen-cooled AC transmission cables* Advances in Cryogenic Engineering vol 24, pp 262–270 1978
49. Lee D Haynes R Skeen D J *Properties of optical fibres at cryogenic temperatures* Mon. Not. R. Astron. Soc. 326, 774–780 (2001)
50. Reed R P Golda M *Cryogenic composite supports: a review of strap and strut properties* Cryogenics vol 37 No 5 pp 233–250 1997
51. Z. Sápi, R. Butler, "Properties of cryogenic and low temperature composite materials—A review", Cryogenics Vol. 111, October 2020, 103190
52. S. Prasanraj, K.P. Dhanbalakrishnan et al, "Advanced composite materials in cryogenic propellant tank", International Journal of Engineering Research & Technology, IJERTV8IS120284, 2019 vol 8 issue 12
53. K.Naito, J. Yang, Y. Kagawa, "Tensile properties of high strength polyacrylnitrile (PAN)-based and high modulus pitch-based hybrid carbon fibers-reinforced epoxy matrix composite, J Mater Sci 2012 47: pp. 2742–2751
54. T. Horiuchi, T. Ooi, "Cryogenic properties of composite materials", Cryogenics 35, 1995, pp. 677–679
55. W. Wei, et al "Cryogenic performances of T700 and T800 carbon fibre-epoxy laminates, IOP Conf. Ser.: Mater. Sci. Eng. 2015 102 012016
56. H. Bansemir, O. Haider, "Fibre composite structures for space applications—recent and future developments", Cryogenics 38 1998 pp. 51–59
57. D. M. Grogan, C.M. Ó Brádaogj et al, "Damage and permeability in tape-laid thermoplastic composite cryogenic tanks, Composites: Part A 78 2015 pp. 390–402
58. N. Liu, B. Ma, F. Liu et al "Progress in research on composite cryogenic propellant tank for large aerospace vehicles, Composites: Part A 143 2021 106297
59. V.T. Bechel, R. Y. Kim, "Damage trends in cryogenically cycled carbon/polymer composites". Composites Science and Technology, 64 2004 pp. 1773–1784
60. J. Hohe, M. Schover, S. Fliegener, "Effect of cryogenic environments on failure of carbon fiber reinforced composites", Composites Science and Technology 212 2021 108850
61. I.G. Tapeinos, D. Zarouchas, O.K. Bergsma, et al, "Evaluation of the mechanical performance of a composite multi-cell tank for cryogenic storage: Part I Tank pressure window

based on progressive failure analysis", International Journal of Hydrogen Energy 44, 2019, 3917–3930
62. J.A. Cerro, "The Membrane Cryotank a Thermal/Structural Concept for Reusable Access to Space", ASCEND event, 2021
63. P. Adams, A. Bengaouer, B. Cariteau et al, "Allowable hydrogen permeation rate from road vehicles", International Journal of Hydrogen Energy 36, 2011, 2742–2749
64. Md. S. Islam et al, "Investigation of woven composites as potential cryogenic tank materials", Cryogenics, vol. 72, Part 1, 2015, pp. 82–89
65. Composite materials handbook, DoD MIL-HDBK-17-1F vol 1 to 5, 2002
66. B. Atli-Veltin, "Cryogenic composite fuel tanks: the mechanical performance of advanced composites at low temperatures", AIAA SciTech Forum 2018
67. T. Aoki, T. Ishikawa, Y. Morino, "Overview of basic research activities on composite propellant tanks in Japan", AIAA-2001-1868
68. T. Aoki, T. Ishikawa, Y. Morino, "Perspective in Basic Research activities on cryogenic composite tanks for RLV in Japan, AIAA 2001-4605
69. T. F. Johnson, D.W. Sleight, R. Martin, "Structures and design phase I summary for the NASA composite cryotank technology demonstration project", 54th AIAA/ASME/ASCE/AHS/ ASC structures, Structural Dynamics and Materials Conference, 2013
70. G. Baschek, G. Hartwig, "Effect of water absorption in polymers at low and high temperatures", Polymer 40 1999 pp. 3433–3441
71. N. O. Brink, "Determination of the Performance of Plastic Laminates under Cryogenic Temperatures", Report No ADD-TDR-62-794, Wright Patterson Air Force Base, Ohio, 1962
72. G. Burkhart, E Klippel, "Strength investigation of FRP material (epoxy resin) in LH2 and LO2
73. D. W. Chamberlain, B.R. Loyed, R.L. Tennent, "Determination of the Performance of Plastic Laminates under Cryogenic Temperatures", Report No ADD-TDR-62-794, Part 2, Wright Patterson Air Force Base, Ohio, 1964
74. K. Dahlerup-Peterson, "Tests of composite materials at cryogenic temperatures: facilities and results, Adv. Cryog. Eng. Materi. 27 1980 pp. 268–279
75. H. Domininghaus, "Die Kunststoffe und ihre Eigenschaften", VDI-Verlag 1986
76. D. Evans, I. Johnson. H. Jones et al., "Shear testing of composite structures at low temperatures", Adv. Cryog. Eng. Materi. 36 1990 pp. 819–826
77. D. Evans, S.J. Robertson, J.T. Morgan, "The long term effects of hydrogen and oxygen on the mechanical properties of carbon fiber reinforced composites", Adv. Cryog. Eng. Materi. 36 pp. 937–941 1990
78. G. Hartwig, "Polymer properties at Room and Cryogenic Temperatures", Plenum Press, New York, 1994
79. J. Hertling, "Ausbreitungsgeschwindigkeit von instabilen Rissen in Polymeren bei tiefen Temperaturen, Dissertation, Universität Karlsruhe 1999
80. R. Hübner, "Zug- und Schereigenschaften von Kreuzverbunden aus kohlefaserverstärkten Polymeren bei tiefen Temperaturen", Dissertation, Universität Karlsruhe 1996
81. R.M. Jones, "Mechanics of composite materials", Hemisphere Publishing Corp. New York, USA 1975
82. M.B. Kasen, "Mechanical and thermal properties of filamentary-reinforced structural composites at low temperatures: 1: Glass-reinforced composites, Cryogenics 33 1975 pp 327–349
83. M.B. Kasen. "Current status of interlaminar shear testing of composite materials at cryogenic temperatures", Adv. Cryog. Eng. Materi. B 36 1990 pp. 787–792
84. L. Lorenzo, H.T. Hahn, "Fatigue failure mechanism in unidirectional composites". ASTM STP 907 1986 pp. 210–232
85. P.K. Mallick, "Fiber-reinforced Composites", Marcel Dekker Inc. New York 1993
86. A. Mayer, K. Pannkoke, "Experimentelle Überprüfung des Ermüdungsverhaltens von Faserverbundwerkstoffen mit unterschiedlichen Probenformen by 77 K mit Flüssigkeit- und Gaskühlung, KfK-Bericht 4887, 1991
87. V.L. Morris, "Advanced composite structures for cryogenic applications", Proc. 14th In SAMPE Symp Company Report: Structural Composites, Pomona, USA 1989 pp 1867–1876

88. T. Okada, N. Nishijima, "Investigation of interlaminar shear behavior of organic composites at low temperatures, Adv. Cryog. Eng. Materi. B 36 1990 pp. 811–817
89. R.P. Reed, P.E. Fabian, "Shear compressive fatigue of insulation systems at low temperatures", Cryogenics 35 1995 pp. 685–688
90. R.P. Reed, M. Golda, "Cryogenic properties of unidirectional composites" Cryogenics 34 1994 pp. 909–928
91. Y. Su, H. Lv, W. Zhou et al, "Review of the Hydrogen Permeability of the Liner Material of Type IV On-Board Hydrogen Storage Tank", World Electr. Veh. J. 2021, 12(3) 130
92. N. L. Newhouse, "Development of Improved Composite Pressure Vessels for Hydrogen Storage", Hydrogen Storage Engineering, Hexagon Lincoln, 2015
93. W. Xu, "Design and analysis of liquid hydrogen storage tank for high-altitude long-endurance remotely-operated aircraft, International Journal of Hydrogen Energy, Vol. 40, Issue 46, 2015, pp. 16578–16586
94. AD 2000-Merkblatt N1, "Pressure vessels in glass-fiber reinforced thermosetting plastics", 2018
95. H. Novak et al., "GFK-Cryostat for a Cryogenic Current Comparator" ICEC29-ICMC 2024 Geneva
96. D. Mariappan, W. Stautner, "Design of Liquid Hydrogen Tanks for Aircraft Applications", GE Research Technical publication 2021090158904
97. Helius Autodesk® software Composite
98. H. Zheng, X. Zeng, J. Zhang, et al "The Application of Carbon Fiber Composites in Cryotank. Solidification" 2018 https://doi.org/10.5772/intechopen.73127
99. US Department of Energy Publication
100. D. Verstraete, "Hydrogen fuel tanks for subsonic transport aircraft", International Journal of Hydrogen Energy 35 2010 pp. 11085–11098
101. J.P. Penland, "Liquid hydrogen fueled Boeing 737-200", Penland diaries, 1973

Cryotankage—Tank Shapes and Airframe Integration

In the following section, we investigate examples of various thought-through design solutions in which way a liquid hydrogen tank can be optimally shaped for different fuselages and airframe structures. Aircraft tanks come in a variety of shapes, e.g., Integral tanks (e.g., wet wing), rigid removable tanks, bladder tanks, tip tanks, external tanks, conformal fuel tanks, or even drop tanks.

Most likely, any medium or long-range aircraft carrier may need multiple fuel tanks, possibly also fulfilling different tasks within an all-electric aircraft. One way of avoiding a large single tank is to split one into more efficient, dedicated storage volumes, this however comes at a challenging manufacturing and reliability cost. Moreover, those tanks could be so-called conformal tanks but then may not benefit from a safe medium to high vacuum environment. To maximize the storage volume for a given tank placement, the ideal configuration would be to construct a conformal tank. A conformal tank can hold up to 20% more hydrogen in a given envelope and pressure than a cylindrical or spherical tank. This is based on the presumption that a conformal tank can utilize over 80% of its envelope volume. This increase in useable storage compared to a conventional cylindrical tank is shown in Fig. 5.1. The ability to construct a lightweight high-pressure conformal tank is presently being investigated by Thiokol Propulsion under a Department of Energy contract [1].

As we go through the various designs with different TRL levels we need to keep in mind that any tank in use in an aircraft needs to be embedded into a cryogenic flow circuit that many early designs omit. The further we move toward a completely non-metallic design the lower the TRL level since any tube connections to the tanks are still challenging in manufacture and assembly [2].

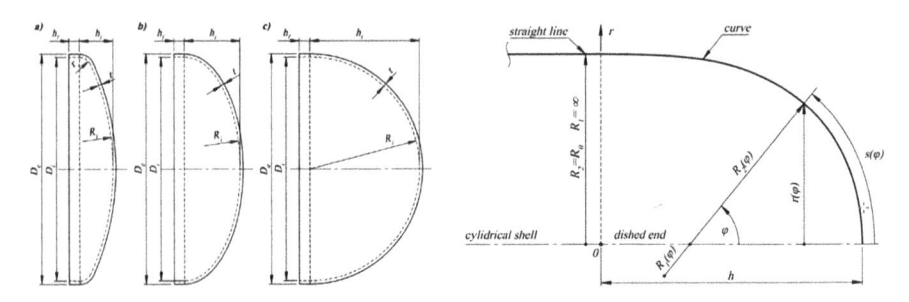

Fig. 5.1 Dished ends according to EN 13445; (**a**) torispherical, (**b**) ellipsoidal, (**c**) hemispherical and cylindrical vessel geometry

5.1 Non-integral vs. Integral Tank Designs

There is most likely no straight answer to whether conformal or non-conformal tanks are more beneficial. The design solution may depend on many parameters, for example, aircraft size, range, and mission profile (hypersonic or subsonic), number of passengers, TRL, or achievable mass weight fraction, as indicated in the previous section, or whether they are for cargo, or military use. Some thoughts for arriving at a decision are shown in Table 5.1.

Brewer [3] points out the differences between both tanks as follows:

> *Nonintegral (non-conformal)* tanks are those which are designed to take only loads associated with containment of the fuel, i.e., pressurization and fuel dynamic loads, plus thermal stresses. They are supported within conventional fuselage skin/stringer/frame structure. On the other hand, *integral (conformal)* tanks are an integral part of the basic aircraft structure; therefore, in addition to the above loads they must also be capable of withstanding all the usual fuselage axial, bending, and shear stresses resulting from the critical aircraft loading conditions.

Integral tanks are also called conformal tanks.

The basic question we need to answer is on what integration route we should choose with some guidelines given in Table 5.1.

Non-integral tanks are directly insulated against the fuselage skin and therefore cannot be easily repaired with respect to insulation defects and other routine maintenance, with only very limited diagnostics means available.

Furthermore, unlike Jet-A or other non-cryogenic tanks, cryogenic tanks need access to cryogenic transfer lines, e.g., as shown in the Cryo-circuit Section.

For example, as mentioned before, tanks can be arranged as single or multiple tanks either above, next to, and below any passenger seating, inside the aircraft fuselage forward or aft of a cabin, enclosed within the wings, within underwing podded tanks, and on the wing tips.

The need for integrating a liquid hydrogen tank with its cryogenic transfer lines and fuel lines, however, severely limits those design options.

5.1 Non-integral vs. Integral Tank Designs

Table 5.1 Integral (conformal) vs -non-integral (nonconformal) tank design trade-offs

LH$_2$ tank design	Non-Integral: dual tank design	Integral: single tank design
Operational	Operating pressure (maintains positive pressure) Refrigeration needs and logistics Guaranteed insulation efficiency in high vacuum (insulation quality/size/thickness/weight/stiffness/maintenance/cost)	Breaking insulation due to temperature gradients (fill, warmup, chilldown) Requires gas purging mechanism to ensure solid air not freezing out at liquid hydrogen walls Bad heat leak, inefficient insulation, high residual gas heat leak, bad vacuum
Transients	Initial chill down / rechill after emptying, tank warmup time Temperature swings during aircraft operation Liquid inventory movement / sloshing / vibration—Heat leak increase	
Supply and storage	Fill level and fill pressure defined	May be different, depending of tank structure
	Tank shapes (adapt to airframe structure), tapered tanks possible	Adapted to airframe structure but may need multicell units
	Tank mass fraction: 0.65 achievable	Can be less than 0.65
	Tank protected when going subatmospheric during operation	Tank cannot go subatmospheric, safety risk, needs interconnect tanks
	Cross feeding of tanks—Transfer of liquids	Different, difficult to inspect
	LH$_2$ delivery line routing	Different, difficult to inspect for gas leaks
Maintenance	No purge required	Requires protective gas atmosphere (Nitrogen / helium purge mechanism)
	Liquid hydrogen pipeline break points in tanks	
	Repair or replacement of components possible, e.g. pump removal, etc.	Repair (local) possible, requires novel leak testing
	Auto-inspect—Remove tank	Cannot be removed easily, if tank rigidly fitted to airframe structure
	Failsafe insulation	Difficulty to inspect cracking or contaminated insulation
	Inspect and automatic repair of fuel tank components	Very difficult for automatic repair
	Pump removal process through manholes	Possible, severe space restrictions
	Removal and inspection of failing control components possible	Difficulty to get to any failing component

5.1.1 Dual Tank—Separate Dual Tank Walls

In the following section, we look at how different tank configurations and structures evolved, independently of the aforementioned definition of integral vs non-integral.

5.1.2 Tank Structural Designs

As Klebanoff [4] succinctly puts it, "If you are going to store LH_2, it is better to store a lot of it" (see also Chap. 4 on tank mass fraction). Naturally, for aircraft, that means storing a maximum amount of liquid hydrogen at lowest weight. This would then call for optimizing the tank structure itself.

Early designs showed that spherical tanks give the best volume to weight ratio, however it may not be the best way to utilize the tight fuselage space. If we design for cylindrical or near-cylindrical solutions on the other hand, we need to decide about head designs and insulation requirements. Even tapered cylindrical tanks could be beneficial as shown in more detail in the following section. Furthermore, the correct head choice not only helps to maximize fill volume but also helps in stiffening the overall tank design. Some rudimentary choice is given by Barron [5] and Brewer [3], for hemispherical, elliptical, and torispherical head sizes as used for cryogenic pressure vessels (Table 5.2).

The dish radius for an ASME torispherical head is related to the outside diameter of the head as follows: (1) for $D_o \leq 1.0668$ m, $R = D_o - 2th$; (2) for 1.0668 m $< D_o < 3.658$ m, $R = D_o - 2th - 0.1524$ m; (3) for $D_o \geq 3.658$ m, $R = D_o - 2th - 0.305$ m.

Fig. 5.1 [6] gives a general geometry of this type of pressure vessels. The question most certainly has to be asked of what shape of end closure for the tanks offers the best combination of lightweight and minimum length for minimum direct operating cost (DOC) for the aircraft operator. According to Brewer [3], hemispherical heads are not recommended for minimum DOC designs.

See also the recent parametric study on tank integration by Huete et al. [7].

Ellipsoidal aircraft tank designs have also recently been analyzed by Winnefeld et al. [8].

Figure 5.2 shows some single tanks as designed in the 1970s. This single-lobed tank comes with a polyurethane foam insulation. Since the insulation is in direct contact at a cryogenic temperature, a purge gas (mainly costly helium) is needed to protect the shell from icing up.

The early Lockheed Starlifter C141 (1962) [9] designs also proposed a sectioned, lobed tank structure. In the past, several different shapes were suggested by engineers to economically and cost-efficiently store liquid hydrogen at low weight,

Table 5.2 Head volume characteristics for cylindrical and non-cylindrical tanks

Hemispherical	2:1 Elliptical	ASME Torispherical
Internal volume		
$V = \pi D^3/12$	$V = \pi D^3/24$	$V = \pi R^3/39.173$
Material volume		
$V_h = 0.5 \pi (D + t_h)^2 th$	$V_h = 0.3450 \pi (D + t_h)^2 th$	$V_h = 0.264 \pi (R + t_h)^2 th$
Outside surface area		
$A_o = 0.5 \pi D_o^2$	$A_o = 0.3450 \pi D_o^2$	$A_o = 0.264 \pi R^2$

5.1 Non-integral vs. Integral Tank Designs

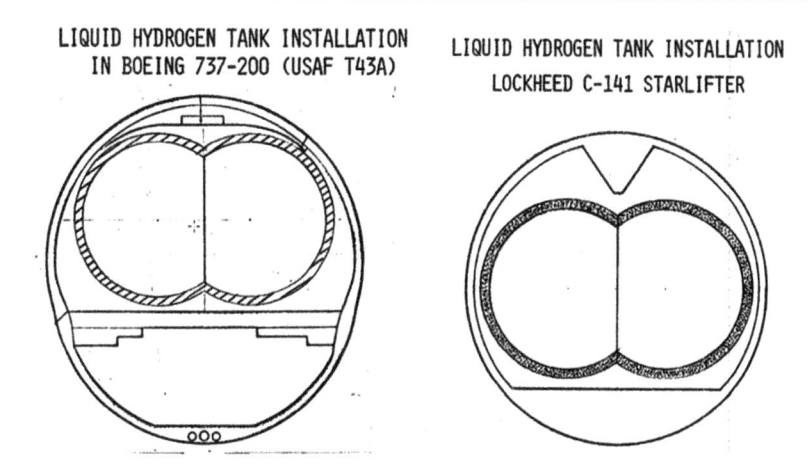

Fig. 5.2 Boeing 737–200 and Lockheed C-141 design studies

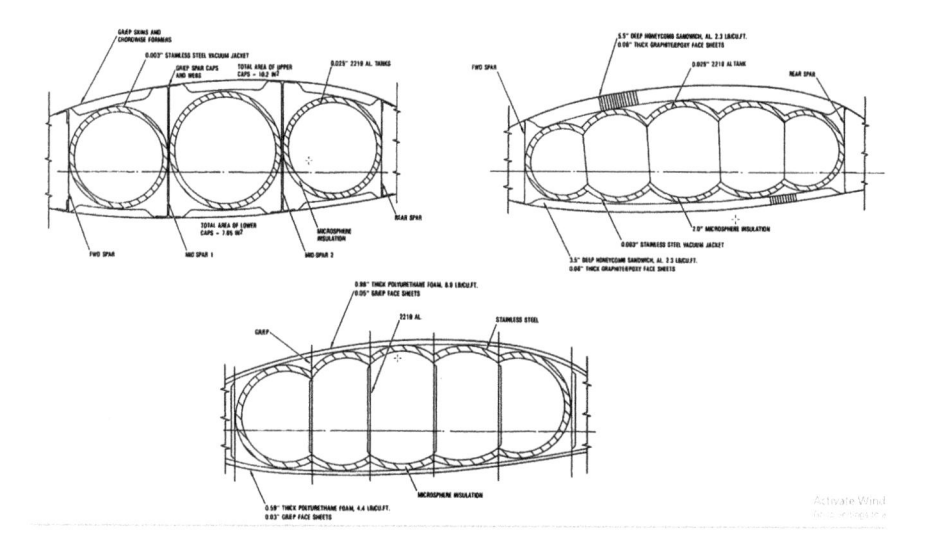

Fig. 5.3 Liquid hydrogen tanks in wings—pillow shapes and spar webs as support members

depending on the specific design target. Not all shapes can be manufactured or assembled easily or would comply with all cryogenic and hydrogen safety standards.

Below are some examples for placing non-integral liquid hydrogen tanks in the wings—pillow-shaped or made of an array of parallel cylindrical tubes [3]. See Fig. 5.3.

As pointed out, since a sphere is the optimal configuration for max volume at minimum weight, a recent design study pushes this shell design even further by increasing the volumetric efficiency combining spherical clusters of liquid tanks, so-called multi-cell composite tanks. See also [9, 11].

Future liquid hydrogen tanks (Fig. 5.4) most certainly need to be conformal or integral with respect to volumetric efficiency and minimum weight fraction. However, those designs are currently not suitable from the cryogenic perspective and at very low TRLs for tanks to work in all operating conditions, especially during many fueling, defueling, and chill-down cycles over the lifetime of an aircraft.

5.1.3 Tank Accommodation Options

When choosing single tanks, many parameters need to be considered apart from the total weight of the tank. The drag for example it takes on the airfoil structure/wings and other components decreases the aircraft efficiency for utilizing liquid hydrogen as a fuel. The first iteration of design concepts therefore looked at conveniently storing liquid hydrogen tanks within a given aircraft design.

In several early Boeing studies (737–200 system) [9] the liquid hydrogen tank was directly fitted into the fuselage (conformal) and placed forward and aft in the aircraft, leaving space before and behind for passenger seating. To achieve this design without having to change the outer aircraft dimensions the number of passengers had to be reduced by 80 and the additional weight was calculated to 1500 kg for the tank, see Fig. 5.5.

Recently Verstraete [10] thought of storage solutions for subsonic transport aircraft of smaller size, shown in the above and Fig. 5.6. However, any optimal storage configuration is rather limited for a given fuselage and changes are required to provide the higher amount of liquid hydrogen storage volume required as compared to Jet-A fuel.

5.1.4 Tank Fuselage Changes

Figures. 5.7 [12] and 5.8 [3] show an early design attempt to change the fuselage so that tanks can be placed in the wing tips itself, rather than in the wings, which helps

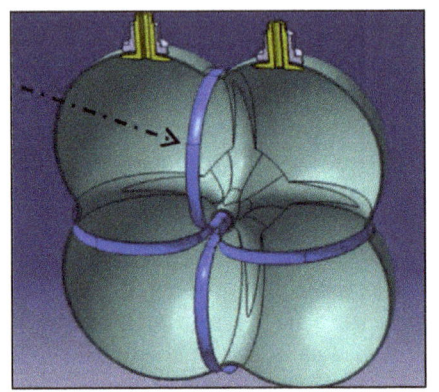

Fig. 5.4 Spherical tanks—study

5.1 Non-integral vs. Integral Tank Designs

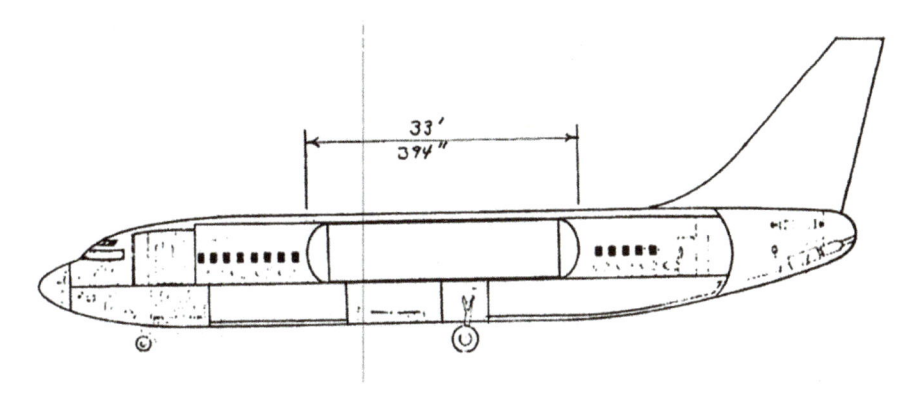

Fig. 5.5 Boeing 737–200 early design study—cylindrical tank

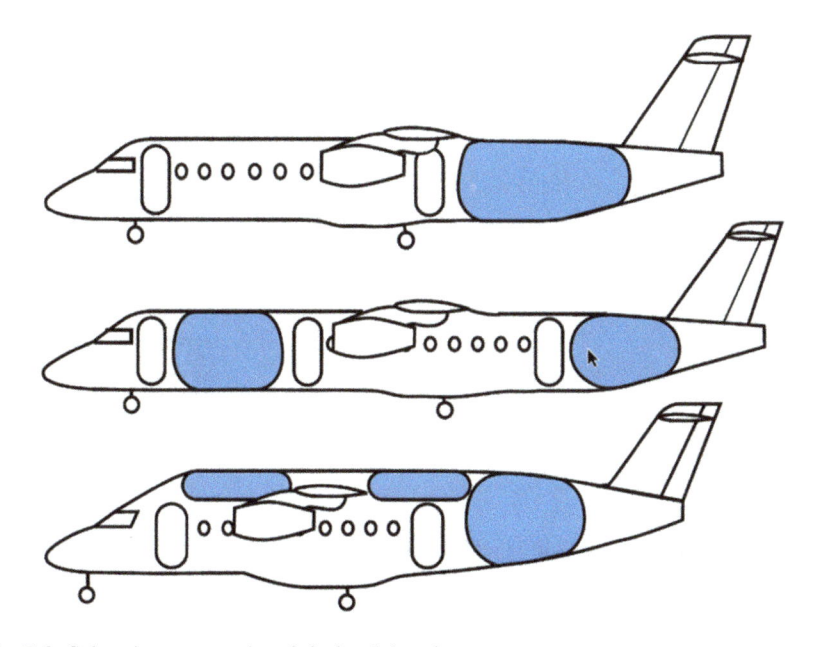

Fig. 5.6 Subsonic transport aircraft fuel tank locations

to maintain a cylindrical tank shape. More recent design studies also aim to place the liquid hydrogen tanks below the wings for some applications. All of those designs come with their intrinsic disadvantages and need to be seen in the context of how liquid hydrogen is used in an aircraft [9].

Another possibility is to change the fuselage structure itself to structurally adapt to a liquid hydrogen tank. For this a liquid hydrogen-fueled aircraft (Mach 2.7) was designed in 1974 for 400 passengers with the liquid hydrogen placed in forward and aft tanks with the 2-story passenger seats located in the aircraft center [3], as shown in Figs. 5.9 and 5.10

Fig. 5.7 Newer design study—conversion of a FD328 Jet

Fig. 5.8 Boeing 737–200—early design study

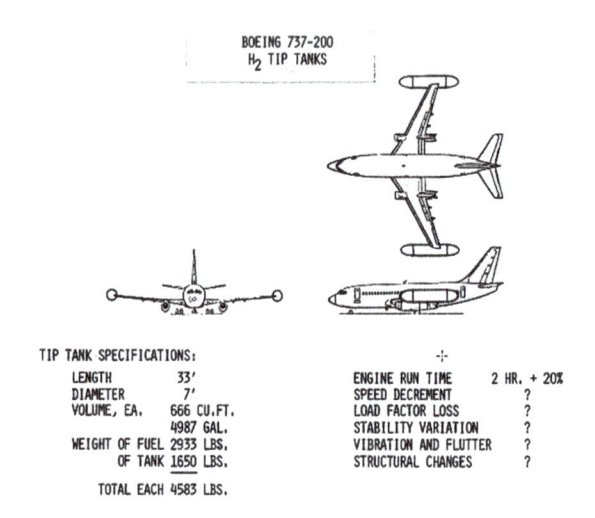

In a further design study tanks were placed in multiple locations wherever room was found to fit into an extended fuselage. Here we can only show a few examples of the envisaged placement of tanks, see Fig. 5.11 [12].

All changes to the outer skin of an aircraft and changes to the fuselage require a thorough aircraft design analysis to ensure drag and other parameters are acceptable for the mission profile.

Figure 5.12 shows an early Airbus cryoplane [13]. See also Henn [14].

With the more recent Enable project (Figs. 5.13 and 5.14) [15] a bulged tank location above the passenger seats was proposed, also considering safety concerns. This project is receiving funding from the European Union H2020 research innovation program under Grant no. 769241.

Yet another approach was thought through by Sefain & Jones [16] with their cryoplane design, with liquid hydrogen storage externally attached to the fuselage and imbedded.

5.1 Non-integral vs. Integral Tank Designs

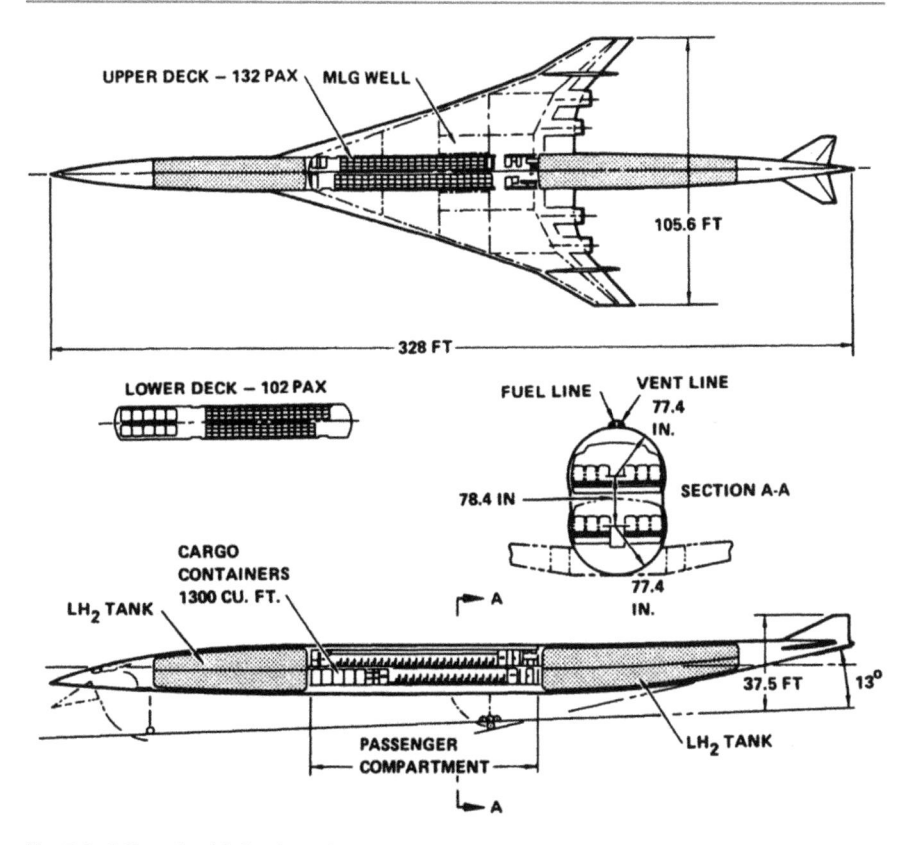

Fig. 5.9 LH$_2$ tank with fuselage changes

A further option was proposed earlier by Coppinger [17] and Verstraete [18] using twin fuselages for launching vehicles into the orbit, using hydrogen-burning GE-90 engines as shown in Figs. 5.15, 5.16 and Fig. 5.17.

Velos Orbiter, Brisbane, for example, teamed up with Boeing on a liquid hydrogen tank-fueled hypersonic launch vehicle [19] (Fig. 5.18).

A further recent, promising, and important effort is given by Airbus on liquid hydrogen storage tank placement in the aft of the aircraft. See Figs. 5.19 and 5.20 [20].

Quite recently, as of 2021, ATI [21] reveals a novel liquid hydrogen tank design for long-haul aircraft. In this design, a "bulged" fuselage accommodates the amount of liquid needed.

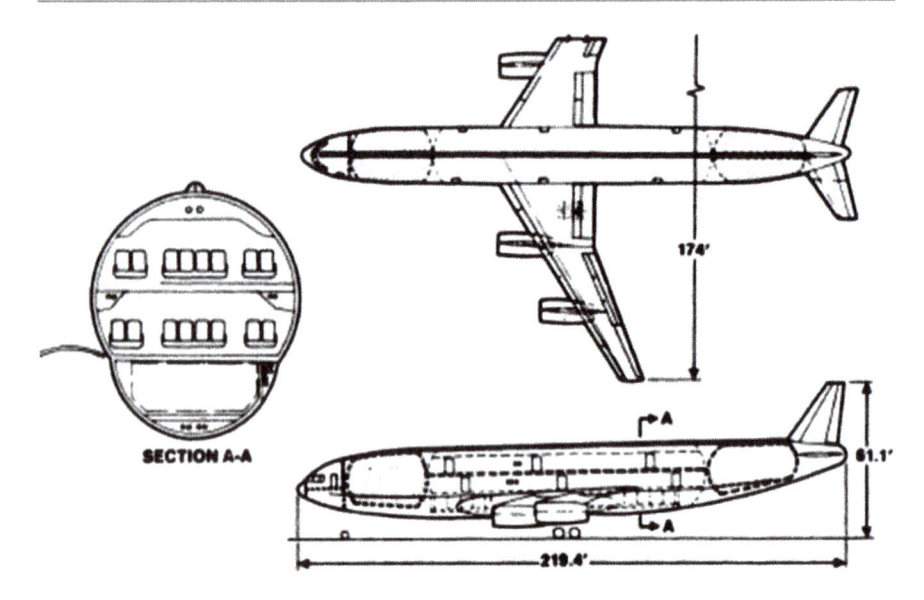

114 5 Cryotankage—Tank Shapes and Airframe Integration

Fig. 5.10 Early design study of a 400-passenger aircraft with liquid hydrogen tanks

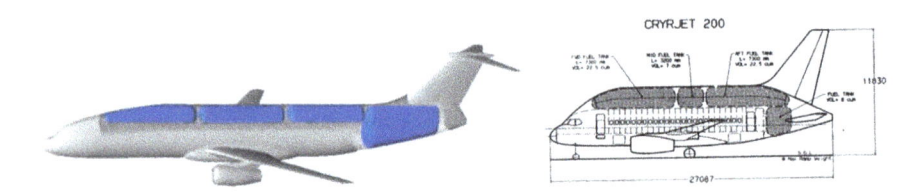

Fig. 5.11 Further LH2 tank placement

Fig. 5.12 Early Airbus cryoplane version

5.1 Non-integral vs. Integral Tank Designs

Fig. 5.13 Enable H$_2$—front view

Fig. 5.14 Enable LH2—design study

Fig. 5.15 Twin fuselage for orbit launch vehicles

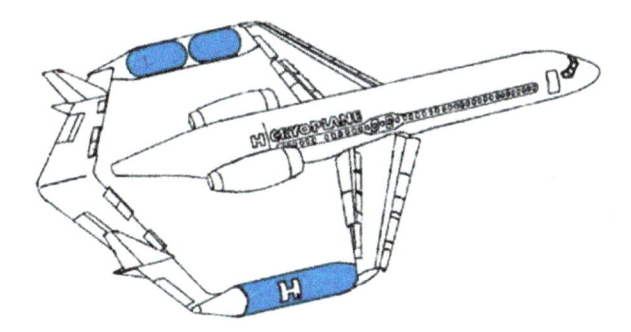

5.1.5 A321 XLR Hybrid-Electric Aircraft with Superconducting Propulsors

A fully superconducting machine model, see Fig. 5.21, is also presented by RollsRoyce [22] with liquid hydrogen tanks placed in the aft and wings (Fig. 5.22).

See also Corduan et al. [23] for hybrid-electric aircraft.

For further details on superconducting motors please see Chap. 9.

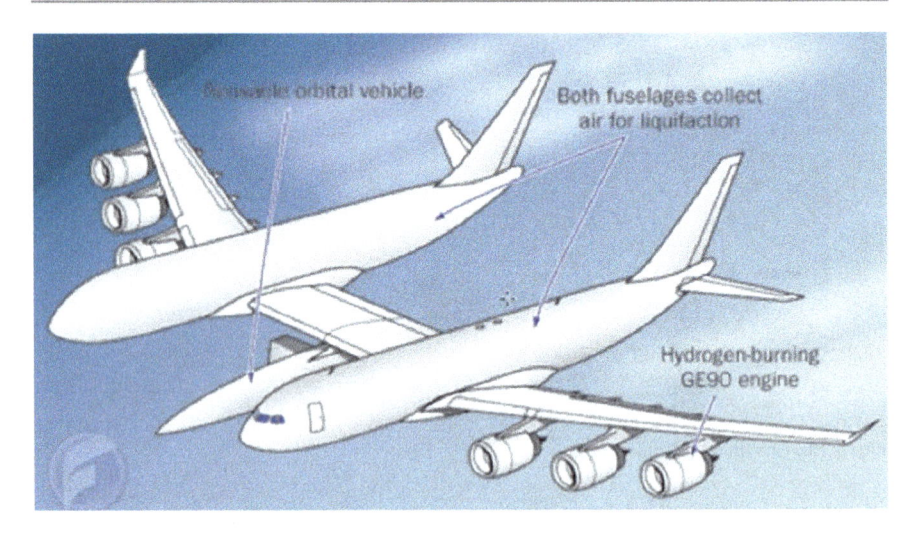

116　　　　　　　　　　5　Cryotankage—Tank Shapes and Airframe Integration

Fig. 5.16 Twin boom—design study

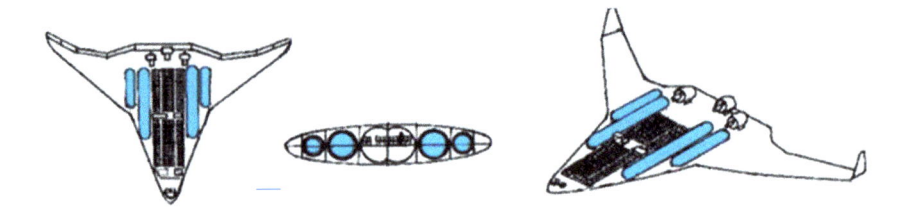

Fig. 5.17 Blended wing body shapes

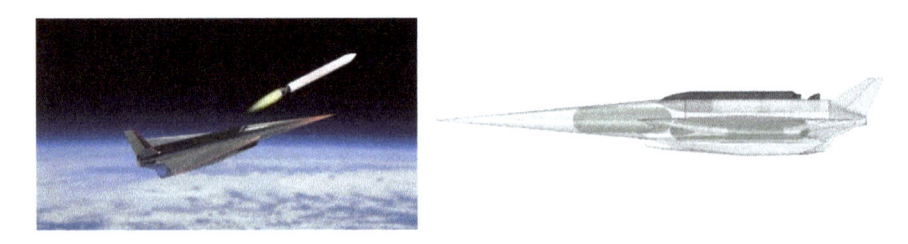

Fig. 5.18 Velos Orbiter LH$_2$ tanks (left) and side view

Fig. 5.19 Airbus ZEROe design

5.1 Non-integral vs. Integral Tank Designs

Fig. 5.20 ATI Long haul aircraft design with bulged forward fuselage liquid hydrogen storage

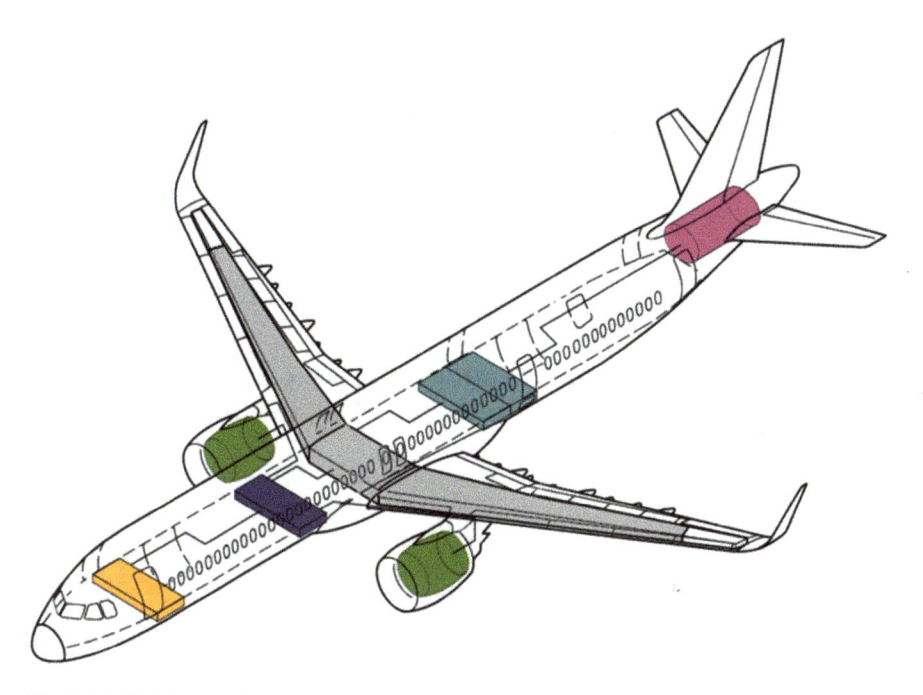

Fig. 5.21 Hybrid electric aircraft A321 XLR

118   5 Cryotankage—Tank Shapes and Airframe Integration

Fig. 5.22 Fully superconducting propulsor for hybrid aircraft

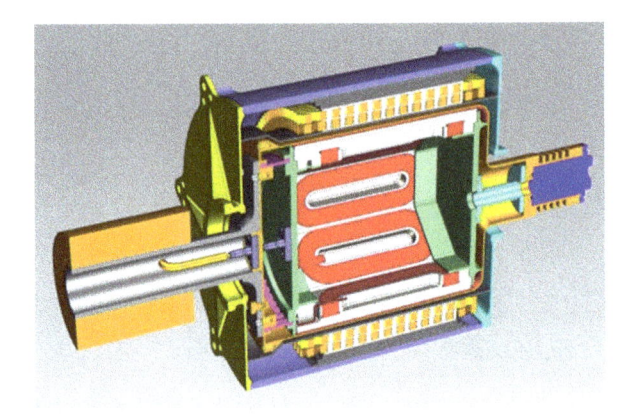

Fig. 5.23 Early design concept of CHEETA cryoaircraft

5.1.6 CHEETA All-Electric Aircraft with Superconducting Propulsors and Superconducting Components

Sustainability through Cryogenic Hydrogen-Electric Aviation: Research of the **C**enter for **H**igh-**E**fficiency **E**lectrical **T**echnologies for **A**ircraft (CHEETA). A wing-storage example that was developed in the early stages of the CHEETA research effort is shown in Fig. 5.23. In the recent, advanced design, CHEETA is an 8-tank cryoaircraft as shown in Fig. 5.24 [24].

For the CHEETA cryoaircraft the fuselage was changed in a drag-minimized way that allows to accommodate 4 cylindrical tanks and 4 tapered cylindrical wing storage tanks. The advantage from the cryogenic perspective is the division of tanks, being placed to the right and left of the airframe. The CHEETA cryoaircraft has to

5.1 Non-integral vs. Integral Tank Designs

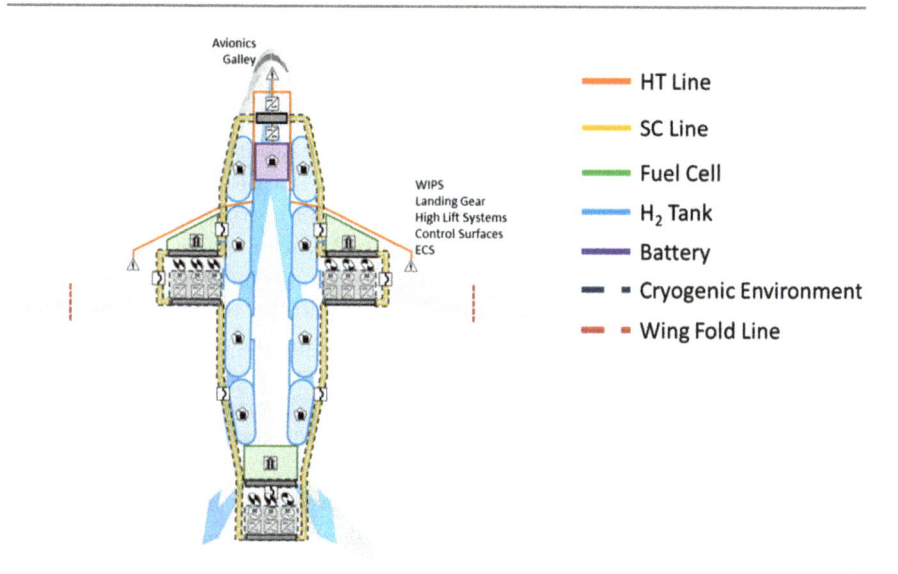

Fig. 5.24 Internal layout of hydrogen-electric power and energy system, CHEETA configuration (8-tank version)

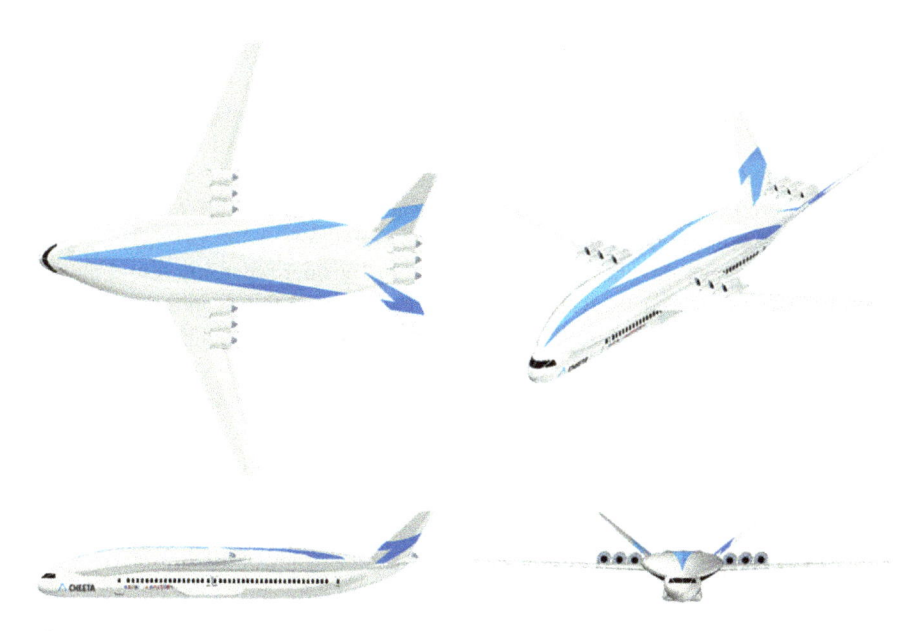

Fig. 5.25 CHEETA aircraft design views

provide space for all components that are required for the all-electric aircraft configuration as shown in Fig. 3.3 of Chap. 3 and Fig. 5.25 [25].

A further advantage is that 4 tanks can be fueled or defueled at the same time, but also provide redundancy in case of pump failures or other.

5.1.6.1 (CHEETA tank configuration) Phillip J. Ansell

When considering the placement of liquid hydrogen tanks into an aircraft configuration, there are several key aspects that limit play a critical role of the installation position and operating logistics. First, the airworthiness standards of fuel tanks for transport aircraft for US certification are discussed in 14 CFR Part 25, under section § 25.963. While these regulations were established for hydrocarbon-based fuels, it is reasonable to anticipate that these standards will remain largely unchanged with future adoption of hydrogen systems. Of note, fuel tanks must be installed to prevent fuel release into the fuselage, or other safety-critical regions of the aircraft. Additionally, fuel tanks must be installed to prevent rupture in the event of a belly landing or other contact of the aircraft with the ground during takeoff or landing operations. Similarly, the tanks must be configured in the aircraft to prevent rupture if subject to failure of landing gear or propulsor components, which is often associated with flying debris.

For these reasons, the aircraft concept studied under the CHEETA program featured hydrogen tank installations configured high on the aircraft for two principal reasons. First, potential puncture of the tanks was avoided under a condition where failed landing gear components or foreign object debris from the runway during takeoff or landing operations were directed into the underside of the fuselage. Second, if any leaks in hydrogen tanks or transfer lines emerged during the aircraft operation, the associated highly buoyant gaseous hydrogen boiloff would not pass through the passenger compartment of the fuselage. Rather, these hydrogen gases would be vented at the apex of an unpressurized center body where the tanks are housed. Moreover, a longitudinal gap in the multitank configuration is observed in the region immediately adjacent to the wing-integrated propulsor bank. This gap was configured to mitigate the probability of tank puncture in the rare event of a fan-blade-off event or other uncontained propulsor failures.

There are numerous other important factors to consider regarding liquid hydrogen placement as well. However unlikely, if a catastrophic failure in the main fuselage structure occurred, additional steps must be taken to ensure that liquid hydrogen spills largely avoid the passenger compartment of the aircraft. As such, for the CHEETA configuration the tanks are installed laterally outside of the region where passengers sit, preventing the potential for frostbite if dripping or spilled LH_2 was produced under extreme circumstances. Moreover, the hydrogen flow pathways never cross immediately overhead of any passengers or crew of the aircraft. The only connections across the aircraft plane of symmetry occur at the aft end of the aircraft and contained within a firewall next to the fore galley of the aircraft. In addition, since hydrogen is gradually removed from the tanks across the duration of a given flight, this flow rate leads to a reduction in the associated aircraft weight. If the local removal of hydrogen mass at the tank installation positions leads to a sufficiently large shift in the overall aircraft center of gravity position, or moments of inertia, such a shift across a flight can lead to stability and control challenges for the aircraft operation.

For more in-depth information, see also Section on "Tank safety" in Appendix 3.

For an update on the most recent activities in the development of liquid hydrogen-powered aircraft see the review by Tiwari et al. [26] and further details given in Appendix 5.

References

1. N. Sirosh, Hydrogen Composite Tank Program, Proceedings of the 2002 U.S. DOE Hydrogen Program Review NREL/CP-610-32405, 2002
2. A.J. Colozza, "Hydrogen Storage for Aircraft Applications Overview", NASA/CR-2002-211867, 2002
3. G.D. Brewer, Hydrogen Aircraft Technology, CRC Press, 1991
4. L. Klebanoff, Hydrogen Storage Technology, CRC Press, 2012
5. R.F. Barron, "Cryogenic Systems", Monographs on Cryogenics, 2nd edition, Oxford University Press, 1985
6. K. Sowiński, The Ritz method application for stress and deformation analyses of standard orthotropic pressure vessels, Thin-walled structures, 162 2021 107585
7. J. Huete, P. Pilidis, "Parametric study on tank integration for hydrogen civil aviation propulsion" International Journal of Hydrogen Energy" vol. 46 issue 74 pp. 37049-37062, 2021
8. C. Winnefeld, "Modelling and Designing Cryogenic Hydrogen Tanks for Future Aircraft Applications", Energies, 11, 2018, 105; pp. 1–23, doi:https://doi.org/10.3390/en11010105
9. J.P. Penland, "Liquid hydrogen fueled Boeing 737–200", Penland diaries, 1973
10. D. Verstraete, "Hydrogen fuel tanks for subsonic transport aircraft", International Journal of Hydrogen Energy 35 2010 pp. 11085–11098
11. I. G. Tapeinos, S. K., R. M. Groves, Design and analysis of a multi-cell subscale tank for liquid hydrogen storage, International Journal of Hydrogen Energy 41, 2016, 3676-3688
12. Airbus Deutschland, "Liquid Hydrogen Fuelled Aircraft—System Analysis", Final technical report, Project no.: GRD1-1999-10014, 2003
13. R. Faß, H2 CRYOPLANE, Flugzeuge mit Wasserstoffantrieb, Airbus Deutschland 2001 https://www.fzt.haw-hamburg.de/pers/Scholz/dglr/hh/text_2001_12_06_Cryoplane.pdf
14. A. Henn, "Entwurf eines wasserstoffgetriebenen Passagierflugzeugs für die lange Mittelstrecke", Diplomarbeit, 1990
15. https://www.enableh2.eu/technologies/ and Vishal Sethi at all IEEE Electrification Magazine / JUNE 2022, Figure 6
16. M. Sefain, R. Jones "*Unconventional Aircraft Configurations*" Task Final Report 2.4-4, Cryoplane project, 2002
17. R. Coppinger, "*ESA study finds in-flight oxidizer collection possible*", http://www.flightglobal.com/articles/2007/07/10/215340/esa-study-finds-in-flight-oxidiser-collection-possible.html: Flightglobal, (2007)
18. D. Verstraete, D. Bizzarri, and P. Hendrick In Flight oxygen collection for a two-stage air launch vehicle: Integration of vehicle and separation cycle design Progress in Propulsion Physics pp 551–568 2009
19. Townhall meeting, Conference, with Airbus and https://spaceaustralia.com/news/hypersonix-and-boeing-join-forces-develop-hypersonic-launch-vehicle
20. https://www.airbus.com/en/newsroom/news/2021-12-how-to-store-liquid-hydrogen-for-zero-emission-flight
21. ATI long haul aircraft, https://aviationweek.com/special-topics/sustainability/ati-unveils-long-haul-liquid-hydrogen-airliner-under-flyzero
22. M. Boll et al, "A holistic system approach for short range passenger aircraft with cryogenic propulsion system" Supercond. Sci. Technol. 33 2020 044014

23. M. Corduan, "Topology comparison of superconducting AC machines for hybrid-electric aircraft" IEEE Transactions on Applied Superconductivity, IEEE Transactions on Applied on Superconductivity vol 30 No 2 2020
24. W. Stautner, P. Ansell, K. Haran, D. Mariappan, "Liquid Hydrogen Tank Design for Medium and Long Range All-electric Airplanes" Invited talk CEC/ICMC 2021
25. W. Stautner, P. Ansell, K. Haran, "CHEETA: All-electric Aircraft takes Cryogenics and Superconductivity 'On board'", *IEEE Electrification Magazine*, June 2022
26. S. Tiwari, M. J. Pekris, J. J. Doherty "A review of liquid hydrogen aircraft and propulsion technologies International Journal of Hydrogen Energy" 57 pp 1174–1196 2024

6

Hydrogen Tank—Cryocircuit, Integration of Components, Instrumentation

Obviously, when we look at the various possible tank designs and as mentioned in the Cryotankage sections, liquid hydrogen tanks are not simple stand-alone tanks. They need to be embedded into an aircraft cryogenic structure with multiple tube connections leading from and deep into the liquid tank. In addition, safety features, as well as measurement and control technology need to be implemented. In the following sections, we will discuss cryocomponents, as well as their integration and process engineering features. The aircraft cryocircuit and the flow of cryogens requiring multiple cryolines have already been discussed Cryogen flow circuit Fig. 3.3 in Chap. 3. Here we discuss the logistics of the hydrogen tank piping and their arrangement in further detail.

Figure 6.1 shows the minimum required piping for a liquid tank, consisting of a hydrogen fill line and two extraction lines, one for liquid hydrogen, the other for gaseous extraction or return flow from the aircraft cryocircuit. For the CHEETA program, the liquid extraction line transfers liquid through the complete cryogenic circuit within the plane fuselage, through the current transfer lines, the motor cooling, the power electronics, and finally to the fuel cell. In addition, one branch of the cryogenic circuit directly feeds into a heat exchanger providing additional gas flow to the fuel cells (see Cryogen flow circuit schematics in Fig. 3.3 in Chap. 3).

To avoid basic tube routing errors, we need to reflect on how to safely implement those.

6.1 Design of Cryogenic Penetrations for Liquid Bulk Tanks

Case study: Design a penetration through a liquid hydrogen tank avoiding pitfalls and design flaws.

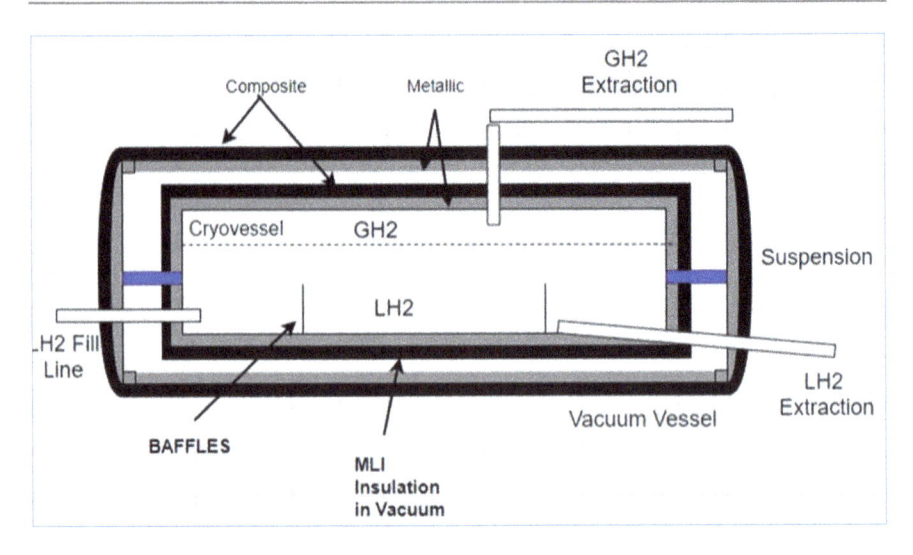

124 6 Hydrogen Tank—Cryocircuit, Integration of Components, Instrumentation

Fig. 6.1 Pipes in liquid tank

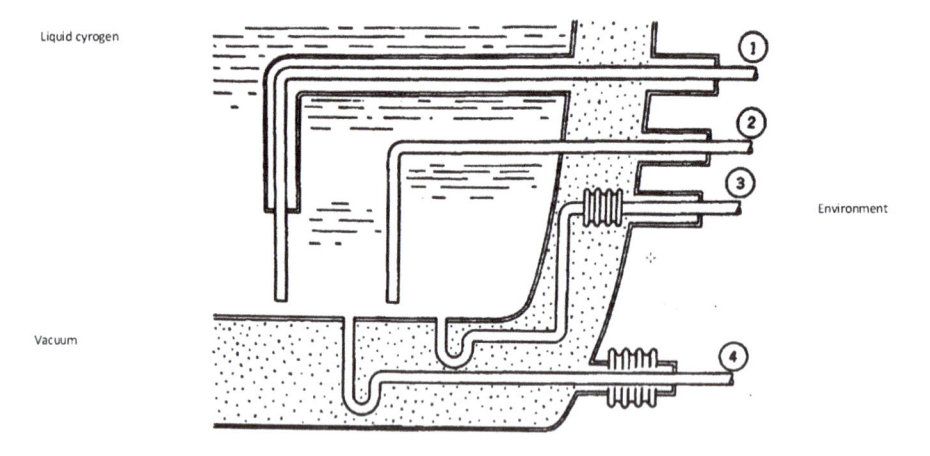

Fig. 6.2 Various cryogenic pipe assemblies for liquid tanks

Cryogenic penetrations are tricky to design for a tank and even more challenging for composite tanks. The following gives some high-level guidelines on how to avoid the most common mistakes for metal tanks [1].

Metal-metal penetrations – Tank piping arrangements (Fig. 6.2)

Pipe 1 is the standard design for extracting liquid from a cryogenic tank (trailer) interfacing to a warm outer environment, mainly used for multilayer insulated vacuum designs. Note that the vacuum jacketed (VJ) tube shares the tank vacuum.

6.1 Design of Cryogenic Penetrations for Liquid Bulk Tanks

Although the inner tube shrinks axially, the outer jacketed tube shrinks by a similar amount keeping the stress level low. The downside is the difficulty in maintenance or disassembly. The downside for aircraft is the fact that liquid can flow into the warm line in some cruising conditions. Heat burden on tank is low however.

Pipe 2 has often been seen being implemented by novices in cryogenics, however even experts sometimes underestimate the grave consequences of this design flaw. An example is shown in Fig. 6.3 where an MRI magnet system ought to have an axial penetration rather than the common vertical turret arrangement. A design driver was the early room height restriction. A horizontal turret therefore gave easier access to the superconducting magnet with its current lead, an at first glance, attractive solution. This pipe attachment version however will not work as learned from implementing it this way.

Figure 6.3 Horizontal multifunction tube on MRI cryo vessel shows the open cryostat at assembly giving access to the SC magnet power connections, ready to weld in the horizontal turret before finally closing off the vacuum chamber. The turret was equipped with diagnostics, current lead and fill port and completely foamed out around those tubes using polyurethan. Initial liquid nitrogen cooldown of the MRI magnet using the fill port failed and completely iced up this horizontal turret. Any efforts in trying to seal remaining small gaps were futile and overrun by the uncontrollable convection currents circulating from the room temperature end to the 4 K helium vessel and back to the warm end acting like a thermal conveyor belt, resulting in a complete thermal short from room to nitrogen temperature. It was not possible to cool down this system in this way.

For an aircraft pipe 2 also has a percolation condition. Condensation of liquid will take place on the top inner surface of the horizontal pipe section within the

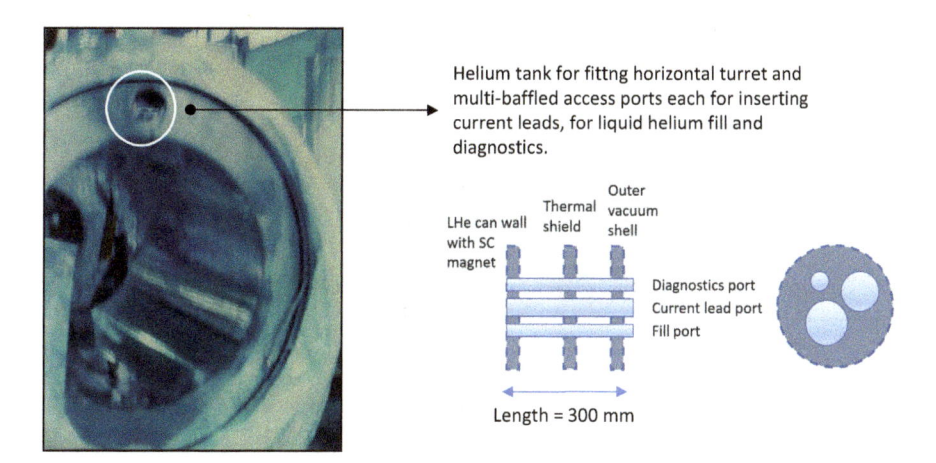

Fig. 6.3 Horizontal multifunction tube on MRI cryo vessel

inner shell. The liquid will then be able to flow into the axial section toward the outer shell where it can evaporate at the warmer end. The only way to prevent this is building another vertical line into the cold end.

Pipe 3 has a reliability problem in that over time the bellow will start to leak without any means of repairing it other than to cut the tank open. This might be acceptable in some instances but not acceptable for aircraft composite tanks. In this solution liquid will always be present in the tubing, creating a high heat load.

Pipe 4 finally has a "stratification tube" or also called "snake" in some configurations that should ensure that the horizontal section always remained filled with static gas. The vertical liquid line has a knee-shaped configuration. Line 4 can also shrink together with the corrugated tube or bellow. The bellow is accessible and can be repaired as needed, a standard procedure for metallic tanks but still problematic for composite tanks.

6.1.1 Inclined Tubes

When working in a confined environment that requires some compactness around the tank as in a typical fuselage, a further option could be an inclined tube arrangement. In practice however that has led to many other issues and sets up convective flow currents causing thermal shorts as discussed in pipe design 2, if not executed properly. This has been investigated in great detail on a liquid hydrogen tank by Langebach [2–4].

Figure 6.4 gives some examples where higher cryogen boil-off occurred due to false tube positioning at 30 and 45 degrees (MRI) and >45 degrees for high-field NMR.

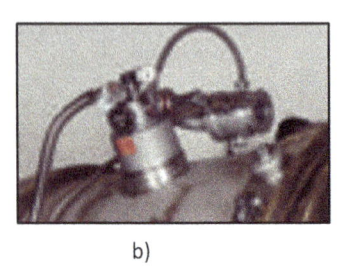

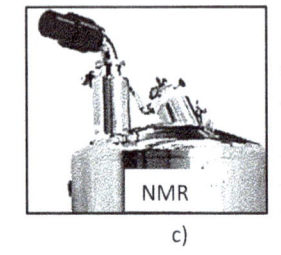

Fig. 6.4 Inclined tubes—MRI (**a, b**) and NMR (**c**)

6.1 Design of Cryogenic Penetrations for Liquid Bulk Tanks

Inclined tubes in cryogenics generally can be implemented in various functions, categorized below:

- Tubes with low pressure (open/closed ends)—dewars, helium, hydrogen, LNG, or LN_2 tanks.
- Tubes with high pressure (closed/closed with reservoirs).
- Tubes with high pressure, low-frequency pulsating flow, e.g., effect of inclined cryocoolers (pulse tube coolers).
- Tubes running at high pressure with high-frequency pulsating flow.
- Cooldown tubes.
- Heat pipes.
- Thermosiphons.
- Multifunction tubes for, for example, current lead insertion, instrumentation ports, siphon fills.

Groundbreaking research work on inclined tubes on a liquid hydrogen tank for the automotive industry that shows a workaround tubing design has been published by Langebach [2, 3]. As a main result of his findings a "snake" tube design as shown in Fig. 6.5 in most cases suffices to control the horizontal convection currents reasonably well from room temperature down to 20 K and lower. For most designs the penalty of a higher pressure drop is acceptable.

Figure 6.6 shows some earlier working designs with alternative coiled fill lines [2].

Figures 6.5 and 6.6 show the tube angle definition and improved results for the "snake" tube as compared with a horizontal tube.

In Fig. 6.7, T_w refers to the warm end, and T_k to the cold end [2].

Figure 6.8 shows research results from Langebach [2] pointing to a jump in transferred heat when inclining a horizontal tube for helium and hydrogen. The lower curve shows the improvement with a given "snake" design as shown in Fig. 6.5. For reduced pressure down to 10 bar the max. amplitude is lower, but the shape of the curve is maintained. Similar results can be seen when operating pulse tube coolers.

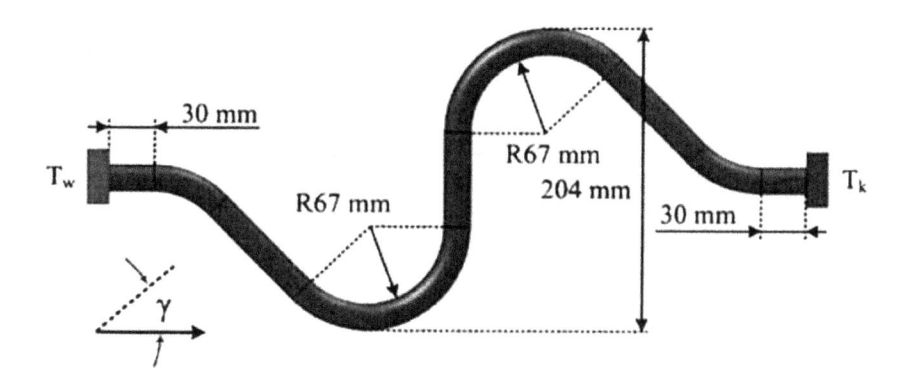

Fig. 6.5 Work around tubing structure after Langebach

128 6 Hydrogen Tank—Cryocircuit, Integration of Components, Instrumentation

Liquid hydrogen tank of 2004 (BMW group)

Liquid hydrogen tank (LLNL and BMW group, co-operation)

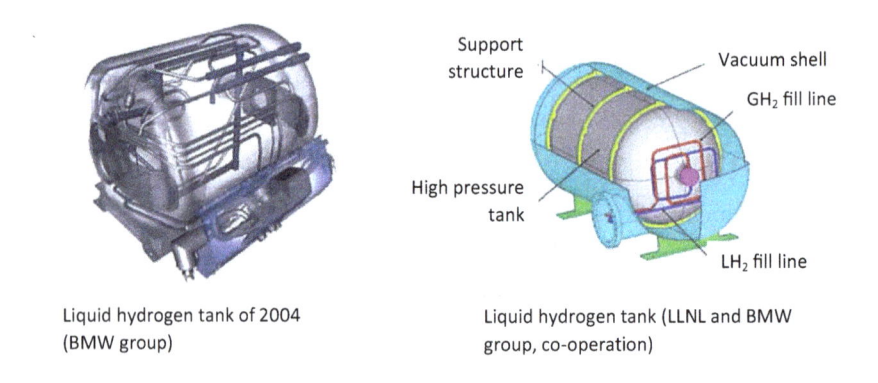

Fig. 6.6 Piping of recent liquid hydrogen tanks

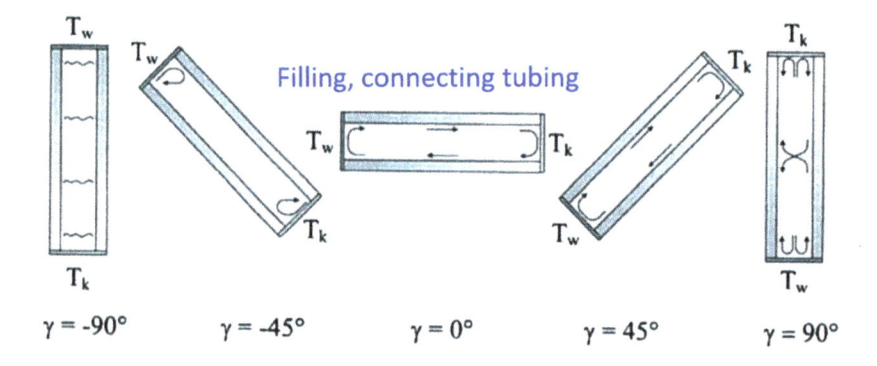

$\gamma = -90°$ $\gamma = -45°$ $\gamma = 0°$ $\gamma = 45°$ $\gamma = 90°$

Fig. 6.7 Tube orientation and definition

Any vapor filled piping connecting a cryogenic liquid warm end to room temperature can occasionally run into cryogenic flow instabilities, so-called thermoacoustic oscillations (TAOs) [5]. This is especially relevant for liquid fill/withdraw, static tubes, and large tank sizes.

In this modified table based on the overview of late Brian Hands [6], the researcher may be warned of events that cannot always be explained thoroughly but once implemented wrongly can have serious design impacts difficult to rectify.

Figure 6.9 maps flow instabilities one may run in occasionally. Please be advised that those instabilities are of concern for many applications and for aircraft since those can appear without warning. Once implemented wrongly those designs are difficult to correct. When designing multiple connected tubing in extended flow circuits with control elements, it may be advisable to back up the safe use of multiple tube circuits using dedicated CFD analysis.

6.1 Design of Cryogenic Penetrations for Liquid Bulk Tanks

a) Helium

b) Hydrogen

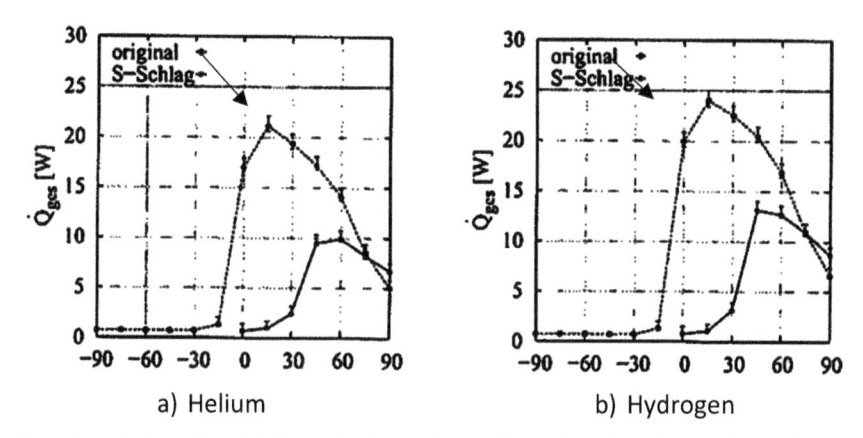

Experimental results of S shaped tube configurations, showing the total heat flux Q_{tot} depending on inclination angle γ for a) helium and b) hydrogen, at 100 bar.

Fig. 6.8 Work around tubing structure—example length 736 mm, extended in comparison to inclined tube of 500 mm

For future aircraft cryotanks bimetallic or pure composite penetrations may be required. We can differentiate as follows:

Bimetallic penetrations—Piping arrangements on a tank with different metallic materials

Bimetallic tubes are usually friction welded together and work reliably. An example would be stainless steel tubes friction welded into high-pressure aluminum gas tanks [7].

Metal/composite penetrations—Piping arrangements on a tank with at least one composite/steel or aluminum end design

As for cryogenics, this technical field is still uncharted territory and subject to further research for commercialization. Glued in tubes in a base of stainless steel are well known in cryogenics but many attempts are needed to arrive at good vacuum tight bonds. Parameters to be concerned about is leakage of any cryogen, and in particular, hydrogen into the vacuum space, leakage from environment into the tube and fatigue at glued interfaces, as well as stress concentration due to material shrinkage, possible hairline crack formation over time, and composite aging. Steel/fiber composite designs are also known as good support structures and capable to support excessive high weight.

Reinforced epoxy penetrations—Piping arrangements on a tank with composite ends

Epoxy penetrations have only occasionally been used so far. For successful implementation the mutual expansion factor is matched as shown in Fig. 6.10 with a solution proposed by Hibbs and Cox [8].

1. *System Oscillations*

Interaction of one part of a plant or component with another. Not well documented and difficult to analyze.

2. *Two-phase forced flow oscillations*

For two-phase flow we must differentiate between forced flow versus natural circulation.

Forced flow, sorted according to **decreased heat flux**:

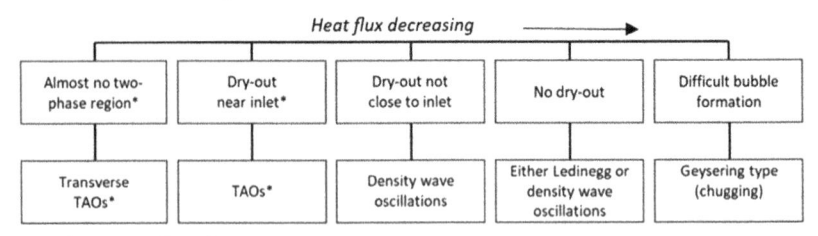

```
                        Heat flux decreasing  ──────────►
┌──────────────┬──────────────┬──────────────┬──────────────┬──────────────┐
│ Almost no    │  Dry-out     │ Dry-out not  │  No dry-out  │  Difficult   │
│ two-phase    │  near inlet* │ close to     │              │  bubble      │
│ region*      │              │ inlet        │              │  formation   │
├──────────────┼──────────────┼──────────────┼──────────────┼──────────────┤
│ Transverse   │   TAOs*      │ Density wave │ Either       │ Geysering    │
│ TAOs*        │              │ oscillations │ Ledinegg or  │ type         │
│              │              │              │ density wave │ (chugging)   │
│              │              │              │ oscillations │              │
└──────────────┴──────────────┴──────────────┴──────────────┴──────────────┘
```

3. *Two-phase natural circulation*

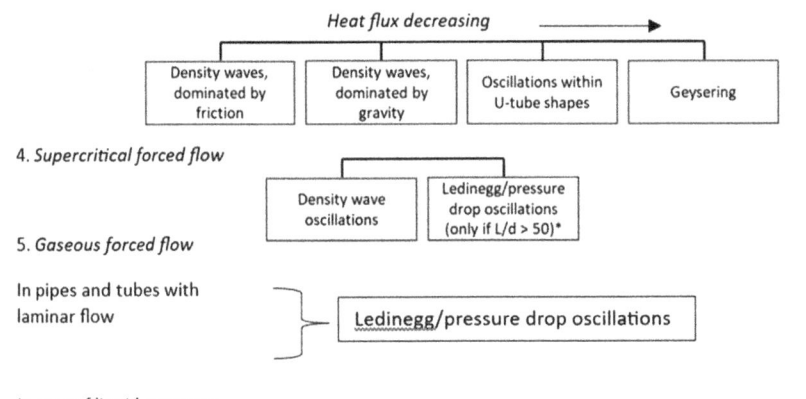

```
                    Heat flux decreasing  ──────────►
┌──────────────┬──────────────┬──────────────┬──────────────┐
│ Density      │ Density      │ Oscillations │  Geysering   │
│ waves,       │ waves,       │ within       │              │
│ dominated by │ dominated by │ U-tube       │              │
│ friction     │ gravity      │ shapes       │              │
└──────────────┴──────────────┴──────────────┴──────────────┘
```

4. *Supercritical forced flow*

```
┌──────────────┬─────────────────────┐
│ Density wave │ Ledinegg/pressure   │
│ oscillations │ drop oscillations   │
│              │ (only if L/d > 50)* │
└──────────────┴─────────────────────┘
```

5. *Gaseous forced flow*

In pipes and tubes with laminar flow — **Ledinegg**/pressure drop oscillations

In case of liquid cryogens:

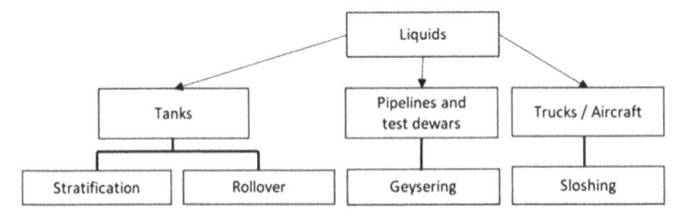

```
                      ┌─────────┐
                      │ Liquids │
                      └─────────┘
             ┌─────────────┼─────────────┐
        ┌────────┐   ┌──────────────┐  ┌──────────────┐
        │ Tanks  │   │ Pipelines and│  │ Trucks /     │
        │        │   │ test dewars  │  │ Aircraft     │
        └────────┘   └──────────────┘  └──────────────┘
        ┌────┴────┐        │                  │
   ┌──────────┐ ┌──────┐ ┌──────────┐    ┌──────────┐
   │Stratifi- │ │Roll- │ │Geysering │    │ Sloshing │
   │cation    │ │over  │ │          │    │          │
   └──────────┘ └──────┘ └──────────┘    └──────────┘
```

* does not normally occur in Cryogenics

Fig. 6.9 Guide to instabilities, after hands

This shows a reliable nonmagnetic liquid helium-tight thermal feedthrough through a fiberglass Dewar wall. The cross section of the thermal connector details a copper ring, filled epoxy, interference fit, separated by a wall of helium and vacuum and is leak tight.

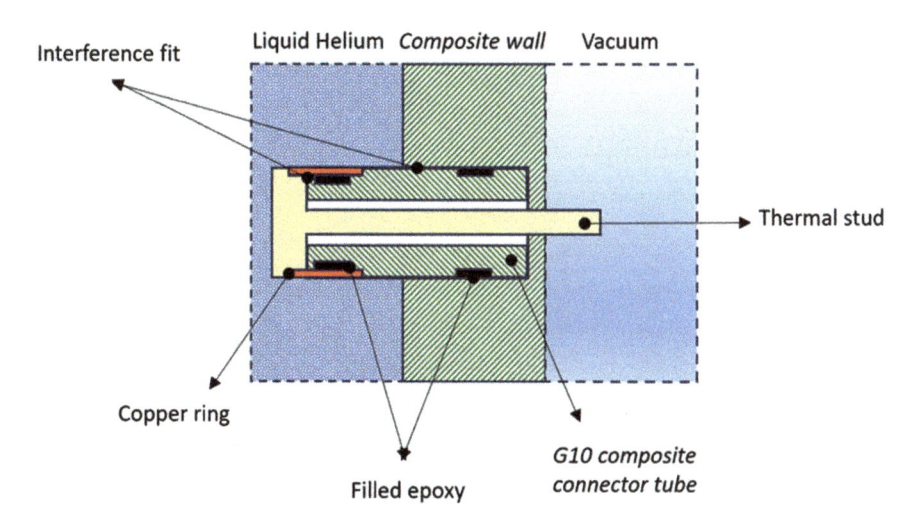

Fig. 6.10 Epoxy penetrations through a composite wall

This is one design approach, but needs to be carefully gauged with respect to operating conditions and investigated and tested for hairline cracks emerging at frequent pressure cycles.

6.1.2 Metal/Composite Bonding

Where contacts between composites and metals are required, for example when connecting flanges or ring-type structures or tubes, the thermal contraction between those components needs to be known and taken into consideration. For any adhesive to work reliably, the bond characteristic "design to failure" is an important parameter. If an outer composite tube is to be bonded to an inner steel or aluminum tube, the bond failure develops due to the contraction of the inner tube. If the composite tube is on the inside, this bond can fail since the outer diameter of the GRP, e.g., G10 or G11 tube shrinks more than steel. Evans [9] points out that those stresses can be reduced by using a sandwich-like structure shown in Fig. 6.11.

6.2 Typical Process Flow Circuit for a Liquid Hydrogen Storage Tank with Instrumentation

A dual-walled composite tank needs to monitor pressure, fill level, temperatures, flow rates, as well as possible tank leakage rates.

A connected process engineering interface is symbolically depicted below, with the attached components as per Table 6.1.

Fig. 6.11 Metal/composite tube bonding example

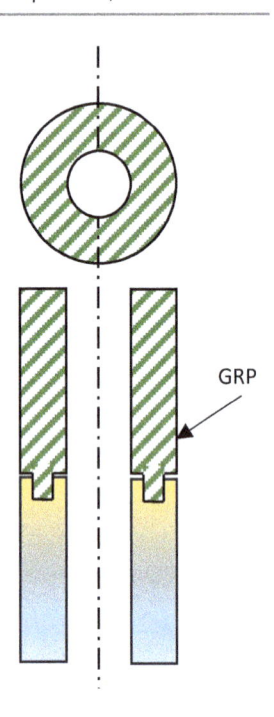

Table 6.1 LH$_2$ tank control elements

A Main suction valve	H Bleed-off valve
B priming valve	I vent valve
C meter inlet valve	J rapid priming valve
D meter cool-down valve	K relief valve (e.g., 50 psi)
E pressurizing valve	L relief valve (100 psi overrated system pressure)
F meter bypass valve	M pressure gauge
G discharge valve	O superconducting motor
P fuel cells	

One of the key elements, the pump, is integrated as per Fig. 6.12. For aircraft, the delivery line either directly feeds propulsors or fuel cells.

For further information, see also Haselden [10] on pump feed logistics.

An alternative, aircraft design is given by Stroman [11] in Fig. 6.13 for the Ion Tiger LH$_2$ system with an immersed pumping arrangement of a turbine feeding the fuel cell. A modified version includes an initial fill tank (not shown).

The strategy here is to divide the tankage into a storage tank and a supply "control" tank, whereas the latter is connected to the process elements. As we extend the cryogenic circuit on the plane, we need to qualify the required piping that feeds the individual components.

6.2 Typical Process Flow Circuit for a Liquid Hydrogen Storage Tank...

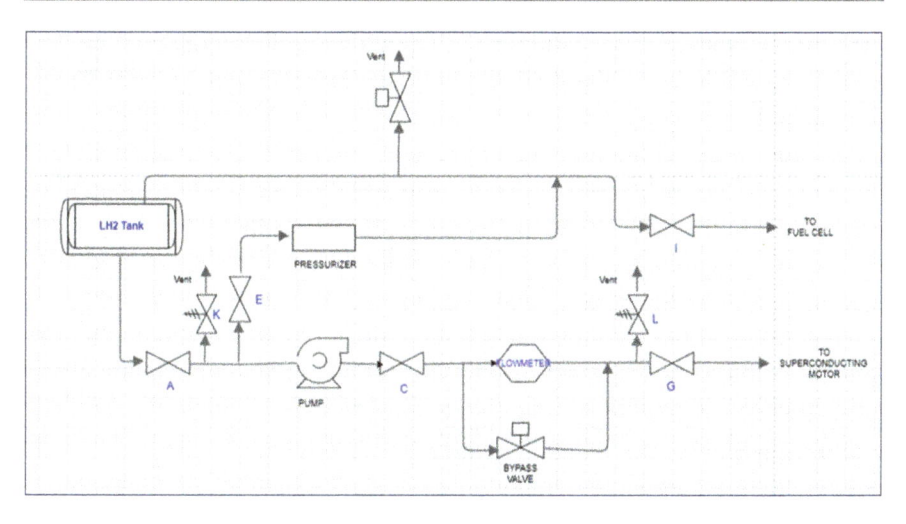

Fig. 6.12 Cryo flow circuit on tank

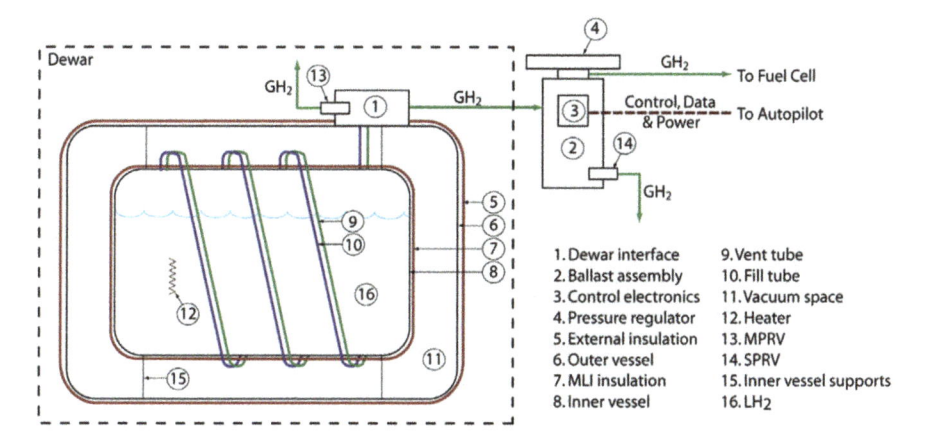

Fig. 6.13 Dual tank liquid withdrawal strategy

6.3 Tank Instrumentation

6.3.1 Fill Level Sensors for Aircraft

Recently, good progress has been made with using capacitive level sensors, similar to routine use on LN_2 tanks or helium reservoirs.

Capacitive sensors rely on the change of the dielectric constants of gaseous and liquid hydrogen. These differences are successfully leveraged in liquid hydrogen storage systems (LH_2). See also Matsumoto [12] et al. for capacitive level meter designs. However, one must take the thermal expansion of liquid hydrogen into

Table 6.2 Dielectric constant of cryogens

Relative dielectric constant ε_r		
Element	Vapor $\varepsilon_{r,v}$	Liquid $\varepsilon_{r,l}$
Oxygen	1.00156	1.484
Argon	1.00175	1.532
Nitrogen	1.00208	1.433
Neon	1.00129	1.174
Hydrogen	1.00387	1.228
Helium	1.00618	1.048

consideration and its resulting capacitance decrease. See also Table 6.2 for dielectric constants of cryogens.

Progress has been reported by the Dresden Team that shows that capacitive sensors theoretically also should work for cryo-compressed hydrogen [13].

A first feasibility study of using superconducting level probes for LH_2 with a medium temperature superconductor (MgB_2) utilizing the liquid hydrogen temperature was executed by Kajikawa et al. [14].

6.4 Cryogenic Transfer Lines

An LH_2-powered aircraft needs cryogenic transfer lines as well as current carrying transfer lines. We differentiate between cryogenic vacuum jacketed (VJ) liquid or gaseous transfer lines, and so-called cable cryostats (CC).

6.4.1 Vacuum Jacketed Lines (VJ)

Figure 6.14 on the left shows some typical rigid VJ lines [1], on the left with their different choices of internal support structure against the vacuum shell [15].

Figure 6.14 on the right [15] combines flexible lines with rigid lines. On the left, some principal control elements are shown that are required for evacuation (pumping port), and to maintain as well as to monitor vacuum conditions in the transfer line. Those control elements require modification before they can be used on an aircraft.

An actual on-ground implementation is shown in Fig. 6.15 [15].

Nexans flexlines (Highflex lines) and semi-flexible Cryoflex transfer lines are key elements in developing new hydrogen uses for energy and mobility decarbonization by enabling the transfer of liquefied hydrogen between storage tanks and end-use tanks [15].

Transfer lines have a long history in cryogenics. A very early rigid transfer line of 1973 is shown in Fig. 6.16 meant for carrying current through an electric cable, cooled with liquid hydrogen.

For that an aluminum conductor with low resistance for 500 kV was used [16].

6.4 Cryogenic Transfer Lines

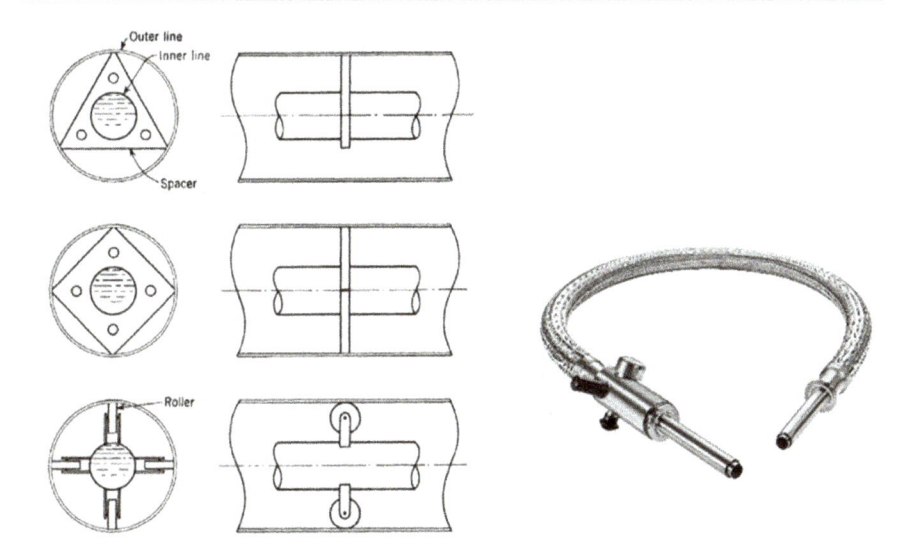

Fig. 6.14 Rigid lines, with ceramic spheres support and flexible lines from Nexans (right)

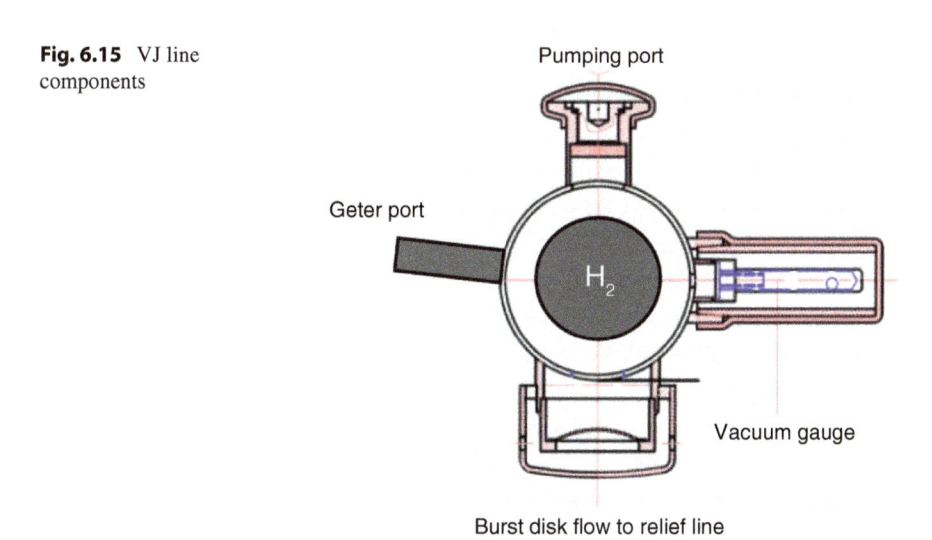

Fig. 6.15 VJ line components

A different, very recent flexible line is shown on the left in Fig. 6.17. This cross section shows two coolant channels. One main coolant channel that can be used for transferring a liquid from A to B or can even include a cable conductor and a low-cost shield coolant channel. This design takes advantage of using one coolant at a higher cryogenic temperature to cool the other at a lower temperature. Those designs were introduced very early in cryogenics, for example cooling liquid helium with a separate liquid hydrogen or liquid nitrogen tank attached to the helium tank [17].

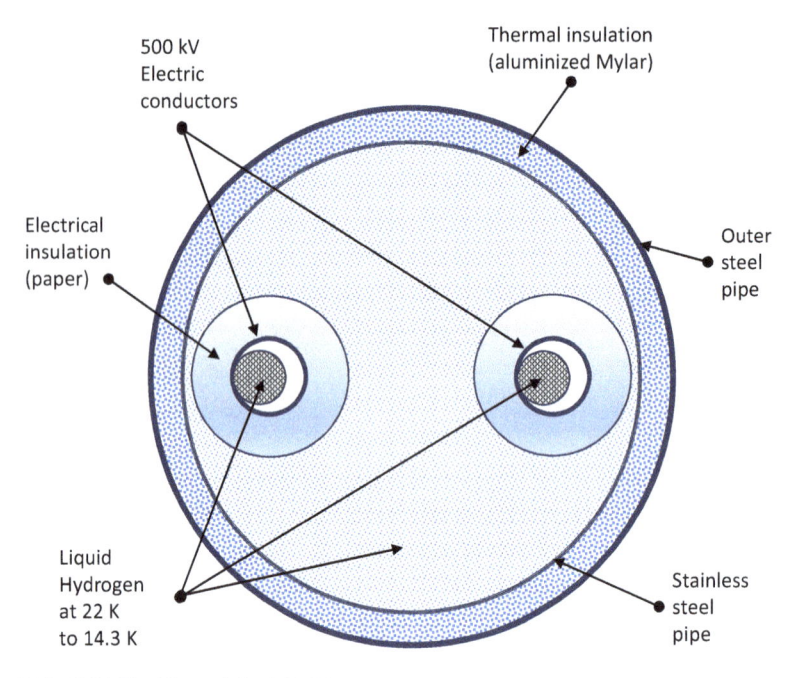

Fig. 6.16 500 kV cable cooled with LH_2

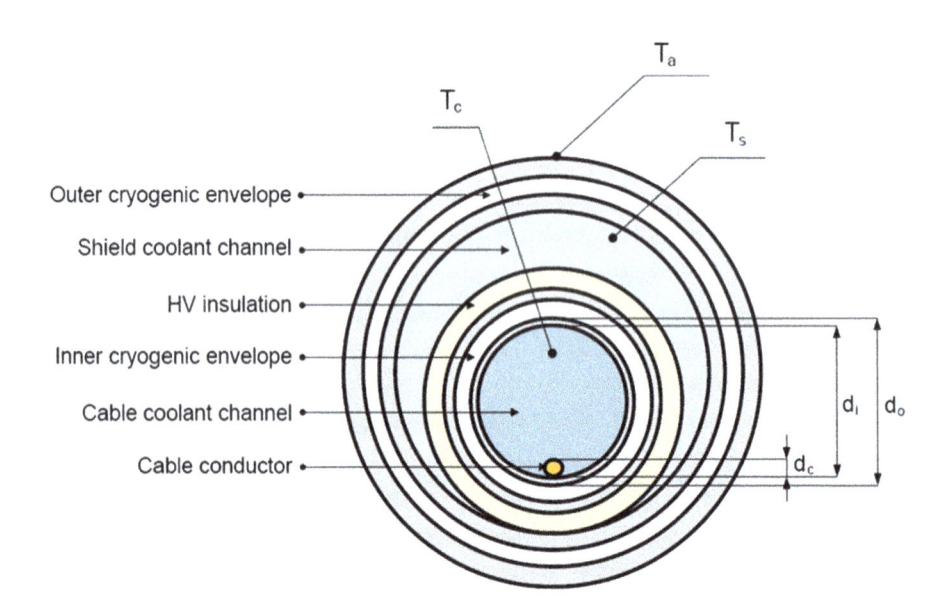

Fig. 6.17 Dual coolant transfer lines

6.4.2 Jacketed Lines Without Vacuum

In 1977, the GE Research Center [18] investigated the use of low-cost glass reinforced epoxy piping for liquid nitrogen-cooled AC transmission cables and showed that, in principle, it should be possible to run cryogens through those pipings with a 40-year life.

Initial technology drivers were cost and the ability of nitrogen to impregnate the cable dielectric like oil in oil paper cables. The resin binders were diglycidylether (or so-called bisphenol A) with 100 parts by weight, methyltetrahydrophthalicanhydride at 85 and Tris (dimethylaminomethyl) phenol with 0.5 parts by weight using Owens-Corning Fiberglass Type 410AAE (Fig. 6.18).

Computed stresses again correlate strongly with different winding patterns.

6.4.3 Cable Cryostats (CC)

Cable cryostats (CC), on the other hand, consist of corrugated tubes in a vacuum environment (shell) and are of interest to carry and transfer current to a consumer. Those lines can also be either rigid or flexible (see Figs. 6.19 [20] and 6.20 [19]).

For aircraft, flexible lines may be of interest. Flexible cables are also known since long, e.g., 55 years since its introduction and since then have seen a few improvements [21].

In addition, those flexible cable cryostat can be fitted with a high temperature superconducting (HTS) cable [19]. An example of a HTS cable cryostat is shown in Fig. 6.21 on the left.

Even though those cables are flexible, the parasitic heat loads are now in the range of 8–10 W/m^2. The losses are higher than with VJs due to their support mechanism and different insulation. A flexible tube, for example, of an outer diameter of 44 mm, with a minimum bending radius of approx. 700 mm, with one bend of

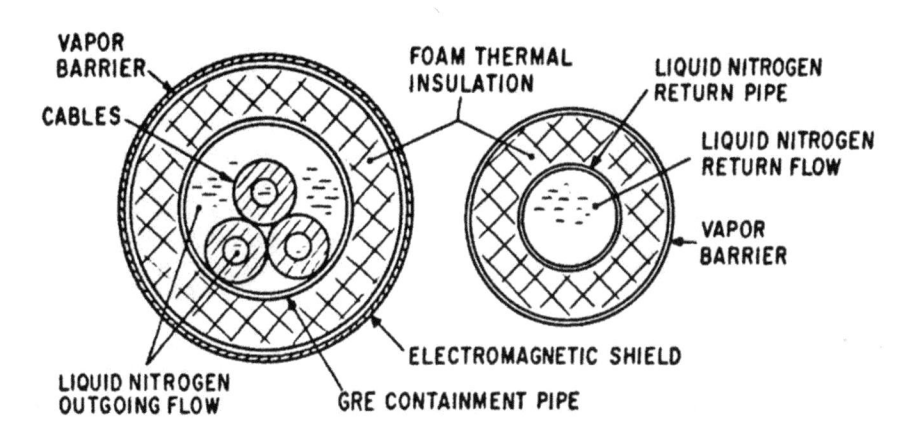

Fig. 6.18 Early AC transmission cryo cable by GE

Fig. 6.19 Nexans cable cryostat (courtesy of Nexans)

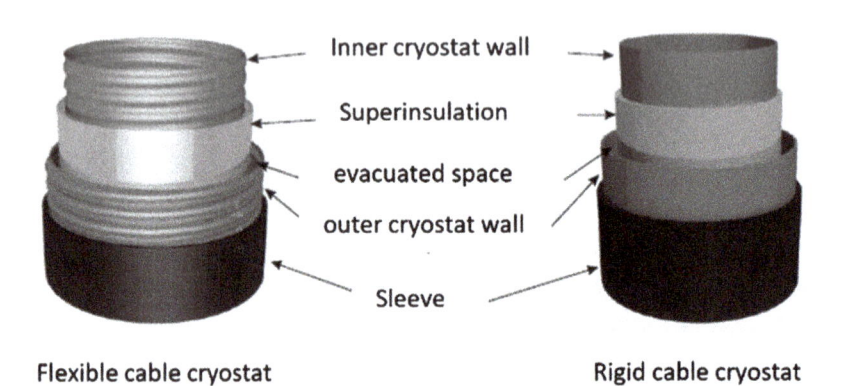

Fig. 6.20 Flexible and rigid cable cryostats

6.4 Cryogenic Transfer Lines

Fig. 6.21 420 kV HTS cable in cable cryostat

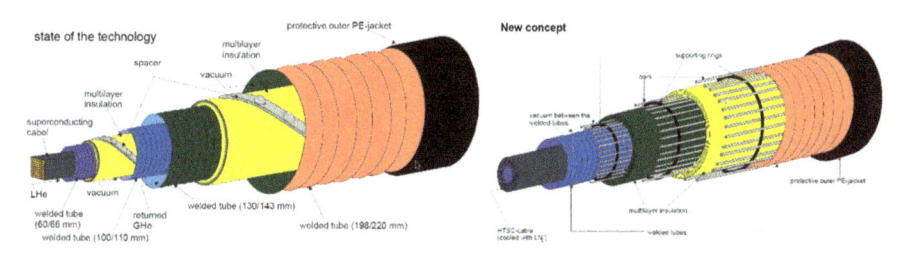

Fig. 6.22 Cable cryostat—state of the art (left) and new design (right)

350 mm, results in a heat leak of 0.6 W/m [22], corresponding to a LH$_2$ boil-off, of about 70 ml per line meter.

Electric wiring around hydrogen tubes and transfer lines must comply with NFPA 50 B 1994 [23] or later guidelines.

Figure 6.22 (left) shows a typical design. The figure on the right shows a new design with a parasitic heat load reduction by almost 70 percent. This is achieved by improving the overall thermal insulation [24]. Karlsruhe Institute of Technology (KIT) proposed a possible strategy to reduce the losses acting on those transfer lines taking them down to 3.5 W/m^2.

For the CHEETA project, the liquid hydrogen flow rate is almost close to zero for use with MgB$_2$-type current leads which requires novel designs for the cable cryostat to permanently remove vapor and to maintain steady state temperature conditions, within the line as shown in Fig. 6.23 [25].

Recently, cable cryostats sparked further interest for transporting cryogenic liquids, and in particular liquid hydrogen over long lengths. In those designs the cooling effect of the coolant is exploited together with the coolant being delivered to a possible consumer. Johnston couplings, invented by Herrick L. Johnston, provide appropriate vacuum-insulated couplings so that connections can be made easily. Those lines are recommended for liquid hydrogen transfer, see also Appendix 7. However, quick connects for VJ lines still do not exist. For some gases, so-called

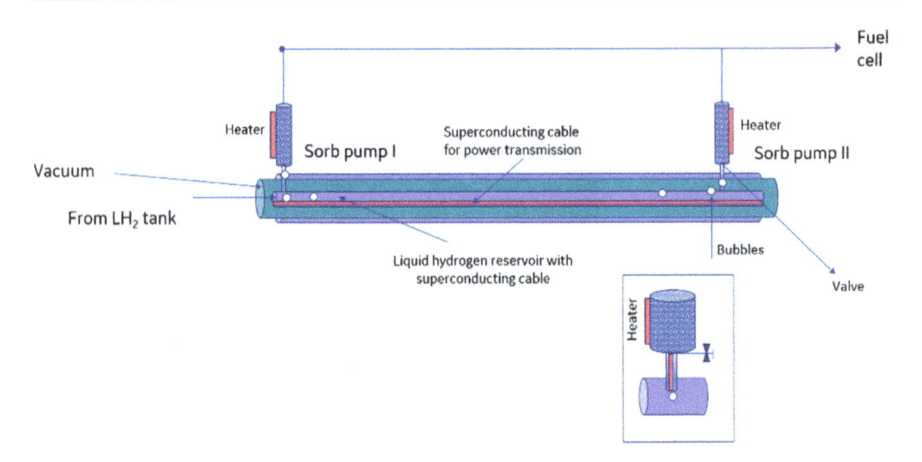

Fig. 6.23 Sorption kits inline with cable

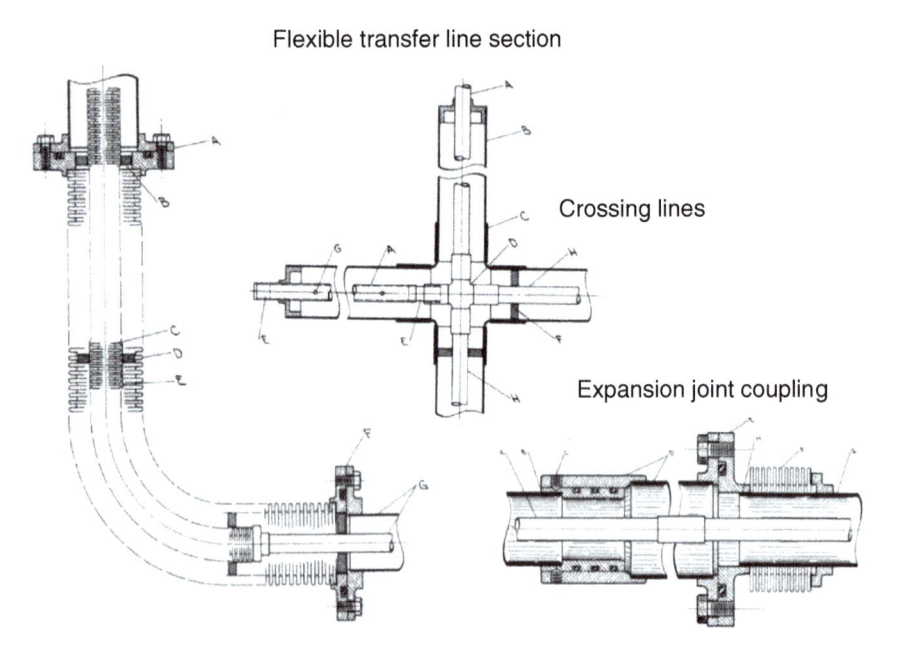

Fig. 6.24 Mixing flexible and rigid transfer lines

"dry break" connections exist up to a nominal diameter of 50 mm by Gather industries, Stäubli, or others.

Martin and Park [26] show the development of LH_2 flow transfer lines, elements known since 1953, with multiple cross points required on an aircraft, as depicted in Fig. 6.24.

6.5 Valves, Safety Valves, and Special Features

Many of the process engineering components have been developed already at an early stage as soon as the first liquid hydrogen tanks were built (see Appendix 4 on hydrogen tanks) and have been continuously developed and improved. Here we can only point to a few qualified components and how they are implemented for space and automotive applications. In 2006, a first attempt in standardizing liquid hydrogen tanks and those associated control elements was published as ISO 13985 [27], that was then reviewed and confirmed in 2015 on "Liquid hydrogen—Land vehicle fuel tanks." This standard interfaces to other cryogenic standards which are listed below for convenience (Table 6.3).

Figures 6.25 and 6.26 show suitably sized samples of flashback arrestors, safety relief valves, and non-return valves to be used in the cryocircuit shown by Witt [28].

Still, there are only a few companies that offer control elements for hydrogen, basically due to the requirement of a very small leak rate and for HEE reasons (see Appendix 1).

Figure 6.27 shows a cryogenic, normally closed (NC) LH$_2$ solenoid valve. Cryogenic solenoid valve manufacturers are, e.g., "SpaceSolutions," and others [30].

Table 6.3 ISO standards

ISO 13984	Liquid hydrogen—Land vehicle fueling system interface
ISO 21010	Cryogenic vessels—Gas/materials compatibility
ISO 21013-3	Cryogenic vessels—Pressure-relief accessories for cryogenic service —Part 3: Sizing and capacity determination
ISO 21014	Cryogenic vessels—Cryogenic insulation performance
ISO 21028-1	Cryogenic vessels—Toughness requirements for materials at cryogenic temperature—Part 1: Temperatures below − 80 degrees C
ISO 21029-1:2004	Cryogenic vessels—Transportable vacuum insulated vessels of not more than 1000 liters volume—Part 1: Design, fabrication, inspection, and tests
ISO 23208	Cryogenic vessels—Cleanliness for cryogenic service

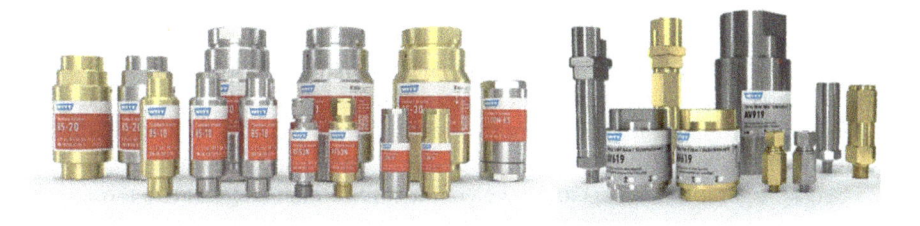

Fig. 6.25 Flashback arresters and safety relief valves for hydrogen use—Courtesy of Witt

Fig. 6.26 Non-return valve for hydrogen—Courtesy of Witt (right)

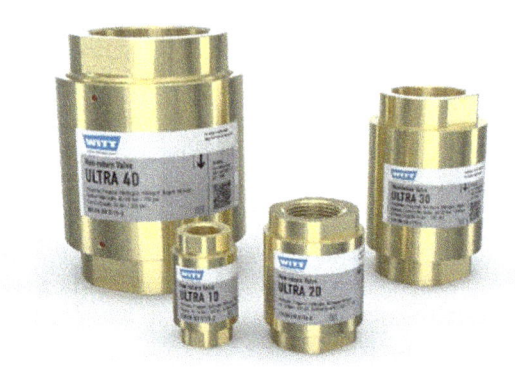

Fig. 6.27 Cryogenic NC LH$_2$ solenoid valve [29]

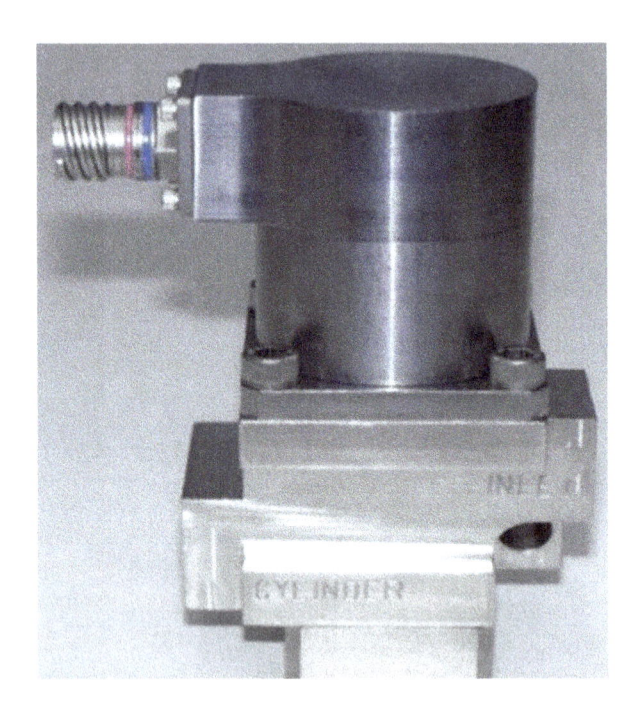

Control solutions also appear at neighboring modalities, one of those is by Bürkert [31], which shows a valve, safely and efficiently feeding a fuel cell stack of the Forze VI Hydrogen Racing team with a compact build. For the anode feed of hydrogen to a PEM fuel cell Bürkert recently developed type 6440 as hydrogen shut-off (and type 6020) as hydrogen pressure control valve. Maximum pressure is 25 barg, for gaseous hydrogen. As we collect information and validate those technologies of those components, one needs to align them to the aircraft specifications, as mentioned. See Fig. 6.28.

The work with hydrogen electrolyzers and fuel cells now sparked some further interest in focusing on this control technology.

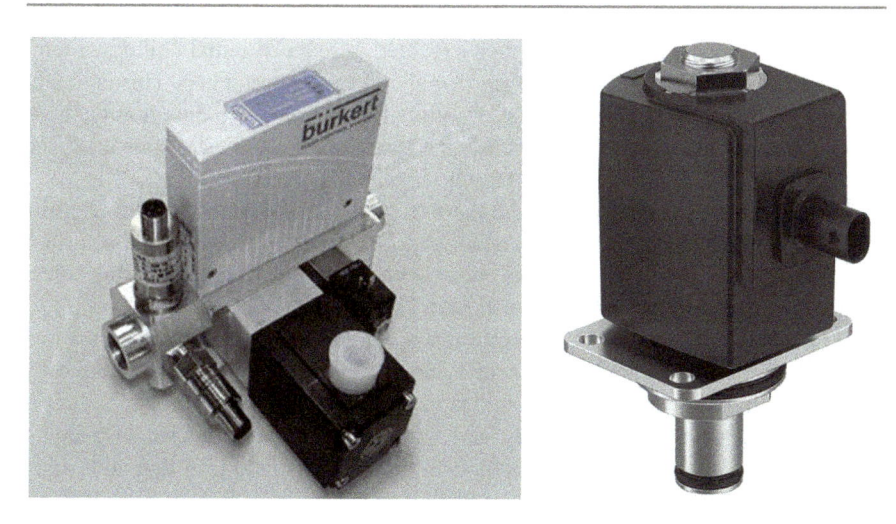

Fig. 6.28 Bürkert valve assembly, Bürkert type 6440 shut-off valve (right), courtesy of Bürkert

However, there is a need to qualify all valves for low leakage to rules and regulations for an aircraft although those are already available and tested for ground tank, or automotive, or other applications.

6.6 Tank Chilldown /Refueling

When first fueling a tank all cryogenic components in the aircraft are at room temperature and need to be cooled down to operating temperature (chilldown). This does not only apply to the hydrogen tank but to all cryolines that need to be operating at 20 K or higher. As the tank is initially warm, introducing a cryogen causes stresses at the internal liner and the composite wall. The chilldown should therefore be done in a way to avoid large temperature differences or hot spots and especially where there are penetrations and feed throughs for sensors, etc. This is where stress concentrations in a composite tank will lead to tank failure according to Hartwig [32] and that is why composite cryogenic tanks, although desirable for many reasons, are not available commercially, with the exception of one company that produces a small thick-walled composite dewar for liquid nitrogen [33] who is now engaged in composite tank manufacture, a Japanese company that developed experimental dewars for experiments in a non-magnetic environment, and Supracon together with GSI for a 100% composite cryostat for a cryogenic current comparator [34].

Although *refueling* (top-ups) is then executed at reduced initial chilldown temperatures, it still requires a tank design that allows for fast refueling. This is less of a problem if there is an aluminum layer (liner) that guarantees a more homogenous temperature profile buildup. In any case, the tank design also needs to be analyzed for fast warmup at minimum stress buildup in the tank shells.

For the thermal response time the thermal penetration lengths and the thermal diffusion time constant need to be known which allows you to estimate the surface penetration time. For G10 fiberglass epoxy with a 2 mm wall thickness this is less than 10 seconds without liner. See also Ekin [35].

With respect to liner-less composite tanks, bigger temperature gradients along the tank circumference can be expected during tank movement, accompanied by significant stress concentrations around penetrations and tank supports.

Figure 6.29 shows an exemplary early design for rapid chilldown of a liquid hydrogen tank, requiring however a built-in precooling shield [36] that adds extra weight to the tank.

Other tank designs may only need a built-in, coiled tube connected to a metallic liner that enables fast chilldown while minimizing the tank mass fraction.

Information is available here from the very early liquid hydrogen work of NASA that first introduced especially built-in coiled chilldown tubes for large tanks.

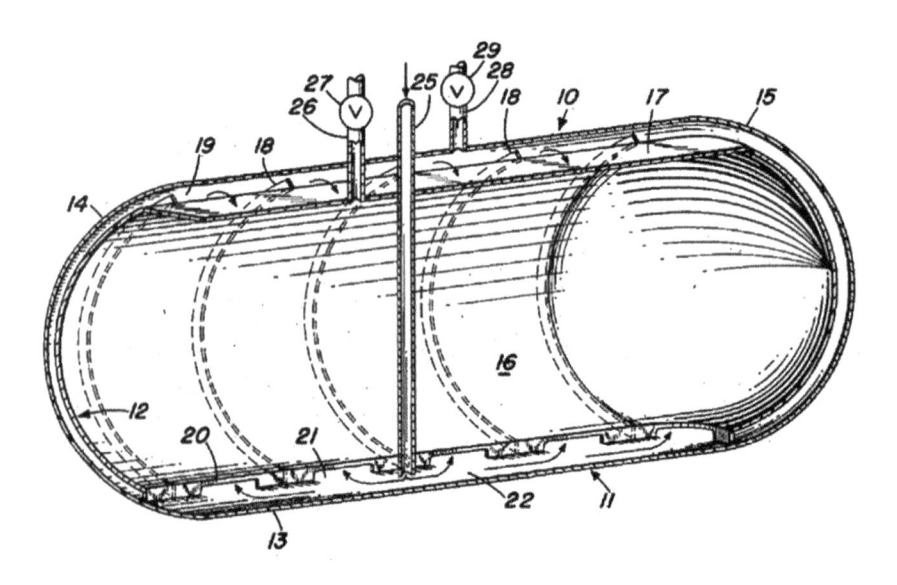

Fig. 6.29 Tank chilldown design example

6.7 Other Designs

Fueling liquid hydrogen has been looked at by space companies and by the automotive industry as shown in Fig. 6.30. Underground filling stations at the airport (see Chap. 11) provide hydrogen in liquid or subcooled form using pumps. For interfacing to the aircraft, nozzle designs exist, some of those can be adapted to suit the fast fill of an aircraft.

Figure. 6.30 shows one set of nozzles with flow meters for the automotive industry [37]. The BMW Hydrogen 7 car was filled with liquid hydrogen.

With respect to tank instrumentation aircraft flow control elements need to be developed. An example of a novel self-calibrating flow meter is shown below initially for the TOSCA KIT program for a helium vapor flow at 20 to 30 K. This flow meter is suitable for use with general cryogenic liquids (in either liquid or gaseous, or supercritical state) and has been developed by Janzen, KIT, and WEKA et al. [38] with helium flow rates of up to 12 g/s, see Fig. 6.31.

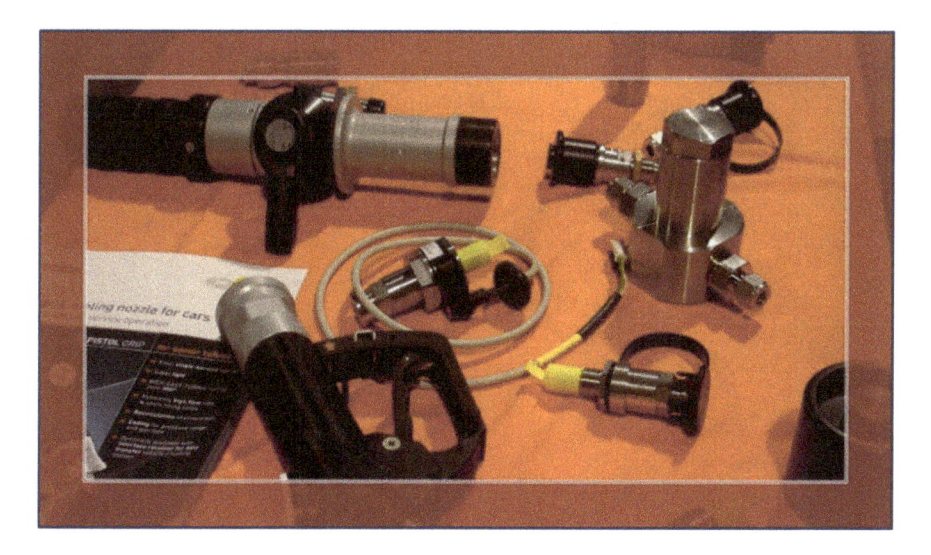

Fig. 6.30 Automotive nozzle system

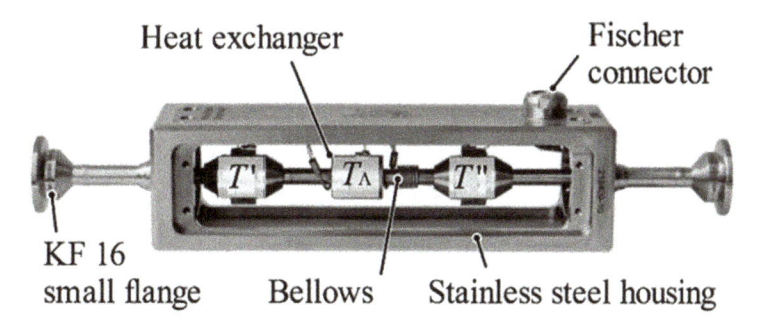

Fig. 6.31 Self-calibrating flow meter

As we will show in Chap. 11, the fueling specification of an aircraft is different from other modalities.

That is why fueling pumps as shown in Chap. 7 are markedly different from any other use, be it stationary at an airport fueling station, or when used in space or in the automotive industry.

Within the fueling station parahydrogen is being supplied at conditions with exothermal reaction already concluded.

In the following section we discuss the design constraints with defueling lines.

6.8 Defueling

For that we connect the extraction line as shown in Fig. 6.1 to a liquid hydrogen pump suction inlet. This leads us to the pump connecting line design for a liquid hydrogen tank for use in an aircraft as shown in Fig. 6.32. As the aircraft follows its mission profile the tank defueling line attached to the pump needs to maintain liquid hydrogen in the line, as shown in Fig. 6.32. At no time, should there be vapor in the liquid pump withdrawal line as discussed in Chap. 7 (Table 6.4).

Ascent and descent angles for the LH_2 supply line need to be considered in the defueling process as indicated in Table 6.2 and Fig. 6.12.

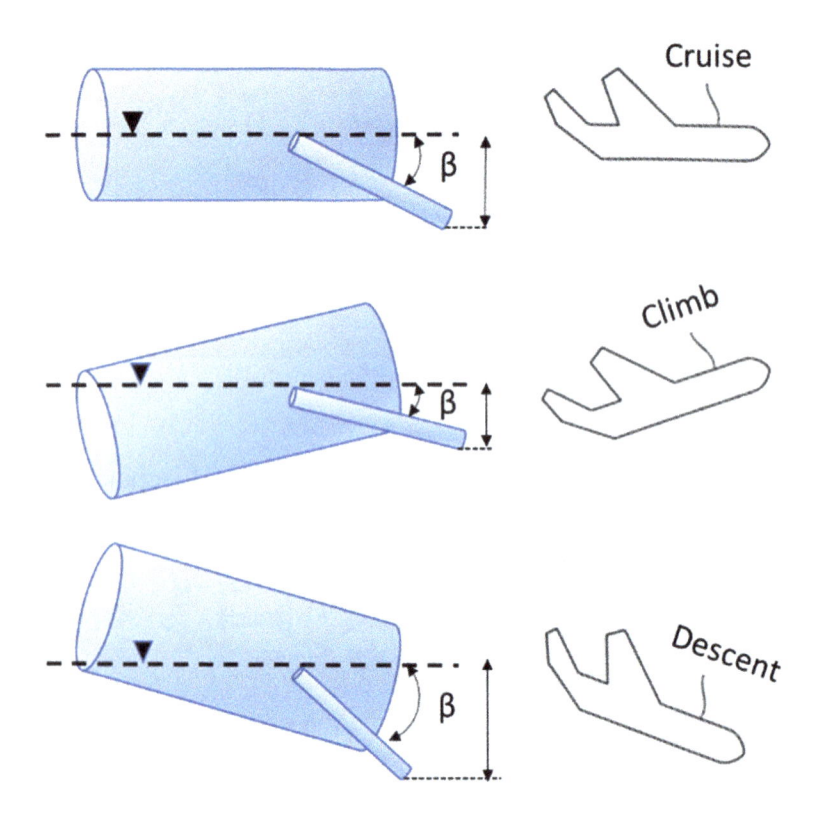

Fig. 6.32 Defueling and mission profile

6.8 Defueling

Table 6.4 Pitch and extraction angles

Mission step	Aircraft pitch angle	LH$_2$ extraction pipe angle
Ground idle	0	25
Takeoff	15	10
Climb	20	5
Cruise	0	25
Descent	−5	30

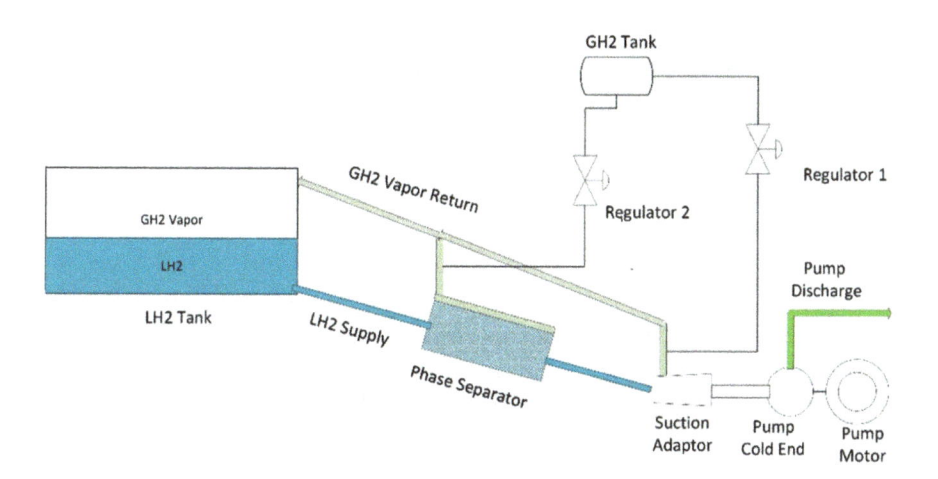

Fig. 6.33 Liquid hydrogen withdrawal from tank (schematics)

Most pumps used for defueling of liquid hydrogen however are used at very high discharge pressures of a few 100 bars. Many pumps thus already exceed a required flow rate for an aircraft and are therefore heavy, bulky, and difficult to maintain as mentioned previously.

The major difference here is that all those pumps currently in use or developed do not deliver liquid at the pump exit, but high-pressure hydrogen vapor and are therefore of inadequate use for all-electric aircraft. The following chapter reveals the landscape on the state of the art of various pumps that are also used in modalities other than aerospace.

6.9 Liquid Hydrogen Transfer Process

6.9.1 Logistics of Single-Tank Operation

Figure 6.33 shows the schematics of a single tank liquid withdrawal using a drainpipe connector to a suction adapter and phase separator with the vapor return siphoned back to the liquid hydrogen tank.

Starting from a single-tank design we can now review future multitank operation in an aircraft, as shown in Figs. 6.34 and 6.35.

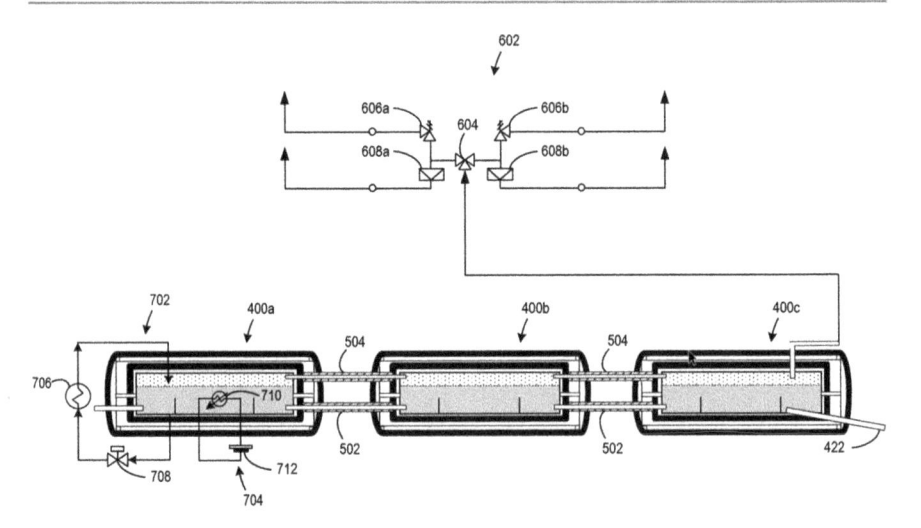

Fig. 6.34 All tanks connected together with single withdrawal line

6.9.2 Logistics of Multitank Operation

Understanding how the process of multitank operation works has been one of the key findings on the CHEETA program with various options covered broadly in a patent application [39]. We can assume the following:

- All 4 tanks at one side are served by one pump giving a fuel redundancy in case of pump failure.
- This requires a novel cryogenic tank design that operates all or a subset of 8 tanks as a single cryogenic vessel.
- Tanks can be filled simultaneously from a single inlet port.
- Tanks can be protected by a single set of pressure relief valves and burst disks.
- LH_2 is extracted simultaneously from all tanks.
- All tanks with the same LH_2 level at the same saturated pressure.
- Pressure is monitored and managed by a single thermosiphoning system.
- Pressure is monitored and managed by a single heater system.
- Cryogenic tank design allows for isolation capability, trimming, and redundancy.

Figure 6.35 shows a further detailed process flow control system capable of running 6 to 8 tanks in an aircraft with built-in redundancy.

Further Reading Fydrych J Chapter 9 "Cryogenic Transfer Lines" in Weisend II J G *Cryostat Design Case Studies, Principles and Engineering* International Cryogenics Monograph Series 2016.

Fuller P D McLagan J N Chapter 9 "Storage and transfer of cryogenic fluids" in Vance R W Duke W NM *Applied Cryogenic Engineering* Wiley New York 1962.

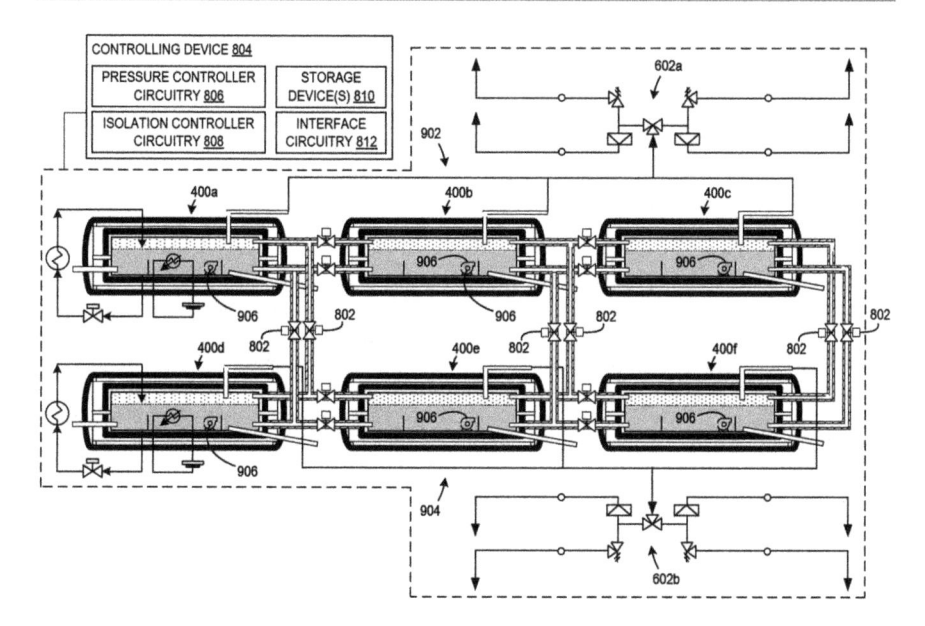

Fig. 6.35 Multitank liquid withdrawal with complete process elements

Further Resources The Gasworld North America Buyer's Guide, issued by Gasworld, Publishing LLC

- Cold Facts International Buyer's Guide issued by Cryogenic Society of America Inc.
- Aerospace & Defense Technology, Tech Briefs Media Group, USA, NY
- H2-View Magazine, Blackwater, Truro, TR4 8UN, UK

Further Reading on Cryogenics a, please see Appendix A6

References

1. Barron R F *Cryogenic Systems* Oxford 2nd edition Oxford University Press 1985 page 379
2. Langebach R *Wärmeeintrag durch geneigte Rohrleitungen in kryogene Speicherbehälter* Phd Thesis 2013
3. Langebach R *Natural Convection in Inclined Pipes - A New Correlation for Heat Transfer Estimation* Fakultät Maschinenwesen, Institut für Energietechnik, Bitzer-Stiftungsprofessur für Kälte-, Kryo- und Kompressorentechnik CEC 2013 Anchorage
4. Langebach R Haberstroh C *Experimental investigation of free convective heat transfer along inclined pipes in high-pressure cryogenic storage tanks* AIP Conf. Proc. 1434, 732 (2012); https://doi.org/10.1063/1.4706985
5. W. Stautner et al "Occurrence of Thermoacoustic Phenomena at 0.8 K, 4 K and above", CEC, Alaska, IOP Publishing, 012038, 2015
6. Hands B A *Cryogenic Engineering* Academic Press 1986 page 133
7. Stautner W *Systems and Methods for storing and distributing gases* Patent US 10670189B2 2020

8. Hibbs A D Cox D W "A reliable nonmagnetic liquid helium tight thermal feedthrough for a fiberglass Dewar" Rev. Sci. Instrum. 62 (9) 1991
9. D Evans, J T Morgan, "Physical properties of epoxide resin/glass fiber composites at low temperature" in G Hartwig D Evans "Nonmetallic materials and composites at low temperatures" 2 Plenum Press Cryogenic Materials Series pp 245-258 1982
10. G.G. Haselden "Cryogenic Fundamentals", Academic Press, 1971, pp. 468
11. R. Stroman, M. Schuette, K. Swider-Lyons, J.A. Rodgers, D.J. Edwards, *"Liquid hydrogen fuel system design and demonstration on a small long endurance air vehicle"*, International Journal of Hydrogen Energy 39 pp 11279-11290
12. Matsumoto K Sobue M Asamoto K Nishimura Y Abe S Numazawa T Capacitive level meter for liquid hydrogen, Cryogenics 2010 https://doi.org/10.1016/j.cryogenics.2010.11.005
13. Funke T Haberstroh Ch Szoucsek K Schott S Kunze K Capacitive density measurement for supercritical hydrogen IOP Conf. Series: Materials Science and Engineering 278 2017 012071. https://doi.org/10.1088/1757-899X/278/1/012071
14. Kajikawa K Tomachi K Maema N Matsuo M Sato S et al Fundamental investigation of a superconducting level sensor for liquid hydrogen with MgB_2 wire Journal of Physics: Conference Series 97 2008 012140
15. Nexans, courtesy of Nexans, priv. communication
16. J.P. Penland, "Liquid hydrogen fueled Boeing 737-200", Penland diaries, 1973
17. S. Klöppel, A. Marian, C. Haberstroh, C. Bruzek, "Thermo-hydraulic and economic aspects of long-length high-power MgB_2 superconducting cables", Cryogenics 113, 2021, 103211
18. Schoch K F Bergh D D *"Glass-reinforced epoxy piping for liquid nitrogen-cooled AC transmission cables"* Advances in Cryogenic Engineering vol 24, pp. 262 –270 1978
19. D. Kottonau, E. Shabagin, W. de Sousa, J. Geisbüsch, M. Noe, H. Stagge, S. Fechner, H. Woiton, T. Küsters, "Evaluation of the Use of Superconducting 380 kV cable", Band 030, KIT Scientific Publishing 2020
20. Nexans, priv. communication
21. K. Kauder „Strömungs- und Widerstandsverhalten in gewellten Rohren", Fakultät für Maschinenwesen an der Technischen Universität Hannover, PhD thesis, 1971
22. Nexans, company leaflet
23. NFPA Guidelines, NFPA 50 B, "Standard for Liquefied Hydrogen Systems at Consumer Sites", 1994
24. H. Neumann, Thermal insulation, KIT publication 2009
25. W. Stautner, Patent pending
26. K. B. Martin, O. E. Park, "A transfer line for liquefied gases", Advances in Cryogenic Engineering, Vol 1, 1954, pp. 95-104
27. ISO 13985:2006, "Liquid hydrogen — Land vehicle fuel tanks"
28. Witt, private communication
29. W. Johnson, T. Tomsik, J. Moder, "Fundamentals of Cryogenics", 25th Annual TFAWS, 2014
30. J.W. Hutchison, ISA Handbook of Control Valves, 2nd Ed. NC: Instrument Society of America, 1976 Control Valve Handbook, 4th ed. Fisher Controls, 2005, Ch. 5. ANSI/ISA–75.01.01–2002, Flow Equations for Sizing Control Valves
31. Bürkert, priv. communication, 2013
32. G. Hartwig, priv. communication, 1996
33. Fabrum solutions, http://www.fabrumsolutions.com/
34. H. Novak et al "GFK-Cryostat for a Cryogenic Comparator" ICEC29-ICMC24 Geneva 2024
35. Ekin J W **Experimental techniques for low-temperature measurements** Oxford University Press 2006
36. US2,882,694, P.C. van der Arend et al, "Cooldown apparatus for cryogenic liquid containers, 1959
37. Fueling means, W. Stautner, conference picture, NHA conference Long Beach 2010
38. A. Janzen, "Experimental validation of a self-calibrating cryogenic mass flowmeter" IOP Conf. Series: Materials Science and Engineering, 278, 012077, 2017
39. US patent application Minas/Stautner 2023/0107610 A1 Onboard liquid hydrogen storage for a hydrogen aircraft

Liquid Hydrogen Pump Overview/Tank Safety

Here we give a short overview on the pump development status and possible designs.

Brewer, very early realized, new pumps may need to be redesigned and built for aircraft proposing a technology development program. Since this never materialized and liquid hydrogen applications are, apart from space or other niche applications, rarely used commercially, the technology readiness level (TRL) of new pumps is still very low. Most of the pumps are by nature centrifugal pumps requiring no dynamic seals. Centrifugal pumps come with higher flow rates but lower compression ratios. Positive displacement pumps do have lower flow rates and higher compression rates. Pump companies are, for example, Cryostar, Linde, Barber Nichols, Nikkiso, Daiho Sangyo, Shinko Industries, and Westport Fuel Systems (see below). Some of those pumps have been developed for LNG, space, and superconducting applications [1].

KIT, JAERI, Atlas, and others, successfully designed, developed, and tested several cryogenic helium circulation pumps for forced flow cooling with liquid helium mainly for fusion applications, like ITER, since the early 1970s, see Fig. 7.1. The early mass flow design for ITER was in the 3 kg/s range at an efficiency of about 70% [2].

Westport Fuel Systems Canada Inc., along with their partners NICE America Research Inc., and ETI Energy Corporation have developed a hydraulically driven, high-pressure submerged cryogenic piston pump with a total length of 2656 mm (other lengths available) for liquid hydrogen. The pump design features a pump socket foot valve actuator for pump replacement [3].

This pump has been tested for a fuel station application with a liquid hydrogen delivery rate of 200 kg/hr. at 450 bar (Table 7.1 and Fig. 7.2).

Barber Nichols [4] also manufactures a range of submersible pumps for liquid ground operation.

These types of pumps are fully submerged into the cryogenic fluid inside of a tank. These are best used in intermittent operation because the motor is fully submerged into the cryogenic fluid and any operating inefficiencies will be imparted

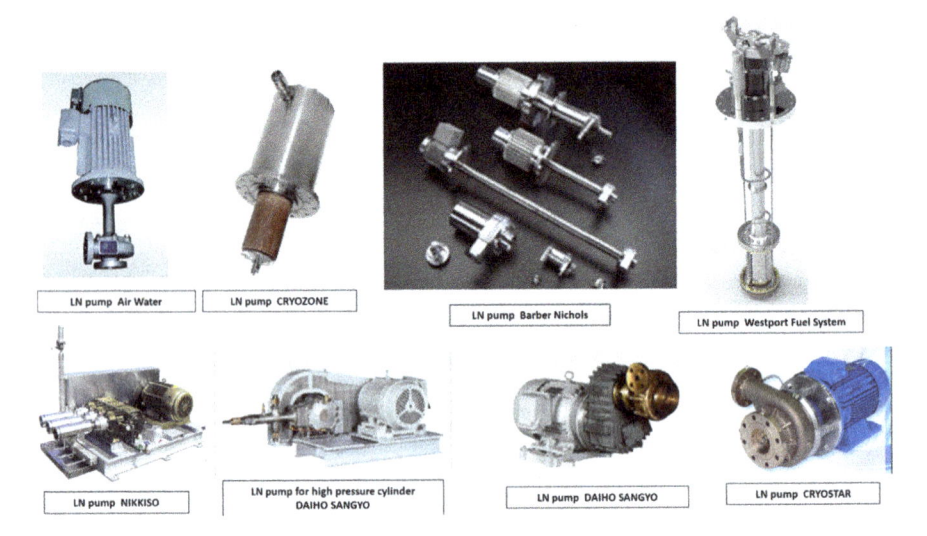

Fig. 7.1 Overview of common pumps for cryogens

Table 7.1 P200H Characteristic data (courtesy of Westport Fuel Systems)

Characteristic	Value
Displaced fuel volume per stroke	3079 cm^3
Maximum rated CPM	21 CPM
Maximum high-pressure gas discharge pressure	450 bar
Maximum allowable hydraulic working pressure	300 bar

into the fluid in the form of heat. The BNHP-40-000 is a submersible pump that is designed for 35 g/s at 0.7 bar (Fig. 7.3).

Vacuum Housing Pumps

These types of pumps reduce heat leak by having a barrier between the motor and the cryogenic fluid. They utilize anti-convection shields to further limit this heat leak. The pump housings are welded into tank and if properly valved off, the rotating assembly can be removed and replaced with spare rotating assembly when overhaul of the pump is needed.

BNHP-41-000 shown in Fig. 7.4 is a vacuum-jacketed, full emission pump that can be used for ground transfer of liquid hydrogen. This full emission pump offers higher flow rate capability and was designed for 1 kg/s at 0.9 bar [4].

A very similar style pump is the BNHP-27-000 designed for 1.7 kg/s of liquid hydrogen at 0.5 bar. This pump is designed with composite materials for components such as the anti-convection shields, shaft, pump support tube, and vacuum housing extension to reduce heat leak from the liquid hydrogen and the motor cavity (Fig. 7.5).

Fig. 7.2 P200H LH$_2$ pump (courtesy of Westport Fuel Systems)

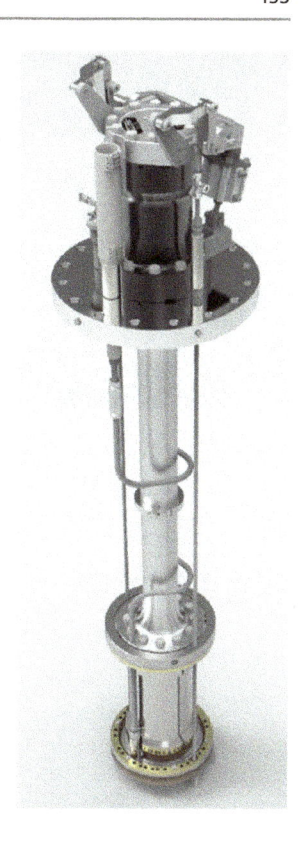

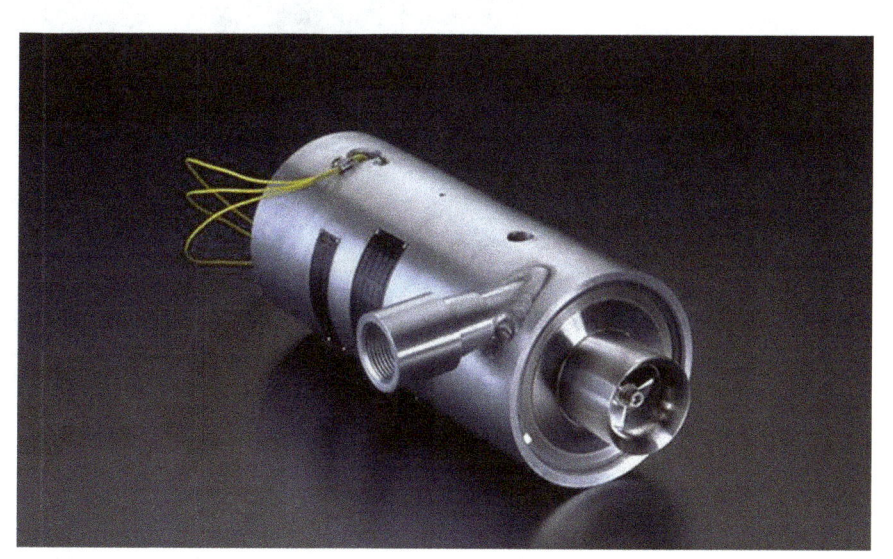

Fig. 7.3 BNHP-40-000 series pump—courtesy of Barber-Nichols

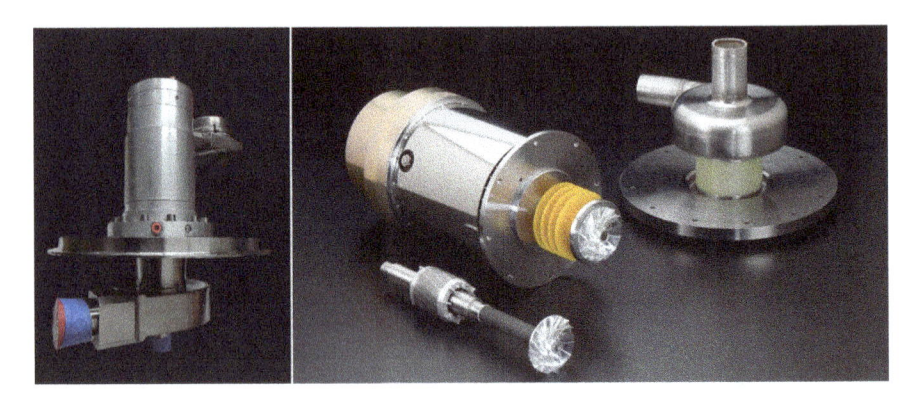

Fig. 7.4 (left) Pumps BNHP-41-000 LH$_2$ and BNHP-27-000 LH$_2$ (right)—courtesy of Barber-Nichols

Fig. 7.5 Pump BNHP-29-000—courtesy of Barber-Nichols

The BNHP-29-000 shown below is a vacuum-jacketed, partial emission pump geared toward lower flow applications. This unit is used to transfer hydrogen for a ground operation test case for an aircraft application.

Table 7.2 gives some typical aircraft pump flow rates during flight operation.

7 Liquid Hydrogen Pump Overview/Tank Safety

Table 7.2 Typical flow rates for engines

Engine	Flow rate (kg/min)	Outlet pressure (MPa)
Turboprop engine	1.2–12	3
Turbofan engine	4.2–30	7
Hydrogen PEM fuel cells	Approx. 30	2.5 bar
Piston pump	< 10	Up to 100.
Centrifugal pump	Up to 1000	< 1.
CHEETA all-electric	Up to 2.6	0.2

For an aircraft, we may need to follow FAR Part 25.955. This section clearly states that (a) Each fuel system must provide at least 100 percent of the fuel flow required under each intended operating condition and maneuver. Compliance must be shown as follows: (1) Fuel must be delivered to each engine at a pressure within the limits specified in the engine type certificate. (2) The quantity of fuel in the tank may not exceed the amount established as the unusable fuel supply for that tank under the requirements of §25.959 plus that necessary to show compliance with this section. (3) Each main pump must be used that is necessary for each operating condition and attitude for which compliance with this section is shown, and the appropriate emergency pump must be substituted for each main pump so used.

At this time, it is understood that several different pumps may be required on an aircraft, with booster pumps only needed, when necessary.

Reciprocating pumps are also shown by Nikkiso [5]. See the following section on the current status of liquid hydrogen pumps.

7.1 Cycling Pumping

The very early work of the Lawrence Livermore National Laboratory of 1976 has already been discussed in Chap. 4 when referring to the testing of low-pressure liquid cryogen tanks. More recently LLNL reported on pump cycling experiments coupled to a prototype thin-lined (1.8 mm) 700 bar composite pressure vessel with 65-liter capacity. The main focus here is on the liquid hydrogen piston pump performance as manufactured by Linde, rated for 100 kgH$_2$/h and 875 bar maximum pressure. However, liquid hydrogen is being pumped into a tank and then pressurized multiple times beyond the pressure vessel design rating. This work has then recently been extended to 700 bar by Petitpas and Aceves in 2018 [6, 7].

The next section looks deeper into the heart of the cryoplane, the pump, as one of the key elements for any liquid hydrogen-driven aircraft and details how the pump interfaces with the liquid hydrogen tank.

The question is on how pumps can be integrated into the airframe with a small and light footprint as compared to those currently available. Obviously, if one could tap into the cryogenic liquid reservoirs a superconducting pump may be feasible.

Excursion: Superconducting Hydrogen Pumps

The first superconducting pump has been developed by GE at the GE Research Center by Darrel and Schoch as early as 1965 [8]. The small pump was designed for use with liquid helium transfer comprising of a superconducting Nb ring, and bellow as a piston element and a superconducting base coil made from Nb wire.

A further approach was thought of by C. Schmidt [9] employing a superconductive electromagnetic drive that includes a stationary superconductive energizing solenoid connected to the pump housing with a superconductive component movable by means of the field generated by the energizing solenoid and attached to the pump pumping member. The pump was meant to drive liquid or supercritical helium in a closed circuit or, more general, for displacing liquid helium.

With the arrival of new types of high temperature superconductors, operating ideally around 20 K, superconducting pumps are becoming even more attractive for use in an hydrogen-powered aircraft where the advantage seems just obvious: Much higher pumping efficiency, paired with unsurpassed compactness due to the shorter pump shaft, resulting in an extreme lightweight pump. Some of these pump components can be made from composite materials.

7.1.1 Cryogenic Magnetic Bearing Technology

As we shall see, a further advantage is the elimination of bearing wear as shown in Fig. 7.15. If those could be made superconducting, this component has a good potential for long endurance as required in an aircraft. Superconducting bearings have also been discussed for example with Walter of Nexans and Nick and Neumüller of Siemens [10].

Contactless magnetic bearings do not produce heat, are not subject to wear, and are an ideal means for maintaining a magnetic gap. A further advantage of the proposed cryogenic magnetic bearing design is its resilience against cavitation since less heat is created during pump operation.

For an aircraft, maintenance intervals (> 30,000 h) are likewise of importance when servicing the pump. Pump bearings are one of the components that experience the highest wear. Any effort of making a contactless bearing is therefore recommended. The latter has been successfully integrated into a superconducting pump for use with liquid nitrogen by Kamioka [11]. For further analysis of superconducting bearings for pumps, please see the more recent work of Kamioka [11].

Currently the most advanced superconducting pump has been designed and tested by Kamioka. Kamioka points out the following advantages of this superconducting pump:

- Runs with high efficiency with synchronous steady operation.
- Delivers a high torque density.
- Has a proven robustness against overload.
- A setting of a short-time rating is available.
- Hunting as well as stepping-out of the pump is avoided.
- Stable rotation is guaranteed with variable speed control.

7.1 Cycling Pumping

Table 7.3 Superconducting pump specifications

Fluid	Subcooled liquid nitrogen
Flow rate	100 l/min
Head	1.0 MPa (2-stage)
Inlet pressure	0.5 MPa gauge
Motor capacity	5 kW, water power 1.7 kW, shaft power 3.3 kW
Motor voltage	3φ, 200 V
Rotation speed	5000 rpm
Pump efficiency	50%
Maintenance interval	More than 30,000 h

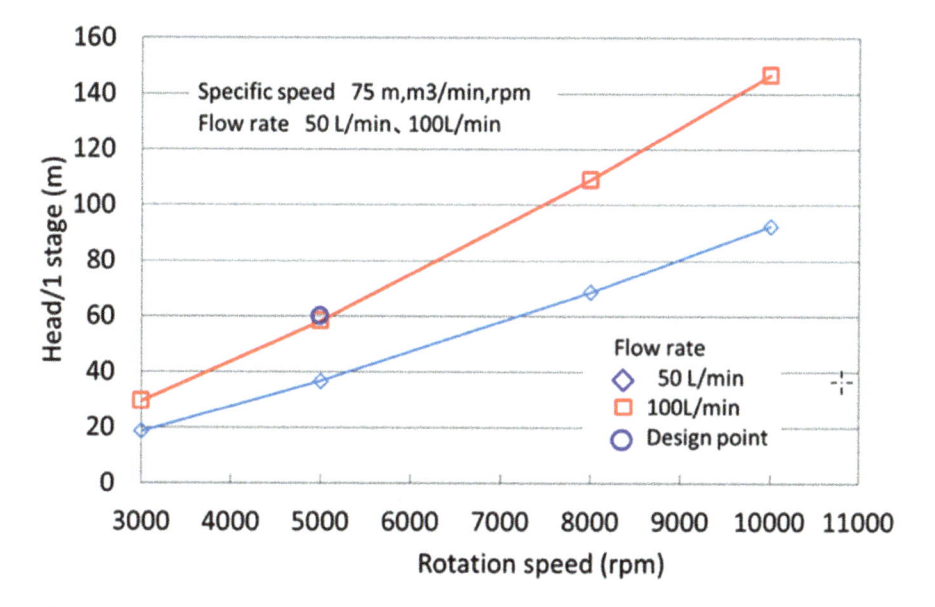

Fig. 7.6 Rotational speed and pressure head

The following Table 7.3 gives design specifications of this superconducting pump [11], Fig. 7.6 the rotational speed, and Fig. 7.7 a cross-sectional view of some of the components.

Alternatively, this liquid nitrogen pump can also run in a liquid helium of hydrogen environment.

This however deserves some close analysis of the thermal load of the superconducting motor caused by AC losses.

For aircraft use, superconducting pumps generally should work with a pressure head of 20 to 70 MPa and a flow rate more than 10 kg/min. Unlike conventional pumps, the superconducting motor efficiency can be as high as 99.9% delivering an overall efficiency of >80%.

Fig. 7.7 Superconducting LN$_2$ pump 1 HTS motor, 2 radial bearing, 3 axial bearing, 4 pump impeller (Courtesy of Y. Kamioka)

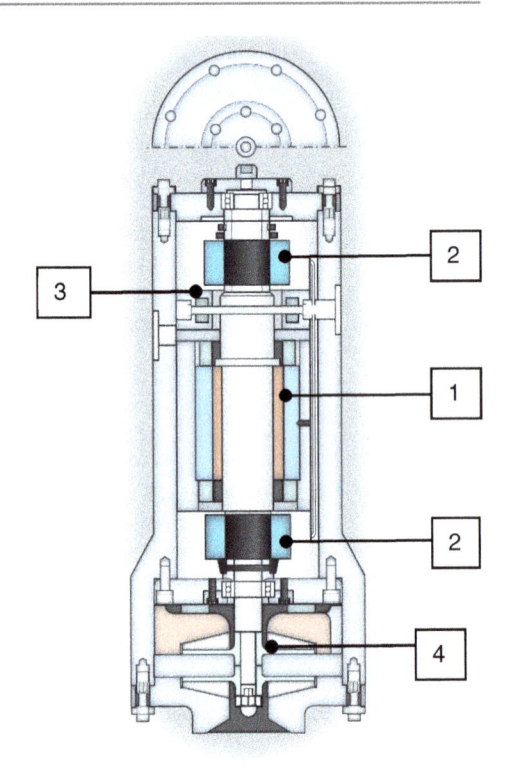

In which way a superconducting pump can be integrated into the circulation loop for cooling HTS cables is shown by Kenta et al. [12] and Kamioka [13]. Ku and Reddi [14] discuss an immersed liquid pump-enabled hydrogen refueling system for medium and heavy-duty fuel cell vehicles.

A superconducting LH$_2$ pump based on the MgB$_2$ as the superconductor has also been developed by Kajikawa et al. [16]. This pump has an experimental flow rate of about 6 liter /min, obtained at 1800 rpm.

Just recently the group around Yim and Hahn [15] presented a superconducting pump study for a 20 kW class axial flux motor for liquid hydrogen, with HTS field windings for 100 MPa adopting the non-insulation (NI) HTS winding technology. The pump is unlike the Linde and Mitsubishi ones of centrifugal type. The advantage is a high torque pump with much simpler structure and increased compactness at reduced weight.

Kobayashi used pulsed-field magnetization of a superconducting (HTS bulk) stepper motor for a centrifugal liquid nitrogen pump as a test bed for a future liquid hydrogen pump [17]. Maximum flow rates were 1.3 l/min for 1350 rpm.

7.2 Liquid Hydrogen Pumps—Current Status

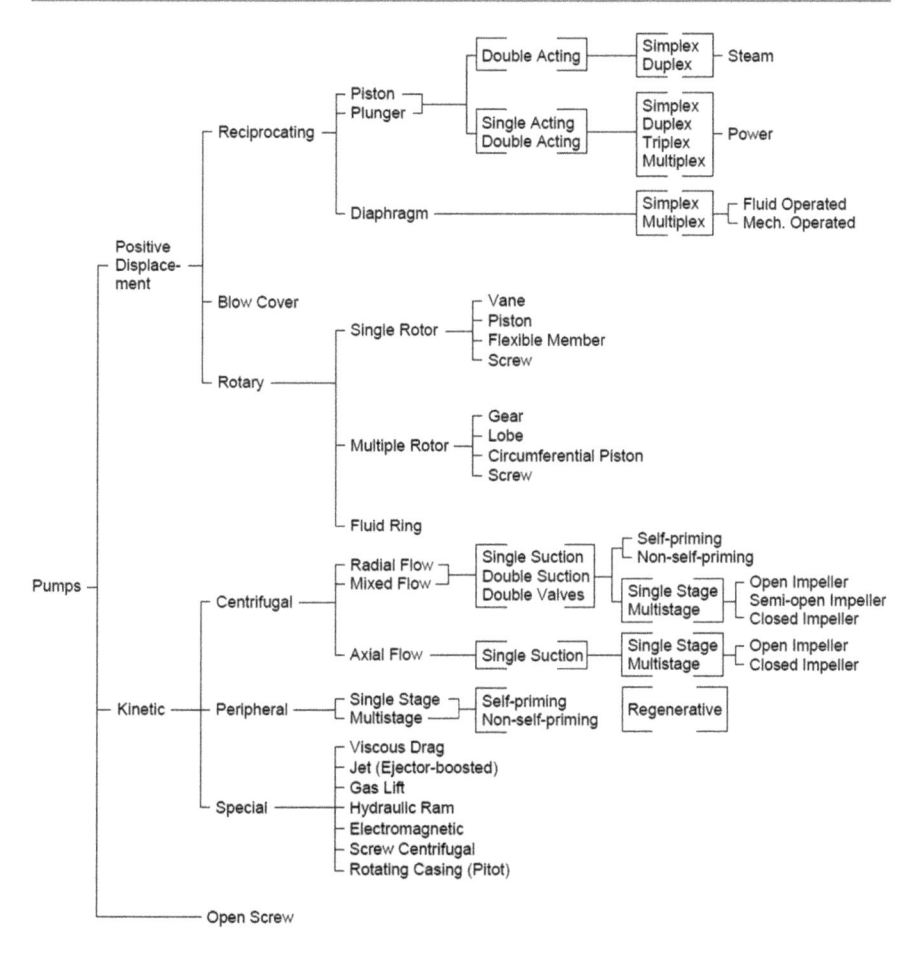

Fig. 7.8 Pump overview

Advanced technology in that field takes advantage of operating the HTS conductor at liquid hydrogen temperature.

Figure 7.8 gives an overview of the pumping landscape [18].

Depending on the targeted specification there will be a shift from piston driven to circumferential radial pumps as shown in the NsDs pump chart in Fig. 7.9.

NsDs pump chart [19].

7.2 Liquid Hydrogen Pumps—Current Status

Liquid hydrogen (LH$_2$) pumps offer by far the least expensive and more convenient way of transferring LH$_2$ from a supply LH$_2$ tank to a high-pressure receiver tank [20]. Accomplishing the same task with a diaphragm H$_2$ compressor will increase the size,

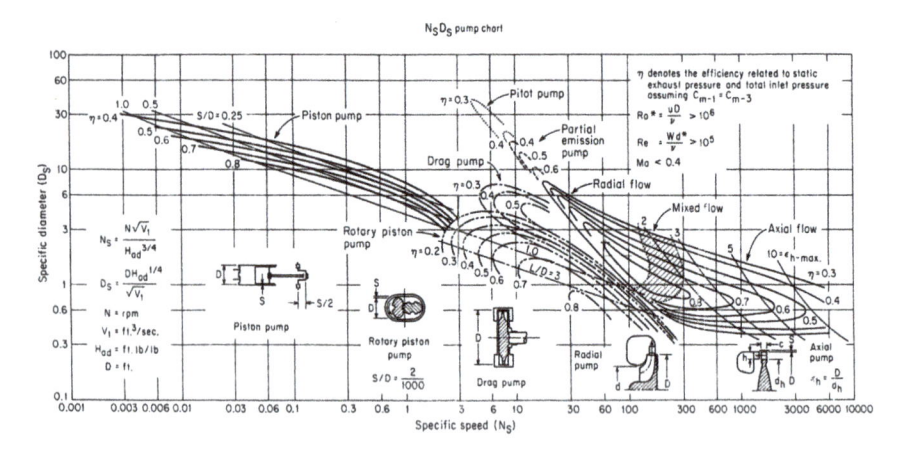

Fig. 7.9 NsDs pump diagram with permission from Barber-Nichols

Fig. 7.10 Nikkiso single-cylinder SGV pump

weight, and cost of the system dramatically. LH$_2$ pumps can be centrifugal single or multi-stage pumps or positive displacement piston pumps. In applications where high compression ratio is required, positive displacement piston pumps are preferred. Typical piston pumps operate at low frequency (275–400 rpm) which results in a physically large and heavy pump for a specific flow rate [21].

The LH$_2$ pump assembly includes a motor, a belt-driven crank drive, and a cold end with a reciprocating piston. A typical LH$_2$ pump with a single-cylinder and three-cylinder design and the associated cold end was shown in Figs. 7.10, 7.11 and 7.12, respectively. The multi-cylinder design offers a compact and reduced component solution for high mass flow rate requirements. The LH$_2$ pump is connected to the LH$_2$ supply tank with two vacuum-jacketed (VJ) lines as shown in Fig. 7.13. The first VJ line is the LH$_2$ supply line which is connected at one end to the LH$_2$ tank via

7.2 Liquid Hydrogen Pumps—Current Status 161

Fig. 7.11 Nikkiso three-cylinder SGV pump

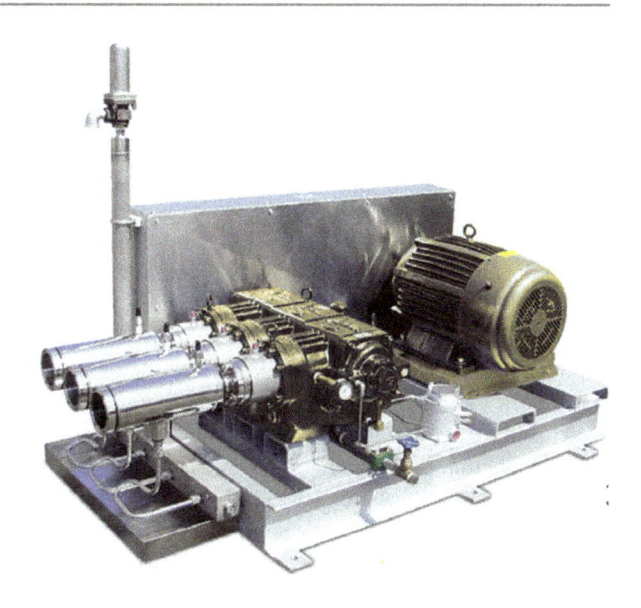

Fig. 7.12 LH2 pump cold end

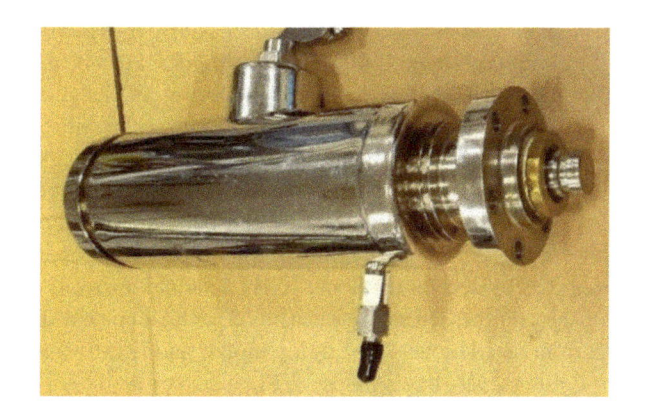

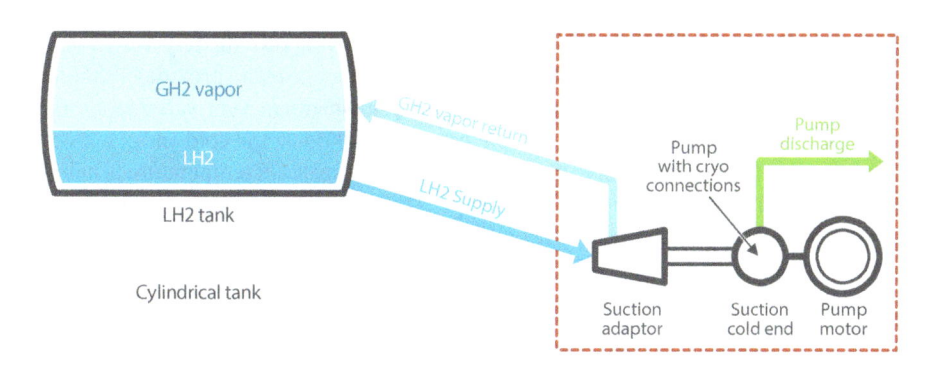

Fig. 7.13 LH$_2$ tank/LH$_2$ pump process diagram

a bayonet connection, and at the other end to the LH$_2$ pump cold end inlet port via the suction adaptor. The suction adaptor includes a conical metal grid filter and a spring-loaded suction adaptor tube. The second VJ line is the vapor return line which is connected at one end to the LH$_2$ tank via a bayonet connection, and at the other end to the LH$_2$pump cold end vapor return port via a flange connection. The pump operating cycle starts with the suction stroke where the cold end piston moves away from the top-dead center (TDC) and liquid enters the compression chamber. When the cold end piston reaches the bottom-dead-center (BDC), the compression stroke begins as the piston reverses direction and the inlet valve closes. The discharge valve is forced to open by the compressed liquid which is discharged in the piping downstream the pump. When the piston reaches the TDC, the discharge valve closes and the piston reverses direction, for the cycle repetition.

The cryogenic liquid in the supply tank will gain heat independent of the quality of its insulation. If the liquid in the tank is in the saturated state, then the actual pressure in the tank will be equal to the vapor pressure. The following two important terms which are critical to the pump operation are defined. The net positive suction head (NPSH) is defined as the difference between the actual pressure and the vapor pressure of the liquid. Cavitation is defined as the bubble formation and collapse in the liquid caused by the localized pressure falling below the liquid vapor pressure. The pump can only be operated successfully if there is sufficient NPSH (~10 Psi) at the pump suction port. A portion of the NPSH can be provided by static pressure, by having the supply LH$_2$ tank supply port at an elevation above the LH$_2$ pump inlet port. The NPSH must be maintained during the entire pumping cycle for efficient pump operation and cavitation avoidance. Operating the pump under cavitation will affect its performance, which includes (i) significant reduction of the mass flow rate, and (ii) shock waves caused by the bubble collapse, resulting in extremely high stresses, pitting, and erosion of the wetted components. Pump cavitation can be detected by several methods: (i) By a sudden increase in the discharge temperature of the compressed mixture, (ii) by a sudden current drop by as much as 50% and current oscillations of the pump motor, and (iii) by a sudden decrease in the pump vibration level shown in Fig. 7.14.

The pump operation starts with opening a supply valve and a vapor return valve to allow thermosyphoning, and cooling of the LH$_2$ supply line, the GH$_2$ vapor return line, and the suction adaptor, typically for a duration of approximately 10–15 min. During this step, the pressure in the supply tank will increase and provide NPSH for the pump to operate. The next step is to open a downstream vent valve along the pump discharge, to flow LH$_2$ through the pump to cool down the cold end. This step is performed for a duration of 2–5 min or until the downstream temperature is below a targeted temperature, typically 100-115 K. The next step is to start the pump and keep the downstream valve open for a short duration since high-pressure hydrogen is vented, approximately 10–30 s. The last step is to close the downstream vent valve which directs the compressed hydrogen to the high-pressure vaporizer and finally to the high-pressure receiver tank. The pump will turn off when the high-pressure receiver tank reaches its targeted value.

7.2 Liquid Hydrogen Pumps—Current Status

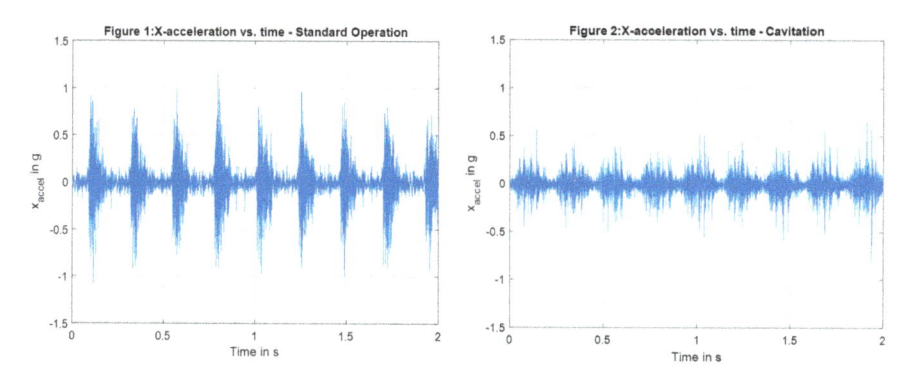

Fig. 7.14 Acceleration measured at LH$_2$ pump under standard operation and cavitation conditions

Fig. 7.15 LH$_2$ pump cold end piston rings

Several sensors can be installed on the pump system to monitor its performance and detect failure. A shaft seal temperature sensor detects hydrogen leakage. A cryogenic hydrogen leak is determined in the case the temperature drops below the ambient temperature range. A discharge temperature sensor monitors the downstream temperature of the pump and can be used to evaluate the pump operation and detect pump cavitation. A crank drive temperature sensor is used to monitor the crank drive temperature and detect low oil condition. A thermistor is used to monitor the temperature of the motor windings and detect a locked rotor condition.

The dry piston (no lubrication) operation of the cold end results in piston ring wear which needs to be replaced periodically (200–3000 h). Typical piston rings shown in Figs. 7.15 and 7.16 are made of bronze-filled Teflon or PEEK. A typical cold end removal and replacement from the LH$_2$ pump assembly can be performed in one hour.

Fig. 7.16 Disassembled piston rings

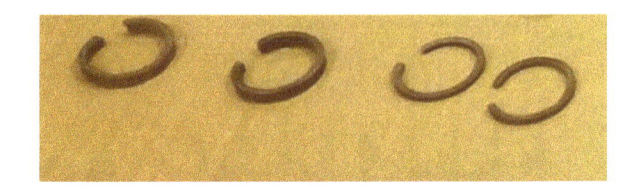

Fig. 7.17 Early phase separator

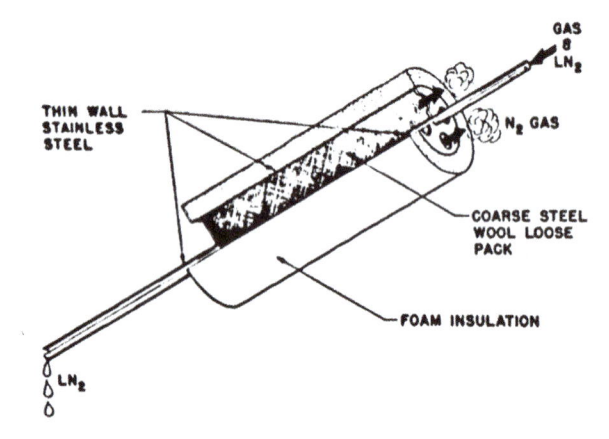

7.2.1 Phase Separator

Referring now to Fig. 7.13 we anticipate long lengths of cryolines through the aircraft connected to the liquid tank as indicated in Chap. 6. During liquid removal and transport through lines, the vapor/liquid fraction in the lines will change, carrying vapor bubbles to the suction pump inlet. Vapor bubbles damage the pump as shown in the previous section due to cavitation.

A remedy against this bubble formation in the pump suction entry line is a so-called phase separator. This component separates the liquid fraction of the cryogen from a very high mass flow, employing gravity.

The very first experimental phase separator was developed by Fretwell at Los Alamos in 1965 [22].

In this early lightweight (470 g) configuration, the phase separator was meant to avoid the flow of warm nitrogen gas into a cold trap.

Figure 7.17 shows a thin-walled stainless tube with top and bottom covers, filled with coarse steel wool, and enclosed by an outer foam insulation (e.g., Armstrong Armaflex). At both ends a stainless steel tube is welded into the cover disks. At the top end, large openings are cut into the disk to allow for nitrogen vapor escape.

Obviously, gravity-assisted phase separating is key to the functioning of the separator. Other designs would be cyclone-type dust separators or today's vacuums or others.

For an optimized phase separator [23] we need to define the following parameters:

- Inlet temperature and pressure to control volume.
- Mass flow rate range.

7.2 Liquid Hydrogen Pumps—Current Status

- Desired yield "y" in %, whereas yield is the mass flow of the liquid to the total incoming mass flow.
- Pressure drop through matrix.
- Packing density/porosity/permeability.
- Hydrogen physical properties.
- Heat transfer characteristics for different materials.

As optimization parameter we need to define

- Efficiency of bubble removal to protect pump.
- Lightweight and compact design.
- Needs to be quickly and easily exchangeable, replaceable.

Matrices that can be considered are steel wool, dimpled ribbon, rolled jelly-type tapes, metal foam, meshes, packed spheres, additively manufactured structures, and sintered disks.

Referring to Fig. 7.18 the pressure drop on 2 choices, sintered disks and meshes for different tube diameters and mesh types is given for the use with liquid hydrogen and liquid nitrogen.

When running tests on matrices with cryogens like helium and nitrogen, we need to scale that pressure drop to the physical properties of hydrogen. So far, calculations reveal a higher pressure drop with hydrogen as compared to liquid nitrogen. In general, the pressure drop through a mesh-type configuration is much lower than using commercial sintered material although additive printed foam could deliver adjustable porosities like meshes. Metal foams and sintered disks are used rarely with liquid cryogens and are mainly used as filters. Meshes are used in cryogenic regenerators, e.g., in cryocoolers but not for liquid/gas fractions. More testing is needed to understand the suitability of those materials.

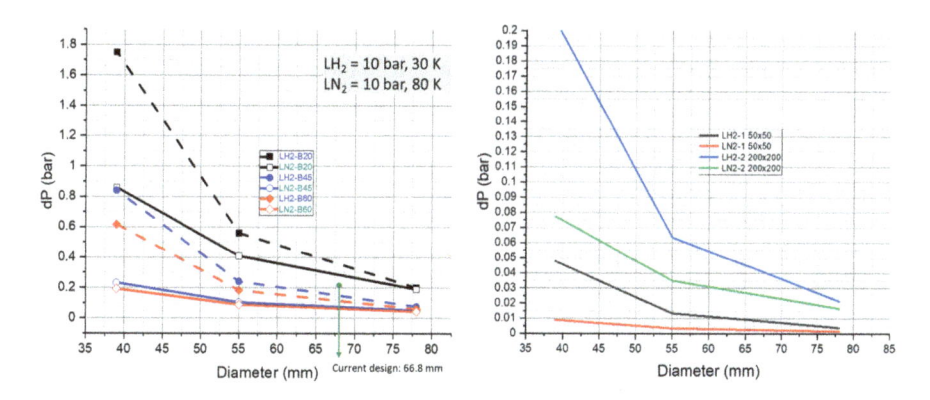

Fig. 7.18 Pressure drop through phase separator matrices (left: SS 316 sintered filter disks, right: typical SS 304 mesh sizes)

Phase separators are also used for space applications with the added complication that they need to work in a microgravity environment. For MRI and NMR systems those separators are recommended whenever there is a transfer of a liquid cryogen from a liquid storage dewar for topping up an external reservoir.

7.3 Tank Safety

As one may recall, gaseous hydrogen was first in use in the airship industry which ended in a negative impact (1937 Hindenburg disaster) for using hydrogen (even though the real cause was never discovered, until recently) with the so-called "Hindenburg syndrome." This public perception on the danger of use of hydrogen hindered the development of safe components that can be used with hydrogen as Edeskuty emphasizes [25]. However, since then good industrial-style efforts have been developed, designed, and manufactured, to tackle the problems when safely handling hydrogen.

Unlike in Chap. 4 "Cryotankage" where we structurally define and design the tank for Faults and Operating conditions, here we look at the need for vent stack, or aircraft louver sizing and safe hydrogen blow-off.

7.3.1 Compromised Vacuum Conditions

Assuming compromised vacuum quality we need to look at the leakage rate of components for hydrogen. The localized permissible leakage is currently limited to 3.56 Ncc/min. [26]. For aircraft this number needs to be much lower and needs to be defined. The permeation limit through a structural shell is set to 150 Ncc/min (flammability within a tight space) according to standards. Again, those need to be validated for all aircraft designs.

Depending on chosen tank insulation type, e.g., closed cell foam is known for long outgassing rates. On the other hand, MLI blankets cannot be compressed by a honeycomb core as proposed in some designs, since those cause virtual leaks which makes it impossible to differentiate from other leak paths.

7.3.2 Sudden Tank Vacuum Failure

Vacuum can slowly degrade over time being detectable with appropriate sensor technology (GE Aerospace Research). Vacuum can also break suddenly due to an external port leakage, or an attached line failure (see Fig. 7.19), or malfunction of one of the safety features. Piezoelectric sensors can also be embedded in the composite matrix for the tank to monitor any crack propagation of hairline cracks in the vacuum shell. Nevertheless, the vacuum shell enclosing the liquid hydrogen tank provides an additional layer of safety against damage of this inner tank.

7.3 Tank Safety

Fig. 7.19 Line failure on a liquid hydrogen pump

In case of a sudden port opening, air ingress accumulates on the inner tank surface and freezes out. Safety hazard: Frosted lines as exemplary shown on a liquid hydrogen pump line [24] cannot be tolerated in cryogenic aircraft circuits.

The heat flux density on an uninsulated, inner liquid hydrogen tank, is of course much higher than of an MLI insulated tank. Experiments with liquid helium dewars have also shown [27] a considerable difference in heat flux density, depending on insulation quality and cryostat build.

A Fermilab estimate on the heat flux caused by a major rupture of an outer shell of a liquid helium 19,000-liter tank of size (diameter 2.13 m, length 6.7 m) with liquid nitrogen cooled shield estimated a heat ingress of 500 kW, corresponding to $1.4 \cdot 10^4$ W/m^2. It is then necessary, to appropriately design the vent stack to safely blow-off the cryogen. With air ingress on 10-layer MLI superinsulated bath cryostat, the heat flux can be as high as up to $0.6 \cdot 10^4$ W/m^2 (for an opening size of 50 mm, as compared to a helium vessel without insulation [28]. For uninsulated tanks this value can get to $1.4 \cdot 10^4$ W/m^2. Although ice build-up may be fast, the thickness of the ice layer most certainly is not built up fast enough to work as an insulation means, supporting a lower heat flux density. Given the values above, an uninsulated tank as discussed in the Section on Cryotankage—Design for Tank Faults in Chap. 4, (CHEETA tank 3, type 2, surface area: 20.8 m^2) 5525 liters of liquid hydrogen would deplete the tank within approx. 10 min, and can take as long as 46 min with MLI insulation (40 layers). Safety aspects for LHe cryostats and LHe transport containers have been well researched by Lehmann and Zahn [29].

It is therefore mandatory to use inner tank insulation techniques that slow down fast tank depletion to give one time to react.

With liquid hydrogen however, as compared to helium, air condensation at a cold insulated surface can pose a risk with respect to explosiveness of the tank itself. See further details in Appendix 3 on Safety.

7.3.3 Inner Tank Rupture

In the unlikely event of an inner tank wall rupture the tank depletes within <10 min, as calculated in the previous section for the tank 3 example. Without appropriate vent stack on the outer vacuum shell this will likely lead to hazardous conditions in the aircraft.

However, excess hydrogen vapor can be routinely vented without incident. Nominal flow rates are 0.11–0.2 kg/s for a single vent tube with a 5 m distance from any transfer line. Multiple vents across prevailing winds spaced 5 m apart can be used. Usually, over-the-road trailers and cars regularly vent hydrogen without any problem. Those generic ballpark figures need to be adapted for aircraft technology.

7.3.4 Ice Formation on Outer Surface of Inner Tank—Refer to Section

As Edeskuty [30] states: "Fuel tanks will nevertheless have to pass rigorous leak testing upon entry into service and throughout their useful life." Brewer [21] also recommends routine visual inspection of the tank structure from inside the tank at least every 8000 h, consistent with airline practice for all primary structure. A very challenging task for cryogenic composite tanks which points to the need of developing a "smart" liquid hydrogen tank.

7.3.5 Lightning Protection

The outer surface of the vacuum shell also needs adequate lightening protection and mitigation and elimination of any electrostatics effect. Lightening protection standards are given in the Federal Aviation Administration Handbook [31].

7.3.6 Stratification, Explosive Boil-Off

Tank liquid stratification can lead to explosive boil off, however this is unlikely to happen in a sloshing aircraft.

7.3.6.1 Tank Failure Diagnostics

GE Research studied latest sensor technology for tank diagnostics and prediction of early tank failure in the composite tank structure, e.g., for indicating composite cracks. Some sensors, e.g., flat polymer piezoelectric sensors, can even be embedded in the laminate. Several ultra-sensitive sensors for hydrogen detection in vacuum have been identified for use at GE. This is important, since during a total breach of vacuum, a tank could empty within 5 to 10 min, depending on tank size and fill level.

7.3.7 Example: The X-33 Project

The X-33 investigation report [32] of a lobed, composite hydrogen tank revealed several problems with liner-less designs in agreement of what has been mentioned earlier in Chap. 4 Sect. 4.12 "Tank Wall-Composites."

The tank was a conformal, load-bearing, composite sandwich structure, consisting of four lobes that provided packaging efficiency for integration into the vehicle (see color figure) (Fig. 7.20).

The report summarizes the main effects that happened during pressurization and gives valuable information for future designs:

Microcracking of the inner facesheet with subsequent gaseous hydrogen (GH_2) infiltration

- Cryopumping of the exterior nitrogen (N_2) purge gas that may have caused ice formation.
- Reduced bondline strength and toughness.
- Manufacturing flaws and defects.
- Infiltration of GH_2 into the core, which produced higher than expected core pressures.

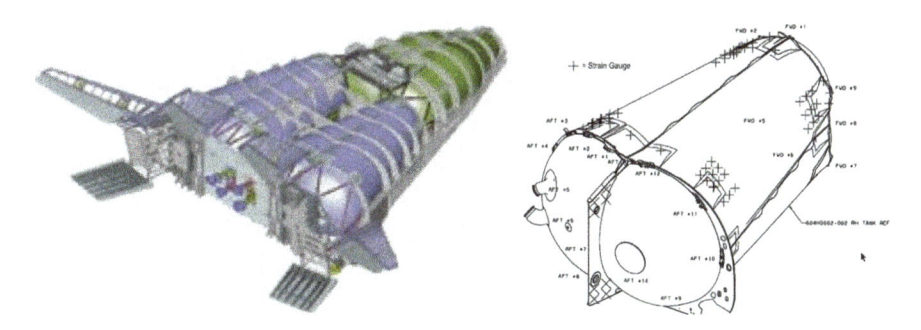

Fig. 7.20 The X33 liquid hydrogen composite tank

More recently however, a collaboration of NASA with Boeing arrived at a simplified non-lobed, cylindrical design that did survive initial pressure tests (see also Fig. 10 in Appendix 4).

References

1. Y. Kamioka, T. Nakamura, H. Hirai, S. Ozaki, K. Kajikawa, S. Imagawa, A. Ishiyama, *A long life and high head liquid nitrogen pump with a superconducting motor and an active magnetic bearing* 3rd IWC-HTS Conference GE Research Center, Niskayuna, NY 2019
2. Neumann H Zahn G *Cryogenic helium circulation pumps for forced flow cooling* www.kit.edu 2009
3. Li J et al., "Liquid pump-enabled hydrogen refueling system for heavy duty fuel cell vehicles: Pump performance and J2601-compliant fills with precooling", International Journal of Hydrogen Energy, Vol. 46, 42, 2021, pp. 22018–22029, https://doi.org/10.1016/j.ijhydene.2021.04.043
4. Barber Nichols, Submersible pump BNHP, private communication
5. Nikkiso, priv. communication, https://www.nikkisoceig.com/product/3-gupd/
6. G Petitpas, J Moreno-Blanco, F Espinosa-Loza, S M Aceves, Rapid high density cryogenic pressure vessel filling to 345 bar with a liquid hydrogen pump, Journal of Hydrogen Energy vol 43 issue 42 pp. 19547–19558, 2018
7. G Petitpas, S M Aceves, Liquid hydrogen pump performance and durability testing through repeated cryogenic vessel filling to 700 bar, International Journal of Hydrogen Energy vol 43 issue 39 pp. 18403–18420, 2018
8. Darrel B Schoch K An electrically pumped liquid helium transfer system Advances in Cryogenic Engineering 1966 J-3 pp. 607–611
9. Schmidt C US 4421464 patent Kernforschungszentrum Karlsruhe 1983
10. Walter H Bock H Frohne Ch Schippl K Nick W Neumüller H-W et al "First Heavy Load Bearing for Industrial Application with Shaft Loads up to 10 kN 7th European Conference on Applied Superconductivity Institute of Physics Publishing Journal of Physics: Conference Series 43 (2006) 995–998. https://doi.org/10.1088/1742-6596/43/1/243
11. Y. Kamioka et al, "Active magnetic bearing for a liquid nitrogen pump", ICEC-ICMC 2018, IOP Conf. Series: Materials Science and Engineering 502, 2019. 012153
12. Kenta T J. "Optimum design of cryogenic pump for circulation cooling of high temperature superconducting cables" Phys.: Conf. Ser. 1293 012070 2019
13. Kamioka [Kamioka Y Kajikawa K Hirai H Ozaki S Nakamura T Imagawa S and Ishiyama A 27th Int. Cryogenic Engineering Conf. and Int. Cryogenic Materials Conf. 2018 (Oxford, UK) E-14-28]
14. Ku A Reddi K Elgowainy A "Liquid pump-enabled hydrogen refueling system for medium and heavy duty fuel cell vehicles: Station design and technoeconomic assessment" International Journal of Hydrogen Energy Vol 47 Issue 61, pp. 25486–25498 2022
15. Yim W Yin J Cha J et al A design study on 20 kW class axial flux motor with HTS field winding for 100 MPa liquid hydrogen pump Cryogenics 139 pp 103829 2024
16. Kajikawa K Kuga H, Inoue T Watanabe K Uchida Y Nakamura T et al, "Development of a liquid hydrogen transfer pump system with MgB2 wires", Cryogenics 52 615–619, 2012
17. M. Kobayashi, M. Komori, "A superconducting stepping motor with pulsed-field magnetization for a pump Cryogenics 47 pp. 101–106 2007
18. after Perry's, Chemical Engineers' Handbook, 7th edition, Pump classification Figure 10-24, redrawn, 1997
19. Https://barber-nichols.com/media/tools-resources/
20. C.F. Gottzmann, "High Pressure Liquid Hydrogen and Helium Pumps", Advances in Cryogenic Engineering, vol. 5, pp. 289–298 1960

References

21. G.D. Brewer, Hydrogen Aircraft Technology, CRC Press, 1991
22. J.H. Fretwell, J.R. Bartlit "An automatic liquid nitrogen distribution system for twenty-five cold traps" Advances in Cryogenic Engineering vol 11 pp. 601–606 1966
23. Patent pending Stautner/Minas Apparatus and systems for separating phases in liquid hydrogen pumps US 2024/0068720 A1
24. Nikkiso, https://www.nikkisoceig.com/product/3-gupd/
25. F. Edeskuty, W. F. Stewart, "Safety in Handling of Cryogenic Fluids", Plenum Press, 1996, page 1
26. L. Klebanoff, Hydrogen Storage Technology, CRC Press, 2012
27. U. Nees, W. Lehmann, "Zur Qualität von Superisolation zwischen Raumtemperatur und 80 K bei Veränderung und Zerstörung des Isolationsvakuums, letzteres durch Belüftung mit atmosphärischer Luft, KFK report, 03.03.07P12A, 1990
28. W. Stautner, Chapter 7, "Special topics in cryostat design", page 213, in J.G. Weisend II, "Cryostat design", International Cryogenics Monograph Series, Springer, 2016
29. W. Lehmann, G. Zahn, "Safety aspects for LHe cryostats and LHe transport containers", Proc. ICEC7, 1978
30. F. J. Edeskuty, "Accidents with cryogenic fluids and what can we learn from them", AIP Conf. Proc. 613, 1743, 2002 and "Safety of liquid hydrogen in air transportation", Hydrogen in air transportation conference, Stuttgart, F.R. Germany, 1979
31. Federal Aviation Administration, Aviation Rulemaking Advisory Committee, Transport Airplane and Engine Issue Area, Electromagnetic Effects Harmonization Working Group, Task 2—"Lightning Protection Requirements", US Department of transportation, Federal Aviation Administration, DOT/FAA/CT-89/22, AIRCRAFT LIGHTNING, PROTECTION HANDBOOK, 1989Federal Aviation Administration Technical Center
32. NASA, Final report of the X-33 liquid hydrogen tank test investigation team, 2000

Hydrogen Detection, Leak Detection, Zero Emission Vehicles (ZEV), Sensor Types

8

8.1 Leak Detection on Hydrogen Aircraft

8.1.1 Introduction

Hydrogen (H2) detection is currently required and implemented in zero emission vehicles (ZEV) and H2 infrastructure sites. A H2 detection system is even more critical on the H2 aircraft where large quantities of fuel are stored and very high mass flow rates are required for aircraft takeoff. A H2 detection system consists of one or an array of H2 sensors which detect the H2 concentration level (C_{H2}) at one or more locations of the vehicle or site and a controller which collects, processes, and transmits this information to the safety system. Since a H2 fuel system is near impossible to seal, the H2 detection system can continuously monitor the H2 level on the aircraft/vehicle or infrastructure site and can inform the safety system. The safety system can make decisions when the H2 level reaches critical values resulting in unsafe operation.

8.1.2 H2 Detection on ZEV

Several H2 vehicles have been in operation mainly in California where there exist more than 50 H2 refueling stations. A popular H2 vehicle is the Toyota Mirai with the first and second generations being sold in 2014–2020 and 2021–2023, respectively. The quantity and location of the H2 sensors depends on the potential leak locations of the fuel storage and fuel distribution systems, the geometry of the volume where H2 is leaked into, and the ventilation of the system. For instance, the 2014 Toyota Mirai was equipped with 2 sensors, the first one in the engine compartment and the second one near the storage tank (Fig. 8.1). It was necessary to increase the number of H2 sensors to three, in the 2021 Toyota Mirai because of changes in the H2 storage and fuel distribution system (Fig. 8.2). The BMW H2 7 car developed in 2007, which stored H2 in the liquid phase was equipped with a safety system which included five H2 sensors at various locations within the vehicle.

Lead Author: Constantinos Minas

The ZEV H2 sensing system monitors and reports to the vehicle safety system the H2 concentration at the sensor locations before vehicle startup and during vehicle operation. Typical responses of the vehicle safety system of the ZEV include increased ventilation and eventually H2 flow shutdown if the previous action is unsuccessful in reducing the H2 level below the critical value.

It is important to realize that there are cases where shutting down the H2 supply valve may not stop the H2 leak. This will occur in the three cases shown in Fig. 8.3 where the H2 leak occurs (i) at the onboard storage tank/H2 supply valve interface (i.e., O-ring failure), (ii) at the storage tank/end plug interface (i.e., O-ring failure), and (iii) at the tank wall caused by a wall crack. These unstoppable leaks will lead to the release of the entire stored amount of H2 in the onboard tank.

Fig. 8.1 2014 Toyota Mirai H2 sensor locations

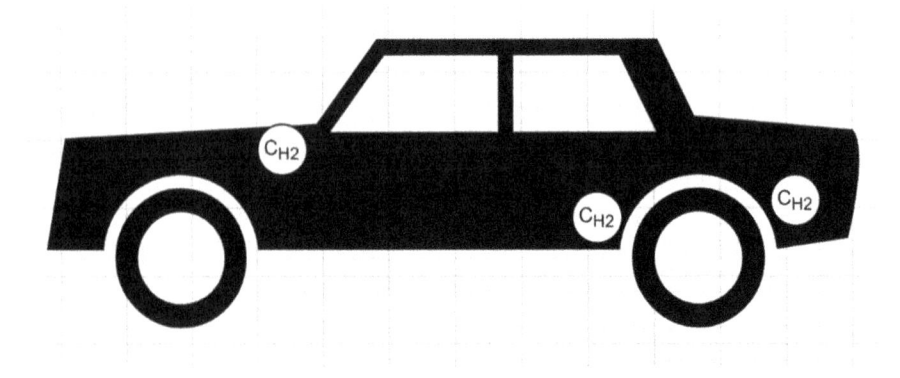

Fig. 8.2 2021 Toyota Mirai H2 sensor locations

8.1 Leak Detection on Hydrogen Aircraft 175

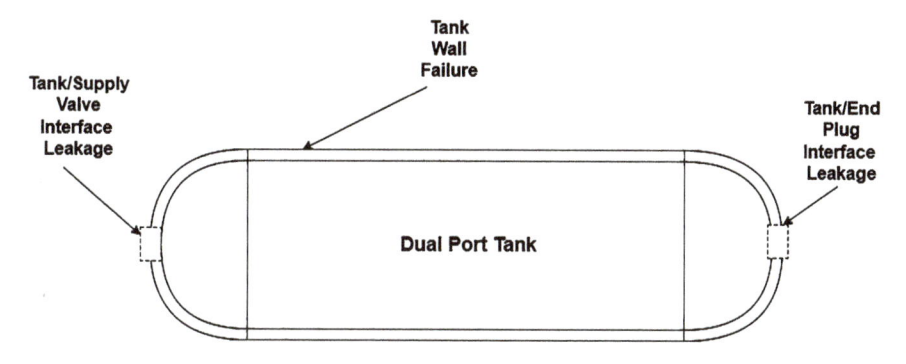

Fig. 8.3 ZEV unstoppable H2 leaks

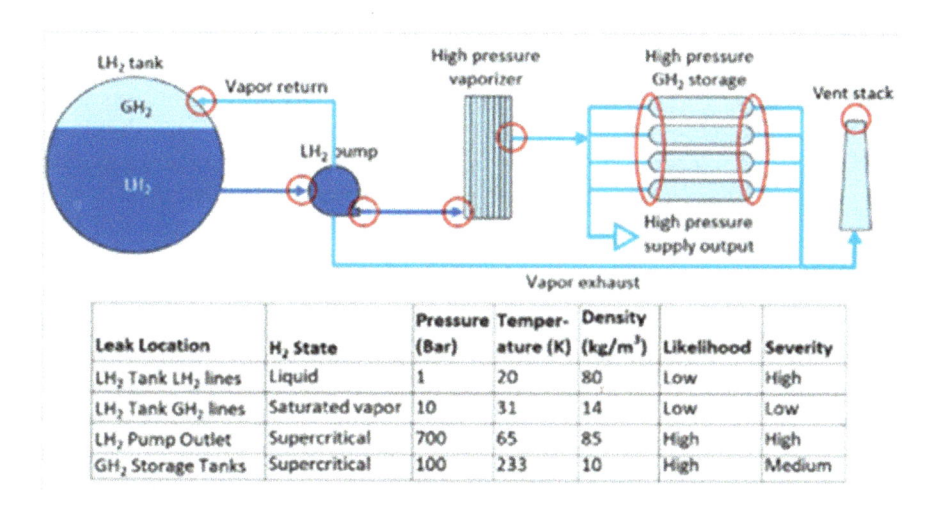

Leak Location	H₂ State	Pressure (Bar)	Temperature (K)	Density (kg/m³)	Likelihood	Severity
LH₂ Tank LH₂ lines	Liquid	1	20	80	Low	High
LH₂ Tank GH₂ lines	Saturated vapor	10	31	14	Low	Low
LH₂ Pump Outlet	Supercritical	700	65	85	High	High
GH₂ Storage Tanks	Supercritical	100	233	10	High	Medium

Fig. 8.4 Layout of an LH2-based H2 delivery with potential leak locations in red ZEV unstoppable H2 leaks

8.1.3 H2 Detection at H2 Infrastructure Sites

Hydrogen infrastructure sites are designed to meet NFPA 2 requirements which can be met with the installation of a single H2 detection sensor. The H2 sensor can only detect a small number of leaks and cannot provide any information on the amount of the leak. A typical layout of an LH2 infrastructure site was shown in Fig. 8.4. Potential leaks can develop at any joint location between the different components. The H2 infrastructure monitoring system includes continuous monitoring of the H2 sensor concentration. Typical responses of the H2 infrastructure safety system are (i) warning of a H2 leak, (ii) LH2 pump shutdown, (iii) compressor shutdown, (iv) partial or total refueling shutdown, and (v) eventually entire system shutdown when the H2 level reaches critical values.

8.1.4 Types of H2 Sensors

H2 sensors of various types are currently being used in ZEVs and H2 Infrastructure sites. These types of H2 sensors include (i) catalytic, (ii) electrochemical, (iii) thermal conductivity, and (iv) metal oxide semiconductor (MOS). The catalytic combustion sensor consists of two beads, which surround a wire which can withstand high temperatures (450 °C). The first bead is passivated without the addition of a catalyst such that it does not react when exposed to H2, acting as a background reference. The second bead is coated with a catalyst to facilitate the reaction with the H2 gas. This bead is placed on a separate leg of a Wheatstone bridge circuit. In the presence of H2, the resistance of the catalytic bead increases, while the resistance of the passivated bead remains the same. The electrochemical sensor consists of a thin layer of electrolyte separating the anode and cathode. When H2 passes through the electrolyte, a reversible chemical reaction takes place, producing an electric current proportional to the gas concentration. This sensor requires the lowest amount of power of all sensor types. This sensor demonstrates high sensitivity, short response time, good reproducibility after calibration, good linearity, stable zero-point, and relatively low cross-sensitivity. A major disadvantage of the electrochemical sensor is the fact that its sensitivity decreases over time due to the loss of the catalytic surface. The thermal conductivity sensor consists of a thermal element such as a thermistor which is heated to a certain operating temperature. When the thermal conductivity of the gas is high, the heat will dissipate more readily from the thermal element and its resistance will decrease. The MOS H2 sensor consists of a heating resistor and a sensitive resistor made of metal oxide layer deposited on a heater which heats the sensor to an operating temperature (200–500 °C). The resistance of the metal oxide layer varies with temperature and the H2 content of the surrounding air. The semiconductor has very low thermal conductivity in clean air. When the gas is present, the higher the concentration the higher the conductivity. A filtering layer can eliminate the influence of interfering gases such as alcohol.

Critical characteristics of the H2 sensors are the range, response time, sensitivity to temperature, and humidity and interference of other gases. Datasheets of these commercially available sensors which are available by Companies like Figaro Engineering, Renesas, and SGX Sensortech show temperature operation in the range of −20 to 40 °C. The aircraft application presents unique challenges caused by the extreme low and high temperatures of the aircraft engine core compartment. Current configurations of automotive-type sensors where the digital board is integrated with the sensing element (Fig. 8.5) cannot survive the high temperatures of the aircraft engine core.

8.1.5 H2 Detection on H2 Aircraft

A H2 detection system is required on both the H2 turbine aircraft where H2 is burned in the combustor and the H2 electrified aircraft where the H2 is consumed in the fuel cells. The amount of H2 needed on the H2 aircraft varies by the size and the

8.1 Leak Detection on Hydrogen Aircraft

Fig. 8.5 (**a**) Figaro TGS6812 sensing element. (**b**) Figaro CGM6812 sensor module

mission of the aircraft. Quantities of onboard stored H2 vary from approximately 400 kg for a small commuter aircraft with a 200 nautical mile mission, to more than 10,000 kg for a wide-body aircraft traveling transatlantic missions. Typical H2 aircraft fuel distribution systems show the on-aircraft H2 stored in the liquid form (LH2). Prototype composite cryogenic LH2 tanks for aircraft are currently under development which presents a novel solution to store H2 at maximum gravimetric storage density and volumetric efficiency [1]. An aircraft LH2 fuel distribution system presents the challenge of leaks of variable temperature, pressure, and density. A typical aircraft fuel distribution system was shown in Fig. 8.6 [2]. Some of the potential states were presented in Table 8.1 which span a broad temperature and pressure spectrum with variable likelihood and severity. For instance, a high-pressure cryogenic leak at the LH2 pump outlet has high likelihood caused by the high pressure and low temperature, and high severity caused by the high density of the leaked H2. On the other hand, a low-pressure cryogenic vapor leak has a low likelihood caused by the low pressure and low severity caused by the low density of the leaked H2.

As mentioned above, commercially available automotive sensors are not appropriate for the aircraft due to the extreme low and high temperatures of the aircraft engine core compartment. Potential solutions to this problem are (i) development of high temperature H2 sensors with high temperature digital boards, (ii) the separation of the digital board from the sensing element, (iii) selection of high temperature cables connecting the sensing element to the digital board, (iv) strategic placement of the digital board and controllers at the lowest temperature locations of the aircraft engine core compartment. Other requirements for the air application include redundancy where all leaks are detectable by at least two independent sensors and response time of a few seconds (Table 8.2).

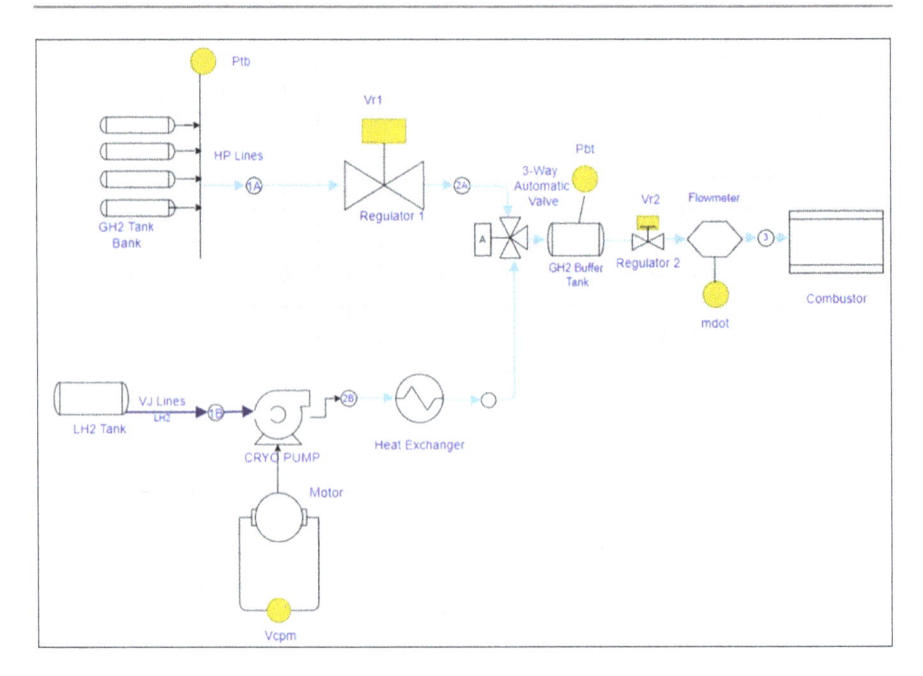

178   8 Hydrogen Detection, Leak Detection, Zero Emission Vehicles (ZEV), Sensor Types

Fig. 8.6 H2 Aircraft fuel distribution system

Table 8.1 H2 sensors in H2 vehicles

Vehicle	Model	Tank technology	Pressure (Bar)	Onboard tank #	Onboard H2 sensors #
BMW	H2–7	LH_2	5	1	5
Toyota	Mirai 2014	GH_2	700	2	2
Toyota	Mirai 2021	GH_2	700	3	3

8.1.6 Example of a H2 Aircraft Detection System

In order to develop a successful design of a H2 aircraft a 2-species CFD model of a representative aircraft engine core is developed. All potential leaks were simulated under various engine conditions like ground idle, takeoff, maximum cruise which include variable air flow, temperature, pressure, and air density. The steady-state and transient CFD analysis was carried out in Ansys Fluent. Potential sensing element locations were evaluated and selected based on criteria like response time derived from the transient CFD analysis and detection ratio (DR) defined by the equation below:

8.1 Leak Detection on Hydrogen Aircraft

Table 8.2 H2 potential states and properties of H2 leaks on H2 aircraft

Leak location	H2 state	Pressure (bar)	Temperature (K)	Density (kg/m³)	Likelihood	Severity
LH2 tank	LH2	1.0	20	79.966	Low	High
LH2 lines		9.9	31	50.036	Low	High
LH2 tank	Saturated vapor	1.0	20	1.2957	Low	Low
GH2 lines		9.9	31	13.989	Low	Low
LH2 pump outlet	Supercritical	100.0	65	35.799	High	High
Heat exchanger outlet	Supercritical	100.0	313	7.321	High	Medium

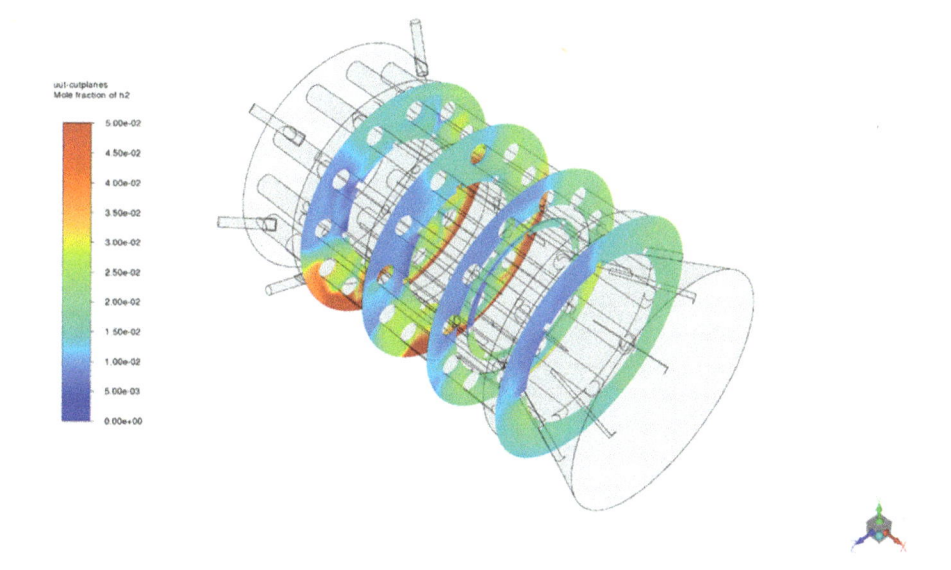

Fig. 8.7 Typical CFD results of H2 concentration spatial distribution

Equation 8.1: Detection Ratio

$$DR = \frac{C_S}{C_{Exit}} = \frac{Local\, H2\, mole\, fraction\, at\, sensor}{Average\, H2\, mole\, fraction\, at\, air\, outlet} \qquad (8.1)$$

A typical value of the Cexit is 20,000 ppm which has a safety factor of 2 over the low flammability limit (LFL) of H2. This ensures that no explosive mixture will be released in the atmosphere by the aircraft. Several results were obtained in the form of H2 concentration (CH2) spatial distribution as shown in Fig. 8.7. It is important to realize that at a small volume in the vicinity of the leak location, the CH2 will be higher than the LFL of H2, which is 4%. And at an even smaller volume in the vicinity of the leak, CH2 will be higher than the low detonation limit (LDT) of H2 which is 18%. The CH2 spatial distribution clearly shows the variability in the engine core and highlights the challenge of identifying the optimal sensor locations.

References

1. Stautner, W., Ansell, P., Haran, K., Mariappan, D., Minas, C "Liquid Hydrogen Tank Design for Medium and Long Range All-Electric Airplanes," *CEC/ICMC21, July 2021*
2. Minas, C., Tang, L. "Hydrogen Fuel System" US Patent 12,092,042 B2

Cooling System Technologies on Superconducting Rotating Machines

9

Uijong Bong and Kiruba S. Haran

9.1 Introduction

Superconducting rotating machines are considered to be an important application area of cryogenics and superconducting technologies, offering the potential to revolutionize energy conversion with remarkable performance metrics such as enhanced efficiency and torque density. Growing interest in eco-friendly propulsion systems for next-generation transportation systems has led to the initiation of various projects incorporating superconducting motors.

The successful development of superconducting rotating machines hinges significantly on cooling system technologies because cooling plays a pivotal role in shaping machine characteristics and performance. Cooling-system technologies on superconducting rotating machines have unique features compared to other superconducting applications, which are mainly stationary DC applications.

- Superconducting rotating machines encompass both "rotary" and "AC" components, with the utilization of superconductors on these components depending on the machine's topology.
- Space for cryogenic systems is limited and must be minimized for higher electromagnetic performance.
- These machines are also generally connected to larger dynamical systems, both on the mechanical and electrical sides, which impose significant transient and fault-tolerance requirements.

Due to these special features, numerous efforts and attempts have been made to enhance the performance in terms of machine design and cooling systems. Figure 9.1

U. Bong (✉) · K. S. Haran
University of Illinois Urbana-Champaign, Electrical and Computer Engineering, Urbana, IL, USA
e-mail: ubong@hinetics.com

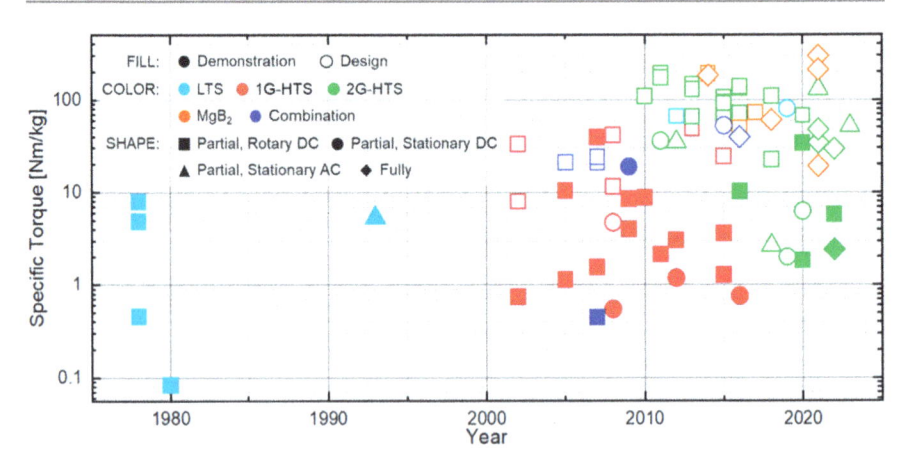

Fig. 9.1 Overall trend of superconducting rotating machine demonstrations and designs [1–4]. In the graph, the x-axis represents the published year while the y-axis is the machine's specific torque [Nm/kg] (=average torque/weight). Filled markers mean demonstration cases, while empty markers are design cases. The color represents superconducting wire materials, and the shape shows the machine's topology

depicts the overall trends in superconducting rotating machines, including their specific torque, type of superconductor used, and machine topologies [1–4].

Figure 9.1 presents several interesting trends in superconducting rotating machines development. Over time, there has been a shift in utilizing superconductors, aligning with their developmental progression from LTS (blue markers) to 1G-HTS (red markers), and subsequently to 2G-HTS (green markers) and MgB_2 (yellow markers). Notably, post-2000s, there is an absence of LTS demonstration cases, indicating a shift toward higher operating temperatures exceeding 4.2 K. Moreover, there is a discernible upward trend in the overall specific torque of both demonstrated and designed machines. This trend may be attributed to the adoption of improved cooling technologies and novel superconductors having higher critical values. Despite several recently published designs with exceptionally high torque density exceeding 100 Nm/kg utilizing 2G-HTS and MgB_2, their practical realization remains pending.

Another notable observation is the attempts at new machine topologies. The predominant one investigated and demonstrated so far is the partially superconducting machine featuring a "rotary superconducting DC field winding" (square markers). While it avoids large heat loads resulting from AC losses in superconductors, the concern is constructing compact and reliable rotary cryogenic systems. Specialized machine topologies employing "stationary superconducting DC winding" (circle markers), such as homopolar induction alternators and clawed pole machines, have been explored to mitigate this challenge. However, achieving high electromagnetic performance is relatively demanding, so research interest is limited despite their attractiveness from the perspective of a simplified cooling structure.

Recent advancements in high-performance superconducting wires and cryogenic technologies have led to the pursuit of topologies utilizing "stationary superconducting AC armature winding" (triangle markers). These include PM-SC machines (permanent magnet on rotor + superconducting AC armature winding on stator) and SC induction machines (squirrel cage rotor + superconducting SC armature winding on stator). Additionally, fully superconducting machines are under active investigation. In contrast to previous topologies, these configurations incorporate AC-operated superconducting coils, requiring comprehensive examination of AC losses and cooling approaches.

Addressing the evolving trends in superconducting rotating machines and their cooling systems, this chapter focuses primarily on two topics. First, we review the cooling system for the most common machine topology, namely those for rotary DC field winding, at a system level. Various approaches, including thermosiphon cooling, forced-flow cooling, and onboard cryocooler cooling (conduction-cooling), are discussed, along with their overarching characteristics. Second, we consider cooling system technologies for stationary superconducting AC armature winding. To realize recently designed superconducting machine performances (including dB/dt) employing such armature winding, aspects such as superconducting wire selection and cryocooler performance are explained through example and calculation.

9.2 Review on Cooling System Technologies for Rotary Superconducting DC Field Winding

Partial superconducting rotating machines with rotary DC field windings have been considered one of the main topologies of superconducting machines. Within this configuration, effective cooling of the rotary DC field windings is critical for ensuring optimal performance, efficiency, and reliability. Over time, a variety of cooling technologies have been developed, each possessing distinct characteristics, advantages, and limitations. These cooling system technologies can be broadly categorized into four approaches: thermosiphon cooling, forced-flow cooling, onboard cryocoolers, and immersion in a cryogenic bath. In this section, we will delve into a comprehensive review of these cooling technologies, exploring their features, benefits, and drawbacks.

9.2.1 Closed-Loop Cooling System

9.2.1.1 Thermosiphon Cooling

Thermosiphon cooling represents a gravity-driven cooling technique that relies on natural convection and phase change to remove heat from the superconducting winding. This method typically involves the circulation of a cryogenic fluid, such as liquid helium or nitrogen, through the winding enclosure. The circulation process is facilitated by the density difference between the warm and cold regions of the fluid. In actual demonstrations, the circulation pipe is connected to the chamber inside the motor through one shaft, and the field winding is cooled through conduction from the chamber. Cryocoolers are often employed as the cooling source, positioned within an additional cold box adjacent to the superconducting motor, with an internal heat exchanger. The pros and cons of thermosiphon cooling are as follows (Fig. 9.2):

184   9 Cooling System Technologies on Superconducting Rotating Machines

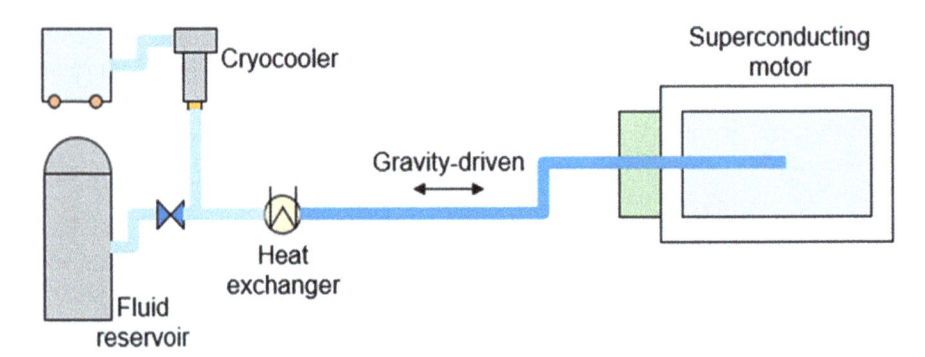

Fig. 9.2 Schematic of thermosiphon cooling for superconducting rotating machine

Pros:
- Relatively simple design requiring no mechanical pumps or moving parts.
- Reduced complexity and cost compared to forced cooling systems.
- Self-regulating property in response to variable heat.

Cons:
- Limited cooling capacity.
- Dependency on gravitational forces, limiting applicability in certain orientations or configurations.

Numerous successful superconducting machine demonstrations with thermosiphon cooling exist, such as the 400 kW machine (2002), 4 MVA generator (2005), and 4 MW motor (2010) developed by Siemens [5, 6]. These machines, utilizing BSCCO wires, achieved rotor cooling down to 27 K with Neon thermosiphon cooling.

9.2.1.2 Forced-Fow Cooling

Forced-flow cooling employs mechanical pumps to actively circulate a cryogenic coolant through the superconducting winding enclosure, enhancing heat transfer. This cooling system provides greater control over coolant flow rates and temperature distribution, making it well-suited for high-performance applications. Similar to thermosiphon cooling, cryocoolers serve as the cooling source, positioned within an additional cold box adjacent to the superconducting motor, featuring an internal heat exchanger. The pros and cons of forced-flow cooling are as follows (Fig. 9.3):

Pros:
- Enhanced heat-transfer efficiency and cooling performance.
- Greater flexibility in controlling coolant flow rates and temperature gradients.
- Improved thermal stability and uniformity within the winding enclosure.

Cons:
- Increased complexity and cost associated with the integration of mechanical pumps.

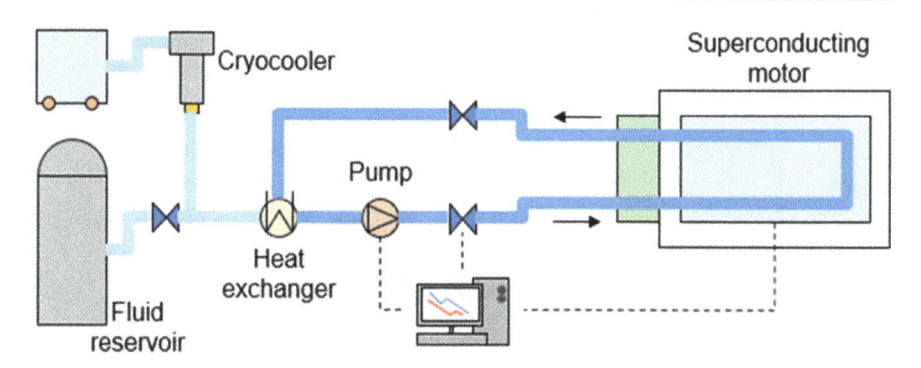

9.2 Review on Cooling System Technologies for Rotary Superconducting DC Field...

Fig. 9.3 Schematic of forced-flow cooling for superconducting rotating machine

- Higher energy consumption compared to passive cooling methods.
- Potential reliability concerns due to the presence of moving parts and mechanical components.

Forced-flow cooling has been implemented in various large-scale demonstrations, including General Electric's 1.5 MVA HTS generator (2004) [7], AMSC's 36.5 MW ship propulsion motor (2010) [8], and Kawasaki's 3 MW HTS motor (2017) [9], utilizing gaseous helium as a common coolant.

9.2.1.3 On-Board Cryocooler

On-board cryocoolers offer a self-contained cooling solution integrated directly within the superconducting machine, eliminating the need for external cryogenic infrastructure. These cryocoolers utilize thermodynamic cycles, such as the Gifford-McMahon or pulse tube, to achieve cryogenic temperatures and maintain superconducting states. This approach, also called *conduction cooling*, involves no circulating coolant in superconducting rotating machines. There are two implementation styles: one utilizes multiple cryocoolers rotating along the superconducting coil, while the other employs one cryocooler rotating along the rotating shaft. The former is suited for large-scale low-speed applications like wind turbine generators, whereas the latter is preferable for relatively small-scale high-speed applications such as aircraft propulsion motors. See also Chap. 10. The pros and cons of on-board cryocooler cooling are as follows (Fig. 9.4):

Pros:
- Simplest cooling structure among closed-loop cooling technologies.
- Decoupling from external circulation system offers greater flexibility.

Cons:
- Limited cooling capacity and performance compared to approaches accompanying external cryogenic structures.
- Potential reliability issues related to the operation of cryocooler components under demanding conditions.

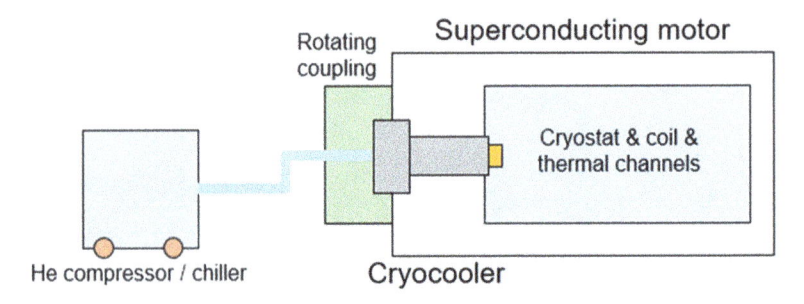

Fig. 9.4 Schematic of on-board cryocooler cooling for superconducting rotating machine

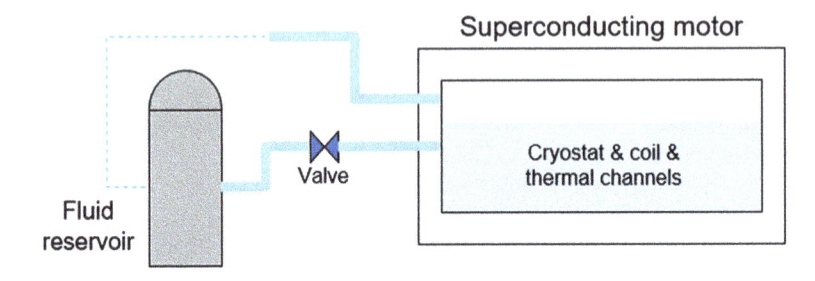

Fig. 9.5 Schematic of bath-cooled superconducting machine

A notable demonstration example of on-board cryocooler cooling is EcoSwing's 3.6 MW wind turbine generator [10], which utilized nine rotating cryocoolers along the superconducting coils. Additionally, NASA Glenn Research Center and UIUC are developing superconducting rotors with an integrated rotating cryocooler [11, 12]. See also "Cryogenic cooling of a hermetically sealed, thermosiphon cooled, 20 MW class superconducting generator for the GE Renewables industry, using 6 (+2 redundant) cryocoolers" [13] (large scale drivetrain cooling).

9.2.2 Open-Loop Cooling System

In certain laboratory-level demonstrations, superconducting motors are submerged into a cryogenic bath to achieve cryogenic temperatures for the superconducting winding. This open-loop cooling concept, known as *immersion in a cryogenic bath*, involves submerging the superconducting winding in a pool of cryogenic fluid, typically liquid nitrogen, to rapidly and uniformly cool it. This technique exploits the high heat transfer coefficient of cryogenic fluids to efficiently dissipate heat from the winding. Because it is open-loop, there is no need for cooling pipes or circulation systems. Immersion in a cryogenic bath can be categorized into total immersion and partial immersion, depending on whether the entire system or only the superconducting part is submerged. The pros and cons of immersing in a cryogenic bath are as follows (Fig. 9.5):

Pros:
- Rapid and uniform cooling of the entire winding assembly.
- High heat-transfer efficiency, enabling effective heat removal even at high loss densities.
- Simplified design and reduced complexity compared to other cooling systems.

Cons:
- Limited scalability and applicability to practical rotating machines.
- Challenges associated with fluid management, including cryogen loss and containment.
- Potential safety concerns related to the handling and storage of cryogenic fluids.

Examples of implementation include early-stage tests or concept verification experiments such as [14, 15].

9.2.3 Comparison of Cooling Technologies

Table 9.1 provides a comparison of the introduced cooling technologies. Each technology offers a unique combination of advantages and drawbacks, necessitating careful consideration of specific application requirements, constraints, and performance objectives when selecting an appropriate cooling solution. For instance, while thermosiphon cooling offers a cost-effective and reliable option for low to medium-power applications, it comes with limited cooling capacity. In contrast, forced-flow cooling provides enhanced cooling performance and control but entails increased complexity and upfront costs. Ultimately, the choice of cooling technology requires a thorough evaluation of trade-offs and compromises, guided by considerations such as cooling efficiency, reliability, cost, and practical feasibility. Engineers can make informed decisions by understanding the characteristics and capabilities of each cooling technique to optimize the performance and reliability of superconducting DC field windings in rotating machinery applications.

9.3 Considerations on Cooling System Technologies for Stationary Superconducting AC Armature Winding

Many superconducting machine engineers are now exploring the feasibility of employing superconducting AC armature winding, given the significant advancements in superconducting wire and cryogenic technologies. By extending superconducting technology to both armature and field windings, substantial performance leaps could be made in terms of efficiency and torque density. Also, even if superconductors are applied to only the armature side, the absence of a rotating cryogenic environment or rotating coupling makes this approach attractive. However, this transition comes with its own set of challenges, especially AC loss. In this section, we delve into the expected performance of superconducting AC armature windings and the associated considerations for cooling system technologies.

Table 9.1 Comparison of cooling-system technologies for rotary superconducting DC field winding [5–14]

	Immersing in a cryogenic bath	Thermosiphon cooling	Forced-flow cooling	On-board cryocooler
Applied capacity	300 W / 400 rpm 5 kW / 150 rpm 10 kW / 150 rpm	400 kW / 1500 rpm 4 MW / 120 rpm 4 MW / 3600 rpm	1.5 MW / 160 rpm 3 MW / 3600 rpm 36.5 MW / 120 rpm	3.6 MW / 15 rpm High-speed applications in development
Characteristics	Used for lab-level motor test Applicable to stator Re-liquefaction can be applied	Many actual application cases Self-regulating to variable heat	Many actual application cases	Multiple cryocoolers: Applied to wind turbine application Integrated one cryocooler: Currently in development
Pros	High cooling power High heat capacity	Relatively simple structure Easy to install & maintain	Large cooling capacity possible Controllable temperature	Conduction cooling type ■ no need for cryogen and circulation system
Cons	Difficult to apply to rotor Relatively a lot of cryogen required Orientation dependency	Difficult to increase cooling capacity Orientation dependency Cooling duct design required	Circulation pump required Cooling duct design required	Cryocooler maintenance issue Rotating coupling for cryocooler required

9.3.1 Challenge with Closed-Loop Cooling System

9.3.1.1 Cryocooler's Coefficient of Performance (COP) and Weight

Of course, superconducting AC armature windings require cryogenic cooling to maintain superconducting states. This can be achieved through closed-loop systems employing cryocoolers, as shown in the previous section. However, the combination of superconducting AC armature windings and cryocoolers presents several challenges.

Figure 9.6 illustrates the COP of commercial cryocoolers across different operating temperatures [16]. Here, Carnot cycle's COP is plotted together for comparison, and as we can see at Carnot cycle's COP curve, the COP of cryocoolers naturally becomes lower at cryogenic temperatures with a thermodynamic principle. Since AC losses are inevitable in superconducting AC armature windings, these lower COP values directly impact system efficiency. This poses a trade-off dilemma for machine engineers: operating temperature should be lowered to maximize torque, but it should be increased to improve efficiency by reducing AC losses and COP^{-1}

9.3 Considerations on Cooling System Technologies for Stationary Superconducting... 189

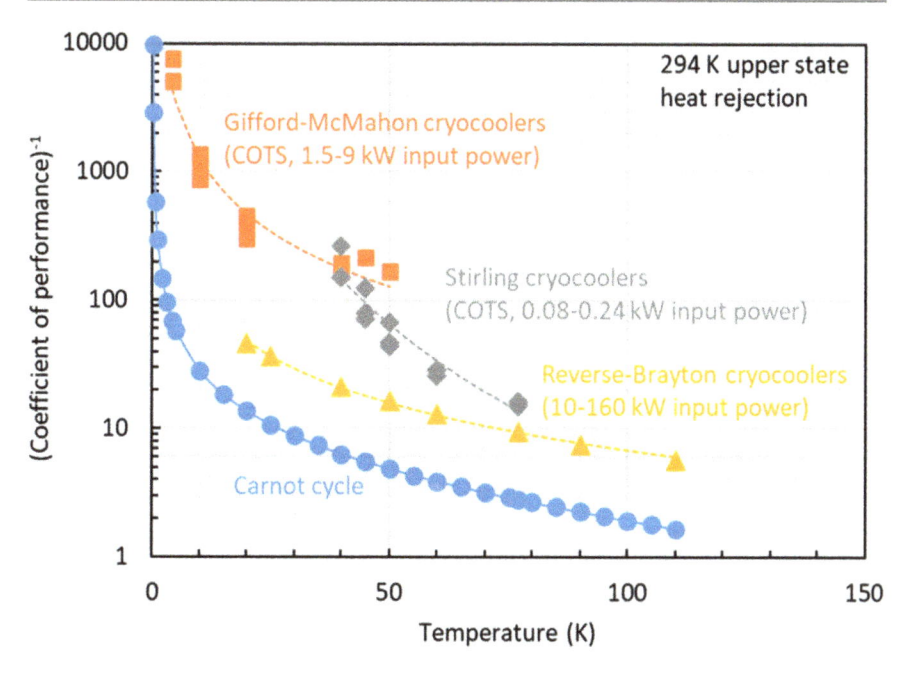

Fig. 9.6 Coefficient of performance of commercial cryocoolers according to their operating temperature. Carnot cycle is plotted together for comparison [15]

value. For instance, numerous superconducting AC armature machines with 2G-HTS wire have been designed at 20 K for enhanced electromagnetic performance, but possible expected COP at 20 K is less than 1/30, so intentionally higher operating temperatures (30 ~ 50 K) are selected in some designs to ensure machine efficiency. Thus, the selection of operating temperature and estimation of AC losses become critical in superconducting AC armature design, as even minor AC loss generation can lead to a significant reduction in system efficiency.

Another consideration is the required mass of the superconducting armature cooling system. As most superconducting machines aim for high power density or torque density, additional weight from the cooling system can pose practical challenges. Figure 9.7 illustrates the cryocooler mass required to achieve 1.5 W of cooling at different operating temperatures [16]. Similar to the COP issue, increasing the operating temperature can improve the cryocooler weight properties. For instance, at five K, the required weight is approximately 100 kg, whereas it reduces to ten kg at 20 K and less than five kg at 50 K. This presents another optimization challenge for machine designers aiming for high power density.

9.3.1.2 Review on *dB/dt* Values of Previous Superconducting AC Armature Windings

A high rate of magnetic field change (*dB/dt*) to the armature winding is typically required to achieve high-power density in rotating machines, even though high *dB/dt* values are associated with thermal losses, including hysteresis and eddy-current

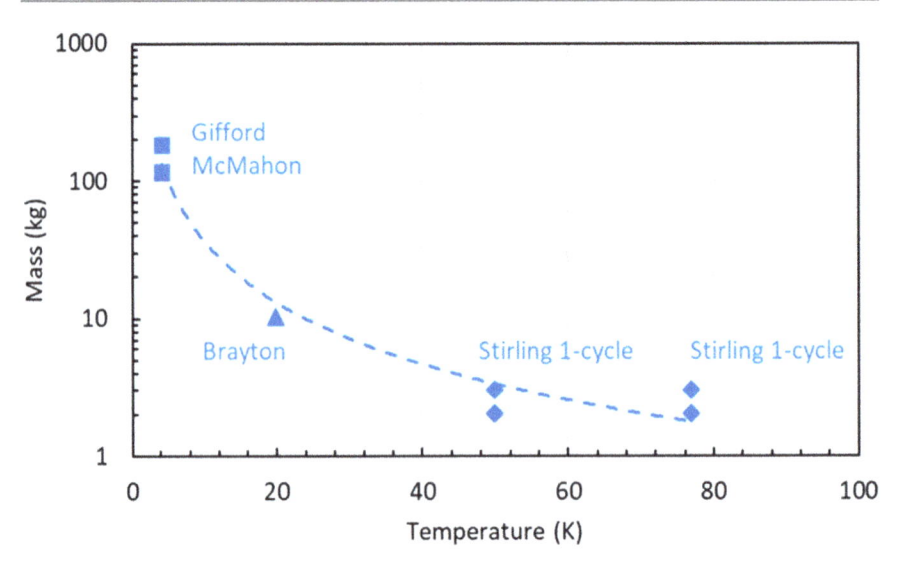

Fig. 9.7 Cryocooler mass required to achieve 1.5 W cooling at different operating temperature [16]

loss. Likewise, the AC loss of the superconducting AC armature winding is dependent on dB/dt, and by reviewing dB/dt values in previous demonstrations and designs, we can estimate typical AC loss level in superconducting AC armature winding. Several machines employing superconducting AC armature windings have been studied, and Tables 9.2 and 9.3 present the development and design cases of superconducting AC armature windings, respectively, along with their corresponding dB/dt values.

As shown in Tables 9.2 and 9.3, superconducting AC armature windings have been applied to PM-SC (permanent magnet in rotor and superconducting AC armature winding in stator) machines and fully superconducting machines (superconducting windings for both DC field and AC armature winding). In demonstrations, BSCCO or REBCO wires have been utilized primarily at relatively high temperatures (around 77 K with liquid nitrogen) and lower speeds (a few hundred rpm), resulting in dB/dt values within <20 T/s. In contrast, in design cases, REBCO wire and MgB_2 wire are principally used, with operating temperatures (20–40 K) lower than those in demonstration cases. These design applications include aircraft propulsion motors with a few thousand rpm and wind turbine generators with around 10 rpm. For wind turbine generators, dB/dt values are less than 10 T/s, while for aircraft propulsion motors, dB/dt values exceed 100 T/s. It is noteworthy that there is a discrepancy in dB/dt values between demonstrations and aircraft propulsion motor designs, suggesting that to achieve designed high performance, dB/dt values exceeding 100 T/s may be required, a value not fully demonstrated in actual motor demonstrations. Thus, in the case of high-speed applications, the actual realization of superconducting AC armature winding poses the main challenge of demonstrating >100 T/s operation at 20–40 K.

9.3 Considerations on Cooling System Technologies for Stationary Superconducting... 191

Table 9.2 Superconducting AC armature winding development cases [17–26]

Year	Group	Topology	SC wire	Temp. (K)	Power (kW)	Speed (rpm)	Freq. (Hz)	Airgap B_{pk} (T)	dB/dt (T/s)
1993	CNRS	PM-SC	NbTi	5	15	750	50	–	–
2007	Fukui univ.	PM-SC	BSCCO	68	400	250	16.67	0.45	7.5
2013	Woosuk univ.	Homo-polar	REBCO	77	5	150	5	0.3	1.5
2014	Tsinghua univ.	PM-SC	BSCCO	82	1.8	290	9.667	0.8	7.733
2014	Cambridge	Fully SC	REBCO	77	0.54	150	5	0.06	0.3
2019	ENEA	PM-SC	REBCO	77	1	600	50	0.2	10
2019	Kyoto univ.	Fully SC ind.	BSCCO	77	12	1500	50	0.4	20
2021	Kyusu univ.	Fully	REBCO	65	1	625	10.4	0.3	3.12
2022	Kyoto univ.	Fully SC ind.	REBCO	77	0.55	200	40	–	–
2024	Kyusu univ.	Fully SC	REBCO	65	10.2	441	7.35	0.69	5.072

Table 9.3 Superconducting AC armature winding design cases (selected) [26–33]

Year	Group	Topology	SC wire	Temp. (K)	Power (MW)	Speed (rpm)	Freq. (Hz)	Airgap B_{pk} (T)	dB/dt (T/s)
2015	DTU	Fully SC	MgB_2	20	10	10	2	2	4
2018	KPS	Fully SC	MgB_2	20	3	4500	150	0.8	120
2020	UIUC	Fully SC	MgB_2	20	2.5	4500	300	0.464	139.2
2021	Sheffield	PM-SC	REBCO	30	10	9.6	2.56	1.3	3.328
2022	MAI	Fully SC	REBCO	21	5	12,000	400	1.23	492
2023	Univ. of Tokyo	Fully SC	MgB_2	20	5.5	5000	166.7	0.8	133.3
2023	Harbin Inst.	Fully SC	REBCO	30	1	6000	400	1.9	760
2023	NTNU	PM-SC	REBCO	40	2.5	5000	333.3	1.09	363.4

9.3.1.3 Expected AC Loss Density

Based on the *dB/dt* values, AC loss density of the superconducting AC armature winding can be computed with analytic eqs. AC losses generated in superconducting wire can be categorized as hysteresis loss (Q_h), transport loss (Q_t), coupling loss (Q_c), and eddy-current loss (Q_e). For each loss, analytic calculation equations have been derived, based on a critical state model, and total AC losses can be estimated by summing these losses. Here, we briefly introduce these analytic equations and discuss the expected AC loss density value according to the *dB/dt*.

$$Q_h = \frac{\mu_0 \pi w_f H_c H_m f}{t_f} \cdot \left\{ \left(\frac{2H_c}{H_m}\right) \ln\left(\cosh\left(\frac{H_m}{H_c}\right)\right) - \tanh\left(\frac{H_m}{H_c}\right) \right\} [W/m^3] \quad (9.1)$$

$$Q_h = \frac{8}{3}\mu_0 \lambda f \left(H_m H_c - \frac{1}{2} H_c^2 \right) [W/m^3] \quad (9.2)$$

Hysteresis loss is one of the dominant loss components generated by the superconductor hysteresis. Equation (9.1) represents hysteresis loss for strip structure [34], which can be used for hysteresis loss of REBCO wire. μ_0, H_m, and f are air permeability, applied magnetic field, and its frequency, respectively. w_f and t_f are thewidth and thickness of the strip, respectively. H_c is given as $\frac{t_f}{2}J_c$, where J_c is thecritical current density. For round-shaped wires (MgB$_2$ and NbTi), eq. (9.2) is used for hysteresis loss calculation [34]. Here, λ means the ratio of the superconductor in the wire. For the hysteresis loss calculation of BSCCO wire, the following eqs. (9.3)–(9.6) can be used [36].

$$m_0 = -\frac{3}{2}\frac{(\alpha+1)}{2\sqrt{1-\alpha^2}} \ln\left(\frac{1+\sqrt{1-\alpha^2}}{1-\sqrt{1-\alpha^2}}\right) \quad (9.3)$$

$$B_p = \frac{\mu_0 J_c w}{\pi} \cdot \frac{\alpha}{2\sqrt{1-\alpha^2}} \ln\left(\frac{1+\sqrt{1-\alpha^2}}{1-\sqrt{1-\alpha^2}}\right) \quad (9.4)$$

$$q = \begin{cases} 4\left(\frac{B_m}{B_p} - \frac{2}{1-m_0}\right)(B_m \geq B_p) \\ 4\left(\frac{2}{1-m_0}\left(\left(1-\frac{B_m}{B_p}\right)^{1-m_0} - 1\right) + \frac{B_m}{B_p}\left(\left(1-\frac{B_m}{B_p}\right)^{-m_0} + 1\right)\right)(B_m < B_p) \end{cases} \quad (9.5)$$

$$Q_h = \frac{2J_c wf}{3\pi} \cdot B_p \cdot q \ [W/m^3] \quad (9.6)$$

In the equations, α is an aspect ratio of the superconductor region in wire, which is given as (thickness, t / width, w). B_m is the magnetic flux density applied to the superconducting wire.

$$Q_t = \frac{\mu_0 I_c^2}{\pi}\left\{\left(1-\frac{I_t}{I_c}\right)\ln\left(1-\frac{I_t}{I_c}\right)+\left(1+\frac{I_t}{I_c}\right)\ln\left(1+\frac{I_t}{I_c}\right)-\left(\frac{I_t}{I_c}\right)^2\right\}[J/cycle/m] \ (9.7)$$

$$Q_t = \frac{\mu_0 I_c^2}{\pi}\left\{\left(1-\frac{I_t}{I_c}\right)\ln\left(1-\frac{I_t}{I_c}\right)+\left(2-\frac{I_t}{I_c}\right)\left(\frac{1}{2}\frac{I_t}{I_c}\right)\right\}[J/cycle/m] \quad (9.8)$$

Transport loss of superconducting wire with strip structure (REBCO) is given as eq. (9.7) [37]. Here, I_t and I_c mean transport current and critical current,

9.3 Considerations on Cooling System Technologies for Stationary Superconducting...

respectively. For other wires having ellipse- or round-shaped wires (BSCCO, MgB$_2$, and NbTi), eq. (9.8) can be applied [37].

$$Q_c = \frac{1}{2\rho_e}(L_p B_m f)^2 \; [\text{W}/\text{m}^3] \quad (9.9)$$

Coupling loss is another dominant loss component generated in normal metals between superconductor filaments. Normally REBCO wires have no coupling loss, because they have a mono-filament structure. The coupling loss of round-shaped wires (MgB$_2$ and NbTi) is expressed as eq. (9.9) [35]. In the equation, ρ_e is effective resistivity between superconductor filaments, and L_p means the twist pitch of the wire. Equations (9.10) and (9.11) represent the coupling loss of BSCCO wire [38]. Again, α is an aspect ratio of the superconductor region in BSCCO wire; t is the thickness of that region.

$$\tau = \frac{\mu_0}{\rho_e\left(1+\frac{1}{\alpha}\right)}\left(\left(\frac{L_p}{2\alpha\pi}\right)^2 + \frac{1}{4}\left(\frac{t}{2}\right)^2\right) \quad (9.10)$$

$$Q_c = \frac{1}{2\mu_0}\left(1+\frac{1}{\alpha}\right)\cdot\frac{\left((2\pi f)^2 B_m^2 \tau\right)}{1+(2\pi f \tau)^2} \; [\text{W}/\text{m}^3] \quad (9.11)$$

Lastly, equations for eddy-current loss are given as eqs. (9.12) and (9.13) [35]. Equation (9.12) is used for tape-shaped wires (BSCCO and REBCO wires). ρ_n and t_a are the resistivity of normal metal in wire and the width of superconductor wire. Eq. (9.13) is for round-shaped wires (MgB$_2$ and NbTi) and d_a means the wire diameter.

$$Q_e = \frac{\pi^2}{6\rho_n}(t_a B_m f)^2 \; [\text{W}/\text{m}^3] \quad (9.12)$$

$$Q_e = \frac{\pi^2}{12\rho_n}(d_a B_m f)^2 \; [\text{W}/\text{m}^3] \quad (9.13)$$

The total AC loss can be estimated by aggregating all individual AC losses. In this analysis, the total AC loss density is evaluated based on the analytical eqs. (9.1)–(9.13), utilizing dB/dt values. The estimation is conducted under the following assumptions:

- Magnetic field strengths ranging from 0 to 2 T and frequencies from 0 to 200 Hz are considered, resulting in dB/dt values within the range of 0 to 400 T/s.
- Various superconducting wires such as REBCO, BSCCO, MgB$_2$, and NbTi are investigated.

- The critical current (I_c) is treated as a function of both magnetic field (B) and temperature (T), denoted as $I_c(B, T)$. The operating current of the wire is expected to be half of the critical current ($I_{op}/I_c = 0.5$).
- For tape-shaped wires, the magnetic field is assumed to be applied perpendicularly to the wide side.
- Operating temperatures of 4 K, 20 K, 50 K, and 77 K are studied.

Table 9.4 provides detailed specifications for each wire used in the calculations, with wire properties compiled from various references [16, 39–43].

Figure 9.8 illustrates the obtained AC loss densities [W/cm^3] as a function of dB/dt [T/s]. The results are differentiated by color, with blue representing REBCO, orange for BSCCO, yellow for MgB$_2$, and purple for NbTi. Overall, it is evident that REBCO exhibits a higher AC loss density compared to other wires, with a density exceeding 1 kW/cm^3 at 100 T/s. Conversely, MgB$_2$ demonstrates the lowest AC loss density—below 10 W/cm^3 at 100 T/s. The high AC loss density in REBCO can be attributed to its mono-filament structure because filamentation effectively reduces hysteresis loss—a predominant component of AC losses. In contrast, MgB$_2$ employs a multi-filament structure, also coupled with higher effective resistivity between filaments, yielding reduced coupling loss compared to other multi-filament wires. Consequently, MgB$_2$ demonstrates the lowest AC loss density among the wires examined, suggesting its suitability for high dB/dt superconducting AC armature winding, particularly in applications such as aircraft.

Variations in operating temperature also influence the results, primarily due to differences in critical current and resistivity of the normal metal component. These temperature-dependent properties produce distinct trends for each wire. However, another critical differentiator across temperatures is the efficiency of the cryogenic cooling system, quantified by the COP. Thus, by dividing the AC loss densities with COP values, the required cooling power density can be estimated.

Figure 9.9 presents the AC loss density divided by COP. COP^{-1} values at 4 K, 20 K, 50 K, and 77 K are assumed to be 5000, 45, 16, and 10, respectively, based on the values depicted in Fig. 9.6. Notably, applying COP renders 4 K operation less

Table 9.4 Wire properties for AC loss calculation [16, 39–43]

Parameters		Unit	REBCO	BSCCO	MgB$_2$	NbTi
Wire dimension		[mm]	4 x 0.1	4 x 0.23	0.32d	0.825d
Filament dimension		[mm]	4 x 0.0001	0.2 x 0.02	0.001d	0.0006d
Number of filaments		[–]	1	69	114	6425
Superconductor ratio		[%]	1	30	11	34
Twist pitch		[mm]	–	5	5	15
Effective metal resistivity	@ 4 K	[nΩ•m]	0.31	0.15	96.4	0.63
	@ 20 K	[nΩ•m]	0.32	0.22	100	–
	@ 50 K	[nΩ•m]	0.85	2.40	–	–
	@ 77 K	[nΩ•m]	2.27	4.00	–	–

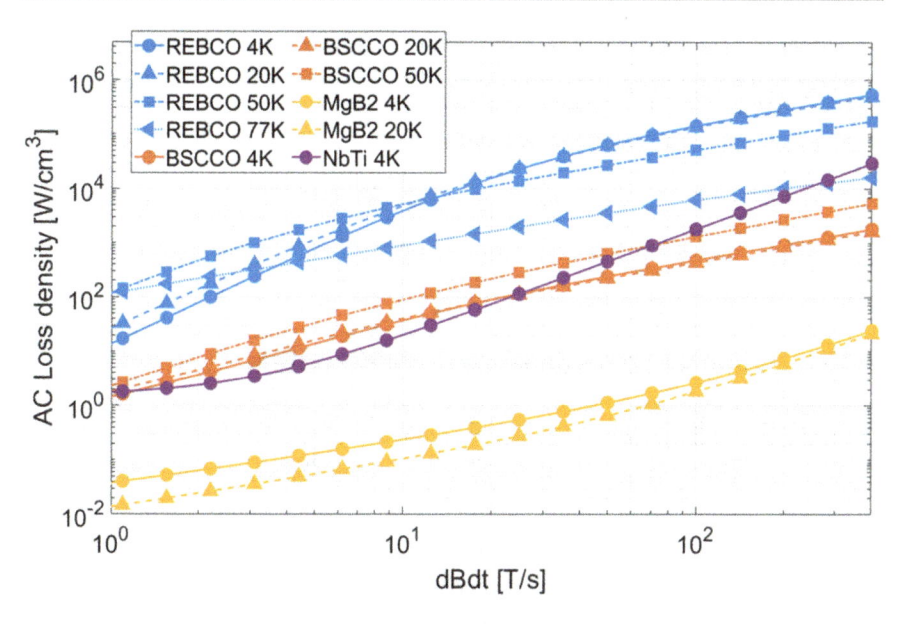

9.3 Considerations on Cooling System Technologies for Stationary Superconducting... 195

Fig. 9.8 Calculated AC loss density [W/cm^3] according to *dB/dt* [T/s] for REBCO (blue), BSCCO (orange), MgB$_2$ (yellow), and NbTi (purple) at different operating temperatures. Both the x-axis and y-axis are in log-scale

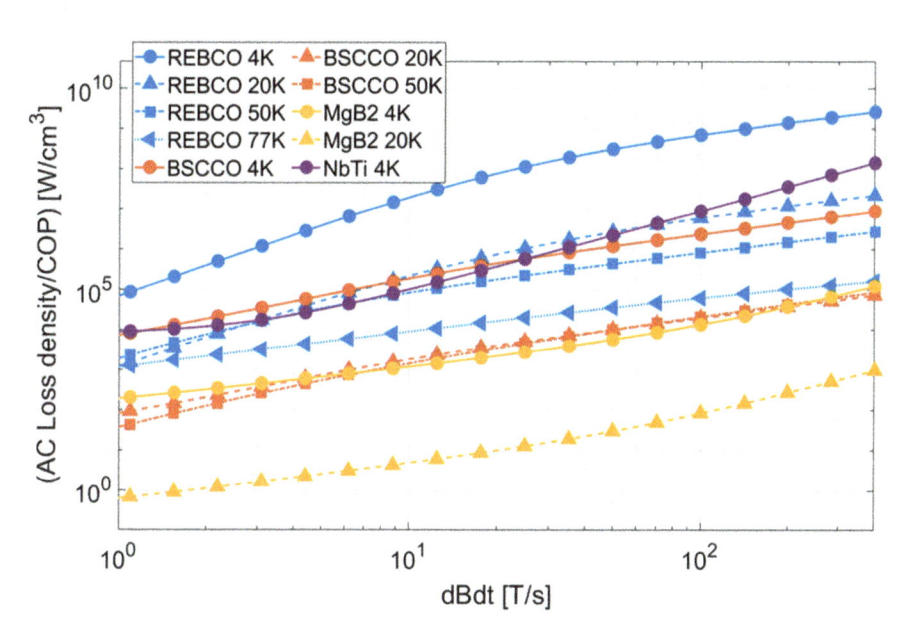

Fig. 9.9 AC loss density [W/cm^3] divided by coefficient of performance at each temperature. REBCO (blue), BSCCO (orange), MgB$_2$ (yellow), and NbTi (purple) at different operating temperatures are plotted together. Both the x-axis and y-axis are in log-scale

appealing due to the relatively large cooling requirement. MgB_2 wire at 20 K emerges as the optimal choice, offering approximately 100 W/cm^3 of cooling power at 100 T/s, while REBCO consistently exhibits higher loss density. Additionally, it is observed that higher temperatures correspond to lower loss density within the same wire when recognizing COP. However, it does not guarantee higher temperatures are better than lower temperatures, as achieving the same ampere-turn at higher temperatures requires larger winding packs due to lower critical current values.

9.3.2 Opportunity in Open-Loop Cooling System

A frequently explored open-loop cooling approach in contemporary research is the *fuel-as-coolant* concept employing liquid hydrogen and hydrogen fuel cells. Here, liquid hydrogen serves a dual role: as a coolant for the cryogenic system encompassing the superconducting windings, and as a fuel for an integrated fuel cell within the system. Liquid hydrogen, characterized by its exceptionally low boiling point (20 K) and high latent heat of vaporization, proves to be an effective coolant for superconductors, offering substantial cooling power to mitigate the significant AC losses from superconducting AC armature windings. Additionally, as liquid hydrogen is stored as fuel for the hydrogen fuel cell, it circumvents the low COP associated with cryocoolers, thereby unlocking new avenues for efficiency and sustainability. The advantages of liquid hydrogen cooling can be summarized as follows:

- High energy density: Liquid hydrogen offers one of the highest energy densities among all fuels, providing ample power for fuel cell operation and cooling requirements.
- Environmental sustainability: Hydrogen fuel cells produce only water vapor and heat as byproducts, offering a clean and sustainable energy solution with minimal environmental impact.
- Efficient cooling: Liquid hydrogen's low boiling point and high latent heat of vaporization enable efficient cooling of superconducting AC armature windings, overcoming the limitations of conventional cooling systems.

In conclusion, cooling system technologies for superconducting AC armature windings have both challenges and opportunities. Considering cryocooler performance, high AC losses caused by high *dB/dt* values present the need for elaborate designs or innovative solutions to optimize efficiency and performance. The liquid-hydrogen cooling concept offers a potential alternative, leveraging the unique properties of hydrogen fuel to enhance cooling efficiency and sustainability.

9.4 Summary

This chapter provides a comprehensive review of cooling-system technologies for both rotary superconducting DC field windings and superconducting AC armature windings. Through an analysis of past demonstrations and designs, we offer forward-looking insights into emerging trends, as well as outline the diverse challenges, opportunities, and considerations to the development of cooling systems for superconducting machines.

In our review of cooling technologies for rotary superconducting DC field windings, we have organized previously attempted approaches into distinct categories, each characterized by its own set of attributes, advantages, and limitations. Techniques such as thermosiphon cooling, forced-flow cooling, on-board cryocoolers, and immersion in cryogenic baths have been utilized by engineers to address the challenge of maintaining superconducting states at cryogenic temperatures under rotating conditions. Understanding the trade-offs and compromises inherent in each method would be a great help for engineers seeking to optimize the performance and reliability of superconducting DC field windings in rotating machinery applications.

Superconducting AC armature windings can reduce auxiliary cooling components for rotation, but they present another challenge—high AC losses. High dB/dt requirements and resulting AC losses that could be difficult to address with cryocoolers, considering their COP and weight of present cryocoolers, and they pose significant hurdles to machine performance. Liquid hydrogen cooling emerges as a promising solution. By leveraging liquid hydrogen's unique properties as both a coolant and a fuel for fuel cells, engineers can overcome the limitations of cryocoolers and enhance their efficiency.

Further Reading Cryocooler integration into an aircraft faces its own challenges and is a relatively new technical field, although the use of cryocoolers on large systems has historically been researched well in various configurations.

An overview on the history and development stages of using cryocoolers for rotating machines on a larger scale is given by Stautner in Chap. 6 **Cryocoolers for Superconducting Generators** in Atrey M (ed.) **Cryocoolers—Theory and Applications** International Cryogenics Monograph Series Springer Publishing 2020.

References

1. Haran, Kiruba S., et al (2017) High power density superconducting rotating machines—development status and technology roadmap. Supercond. Sci. Technol. 30(12):123002.
2. Chow, Calvin CT, Mark D. Ainslie, and K. T. Chau (2023) High temperature superconducting rotating electrical machines: An overview. Energy Rep. 9:1124-1156.

3. Y. Liu (2018) Design of a superconducting DC wind generator. Dissertation, Karlsruhe Institute of Technology.
4. Bong, Uijong, et al (2021) Investigation on key parameters of NI HTS field coils for high power density synchronous motors. IEEE Trans. Appl. Supercond. 31(5):1-5.
5. L. Tomkow, I. Harca, K. Machaj, A. Smara, T. Reis, and B. Glowacki (2004) Thermosyphon Cooling System for the Siemens 400kW HTS Synchronous Machine. AIP Conf. Proc. 710:859–866.
6. Nick, W., et al (2010) Development and construction of an HTS rotor for ship propulsion application. J. Phys. Conf. Ser. 234(3) .
7. Urbahn, J. A., et al (2004) The thermal performance of a 1.5 MVA HTS generator. AIP Conf. Proc. 710(1).
8. Gamble, Bruce, Greg Snitchler, and Tim MacDonald (2010) Full power test of a 36.5 MW HTS propulsion motor. IEEE Trans. Appl. Supercond. 21(3):1083–1088.
9. Yanamoto, Toshiyuki, et al (2017) Load test of 3-MW HTS motor for ship propulsion. IEEE Trans. Appl. Supercond. 27(8): 1–5.
10. Song, Xiaowei, et al (2019) Ground testing of the world's first MW-class direct-drive superconducting wind turbine generator. IEEE Trans. Energy Convers. 35(2):757–764.
11. Jansen, Ralph H., et al (2019) High efficiency megawatt motor preliminary design. 2019 AIAA/IEEE Electric Aircraft Technologies Symposium (EATS).
12. Xiao, Jianqiao, et al (2022) A spoke-supported superconducting rotor with rotating cryocooler. IEEE Trans. Magn. 58(8):1–5.
13. Stautner W, et al "Cryogenic aspects of a 20 MW class superconducting generator for the Renewables industry" invited talk 2024 *IOP Conf. Ser.: Mater. Sci. Eng.* 1301 012048. https://doi.org/10.1088/1757-899X/1301/1/012048
14. J. Lee, S. Park, Y. Kim, S. Lee, H. Joo, W. Kim, K. Choi, and S. Hahn (2013) Test results of a 5 kW fully superconducting homopolar motor. Progr. Supercond. Cryog. 15(1):35–39.
15. U. Bong (2022) Applicability of no-insulation high-temperature superconductor field winding to superconducting synchronous motor. Dissertation, Seoul National University.
16. Balachandran, Thanatheepan, Timothy Haugan, and Kiruba Haran (2022) 4 Superconducting Machines and Cables. In: Kiruba Haran (ed) Electrified Aircraft Propulsion: Powering the Future of Air Transportation. University Printing House, Cambridge.
17. Tixador, P., C. Berriaud, and Y. Brunet (1993) Superconducting permanent magnet motor design and first tests. IEEE Trans. Appl. Supercond. 3(1):381–384.
18. Sugimoto, Hidehiko, et al (2007) Development of an axial flux type PM synchronous motor with the liquid nitrogen cooled HTS armature windings. IEEE Trans. Appl. Supercond.17(2):1637-1640.
19. Lee, Ji-Kwang, et al (2011) Electrical properties analysis and test result of windings for a fully superconducting 10 HP homopolar motor. IEEE Trans. Appl. Supercond. 22(3):5201405–5201405.
20. Qu, Timing, et al (2014) Development and testing of a 2.5 kW synchronous generator with a high temperature superconducting stator and permanent magnet rotor. Supercond. Sci. Technol. 27(4):044026.
21. Huang, Zhen, et al (2013) Trial test of a bulk-type fully HTS synchronous motor. IEEE Trans. Appl. Supercond. 24(3):1–5.
22. Messina, Giuseppe, Edoardo Tamburo De Bella, and Luigi Morici (2019) HTS axial flux permanent magnets electrical machine prototype: Design and test results. IEEE Trans. Appl. Supercond. 29(5):1–5.
23. Nakamura, Taketsune, et al (2019) Load test and variable speed control of a 50-kW-class fully superconducting induction/synchronous motor for transportation equipment. IEEE Trans. Appl. Supercond. 29(5):1–5.
24. Sasa, Hiromasa, et al (2021) Experimental evaluation of 1 kW-class prototype REBCO fully superconducting synchronous motor cooled by subcooled liquid nitrogen for E-aircraft. IEEE Trans. Appl. Supercond. 31(5):1–6.

References

25. Nakamura, Taketsune, et al (2022) Experimental and theoretical study on power generation characteristics of 1 kW class fully high temperature superconducting induction/synchronous generator using a stator winding with a bending diameter of 20 mm. IEEE Trans. Appl. Supercond. 32(6):1–5.
26. Miyazaki, H., et al (2023) Fabrication and Test of a 400 kW-Class Fully Superconducting Synchronous Motor Using REBCO Tape for an Electric Propulsion System. IEEE Trans. Appl. Supercond. 34(5):5200506.
27. Song, Xiaowei, et al (2015) Design study of fully superconducting wind turbine generators." IEEE Trans. Appl. Supercond. 25(3):1–5.
28. Kalsi, Swarn S., Kent A. Hamilton, and Rodney A. Badcock (2018) Superconducting rotating machines for aerospace applications. 2018 Joint Propulsion Conference.
29. Balachandran, T., et al (2020) A fully superconducting air-core machine for aircraft propulsion. IOP Conf. Ser. Mater. Sci. Eng. 756(1).
30. Xue, Shaoshen, et al (2020) Stator optimization of wind power generators with high-temperature superconducting armature windings and permanent magnet rotor. IEEE Trans. Appl. Supercond. 31(2):1–10.
31. Dezhin, Dmitry, and Roman Ilyasov (2022) Development of fully superconducting 5 MW aviation generator with liquid hydrogen cooling. EUREKA: Phys. Eng. 1:62-73.
32. Terao, Yutaka, et al (2023) Electromagnetic analysis of fully superconducting motors employing dilute gas rotor and liquid hydrogen stator cooling structure. J. Phys. Conf. Ser. 2545(1).
33. Wang, Rui, et al (2023) Preliminary design optimization of a fully superconducting motor based on disk-up-down-assembly magnets." Supercond. Sci. Technol. 36(5):054003.
34. Mellerud, Runar, et al (2023) Design of a Power-Dense Aviation Motor With a Low-Loss Superconducting Slotted Armature. IEEE Trans. Appl. Supercond. 33(8):5204013.
35. Iwasa, Yukikazu (2009) Case studies in superconducting magnets: design and operational issues. Springer science & business media, New York.
36. Ten Haken, Bennie, Jan-Jaap Rabbers, and Herman HJ Ten Kate (2002) Magnetization and AC loss in a superconductor with an elliptical cross-section and arbitrary aspect ratio. *Physica C.* 377(1):156–164.
37. S. Kalsi, et al (2021) Motors Employing REBCO CORC and MgB2 Superconductors for AC Stator Windings. IEEE Trans. Appl. Supercond. 31(9):5206807.
38. Banno, N., and N. Amemiya (1999) Analytical formulae of coupling loss and hysteresis loss in HTS tape. Cryogenics. 39(2):99-106.
39. SuperPower Inc. 2G HTS Wire Specification. https://www.superpower-inc.com/specification.aspx Accessed 11 Apr 2024
40. Sumitomo Electric. Bi-2223 Superconducting Wire. https://sumitomoelectric.com/super/wire Accessed 11 Apr 2024
41. Zhou, Chao, et al (2012) Inter-filament resistance, effective transverse resistivity and coupling loss in superconducting multi-filamentary NbTi and Nb3Sn strands. Supercond. Sci. Technol. 25(6):065018.
42. Sumption, M. D., et al (2019) Ac loss of superconducting materials-refined loss estimates of mgb2 wires for superconducting motors and generators. 2019 AIAA/IEEE Electric Aircraft Technologies Symposium (EATS).
43. Oomen, M. P., R. Nanke, and M. Leghissa (2003) Modelling and measurement of ac loss in BSCCO/Ag-tape windings. Supercond. Sci. Technol.16(3):339.

10 Rotating Vacuum Heat Transfer, Rotating Cryocoolers, Slip Rings, Rotating Bearings, Ball Bearings

10.1 Excursion: Rotating Heat Transfer for Motors

The following description of technology is directly related to cooling a rotating superconducting field coil. While there are, as we have seen in Chap. 9, several technologies available for different boundary conditions, efficiencies, and operating conditions, like the self-driven, pumped gas approach of Morris [1], Hofmann [2], and Schnapper [3], the AMSC approach [4], the Siemens stationary thermosiphon within a rotating tube [4], or of Callaghan Innovation [5], in this section we discuss further, ambitious, and novel approaches.

Case Study
Provide heat transfer from a rotating to a stationary cylindrical superconducting component without the use of a liquid transfer coupling on the main axis of rotation (shaft) as for examples shown by Xu [6].

Introduction
Hydrogen-electric aircraft technologies require electric propulsors to achieve the goal for zero emission. Those electric propulsors are preferably superconducting with high current density, resulting in an increased power density.

Task

- Develop a feasible cryogenic cooling concept for indirect cold mass cooling in the above 20 K or higher temperature range, depending on conductor choice.
- For several reasons, one would not bath-cool the propulsor (direct cooling) but prefer an indirect cooling approach where field and armature windings are not in direct contact with a cryogen, e.g., hydrogen.
- The stator of this motor is exposed to the rotating magnetic field of the field coils that rotate at, e.g., 4500 rpm for the CHEETA design initiating eddy currents in

the armature structure. Those AC losses need to be transferred to a cooling medium. In the proposed configuration a helical cooling coil is mounted on the inner surface of the stator. The cooling coil is configured such that liquid hydrogen can pass through the stator. We call that an armature winding cooled by highly efficient liquid hydrogen forced-flow boiling. The heat load generated from the armature due to those AC losses is quite substantial and may be around 2.3 kW [7].

Figure 10.1 shows a high-level drive train integration including fuel cells with battery back-up, driving a fan. In which way heat can be transferred in a cryogenic vacuum environment is depicted in Fig. 10.2 showing some generic design principles:

10.1.1 Rotating Cryocoolers

Depending on motor operating conditions and cryocooler specification a cryocooler can be mounted directly onto the field winding of a high temperature superconductor rotor with racetrack coils [8, 9], as shown in Fig. 10.3a. Figure 10.3b shows the temperature profile of the experimental operation obtained from the rotating cryocooler test facility (at the University of Illinois, Hinetics) with respect to the cryocooler rotational speed for a typical HTS operating temperature range of 40–45 K [10].

General issues with this novel design approach are related to cryocooler performance (Stirling type or other), unknown mean time between failure for rotational operation and consequently, in-field cryocooler replacement, as well as dependency on a fluctuating temperature window, that changes during various aircraft operating conditions, to name but a few aspects. This is subject to further research. Figure 10.4 shows the schematics and the assembly in the drive test facility [10] for the results in Fig. 10.3.

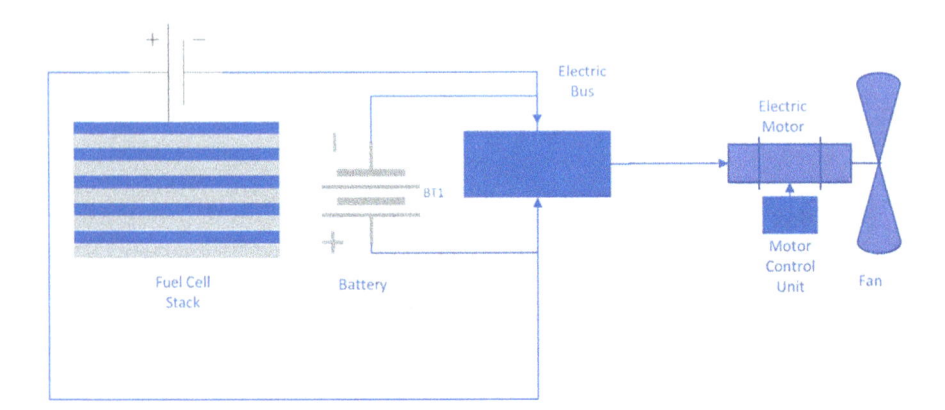

Fig. 10.1 Bare-bone drive train integration

10.1 Excursion: Rotating Heat Transfer for Motors

```
┌─────────────┐      ┌─────────────┐      ┌─────────────┐
│  Rotating   │ ⇐    │  Rotating   │ ⇒    │   Thermal   │
│ cryocoolers │      │ Cryobearings│      │  slip rings │
└─────────────┘      └─────────────┘      └─────────────┘
```

Rotating cryocoolers fitted to the field winding – cooler MTBF, maintenance

Current research activities at:
- University of Illinois at Urbana-Champaign, Urbana, Illinois
- University of Auckland

Cryobearings, and other – No heat transfer possible

⇓

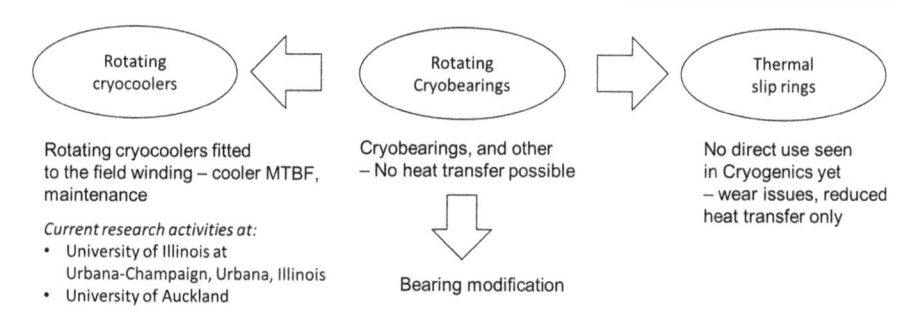

Bearing modification

No direct use seen in Cryogenics yet – wear issues, reduced heat transfer only

Fig. 10.2 Rotating heat transfer design options

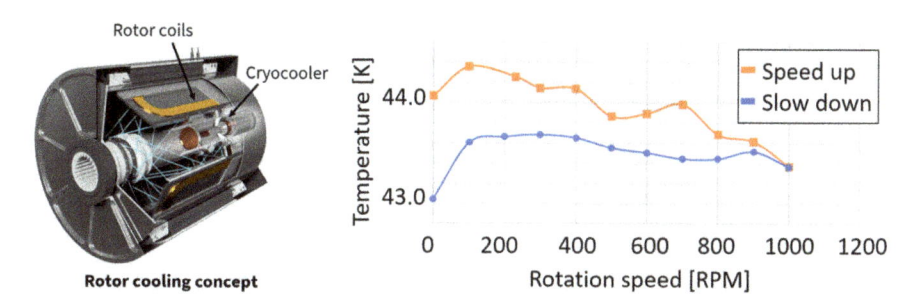

Fig. 10.3 (a) Rotor cooling concept with rotating cryocoolers and Kevlar suspension elements. (b) Temperature fluctuation vs rpm as obtained from a rotating cryocooler test facility at the University of Illinois, Hinetics

The use of a cooler driven by a linear compressor offers a further design improvement. The absence of a piston in a pulse tube cooler for example is beneficial for long-term operation. Although pulse tube coolers substantially show a cooling power degradation when implemented horizontally, PTRs for this particular use run at high operating frequency with only minor power degradation. The advantage of the PTR is the higher cooling power combined with a small footprint and increased torque. There is also a less documented, additional benefit in that PTRs can deliver higher cooling powers during cooldown when optimized for initial higher thermal burden.

10.1.2 Thermal Slip Ring/Brush Design

Figure 10.5 shows a typical, known design with slip rings and brushes for transferring electrical power. The same principle applies for thermally transferring heat as a thermomechanical device from a rotating shaft to an outer surface. Common slip ring brush materials are copper, silver, gold, aluminum, stainless steel, ceramics, or carbon (e.g., with company Schunk), whereas metals need to be plated with a conductive material like tin or other. To provide insulation and reduce wear on the

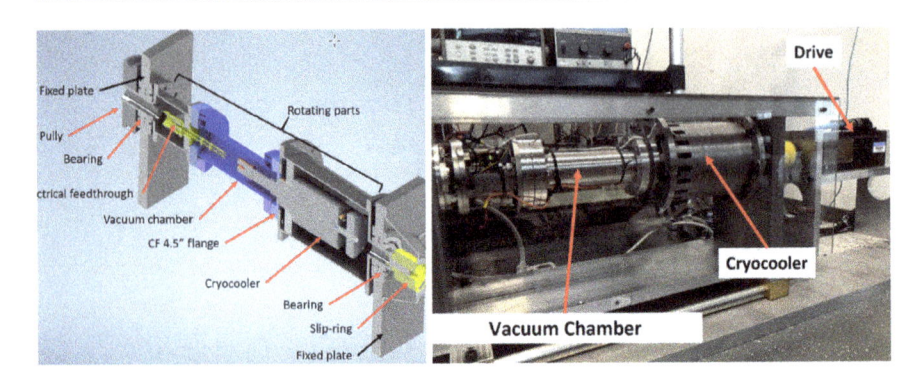

Fig. 10.4 (**a**) Design and (**b**) implementation of a spinning cryocooler

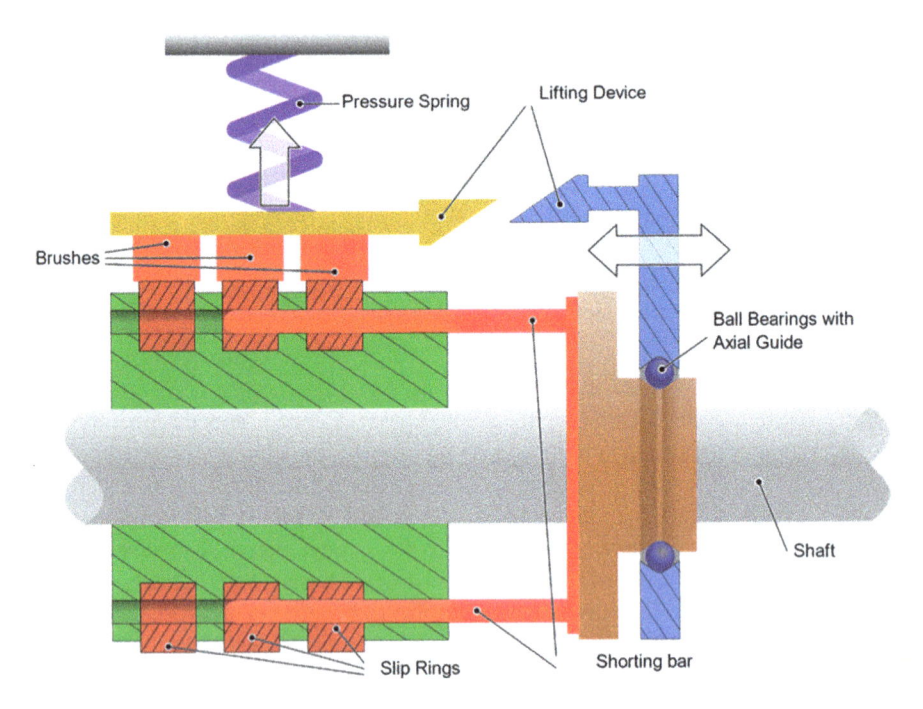

Fig. 10.5 Simplified configuration of thermal slip rings/commutators with brushes

brush, ceramic material can be used. Brushes are usually fixed on rotating copper slip rings with a brush block assembly, usually on the stationary component.

Generally, various types of slip ring/brush designs are possible as shown in Fig. 10.5 [11, 12].

Main issues with this design are usually related to maintenance of the wearing parts over time at various temperatures and operating conditions, as well as fatigue in the brushes related to performance degradation (increased thermal contact resistance) and generally, brushes coping with vibration effects in an aircraft.

10.1.3 Bearings and Their Derivatives

Cryogenic bearings on the other hand already found their place in certain applications where a rotational component is required. Those are used, for example, in cryogenic pumps, or for aerospace/space missions, or even for launched infrared spectrometers. Typical cryogenic bearings are made by RBC Transport Dynamics, Carter Manufacturing and SKF, for example, the Cryogenic 6206 ABEC 3 with Si3N4 ceramic balls with the cage made of PEEK.

Design considerations when using bearings for aircraft are:

The bearing should be of lightweight and high strength. Since bearing components are subject to intense vibration over lifetime, fatigue life needs to be considered. For today's aircraft self-lubricated aero bearings work in the range from 220 to 610 K (third stage of a turbine engine) but for a cryogenic aircraft motor those bearings are subject to high thermal gradients. Material phase transformations toward embrittlement can occur and are of particular concern for some bearing components when going through those large temperature cycles. For coping with abrasive wear, some high-tech cryogenic nano lubricants greatly mitigate this risk and deliver constant wear rates over many thousands of cycles. More research is required here.

Let us assume the following cryogenic environment as shown in Fig. 10.6: The outer field coils need to be maintained, e.g., at 23 K and a bearing structure needs to be designed such that it acts as an efficient thermal link between those fields and stationary armature coils where we need to remove heat of, e.g., 1 W, caused by eddy currents in the field coils. For motor stability reasons we locate a suitable

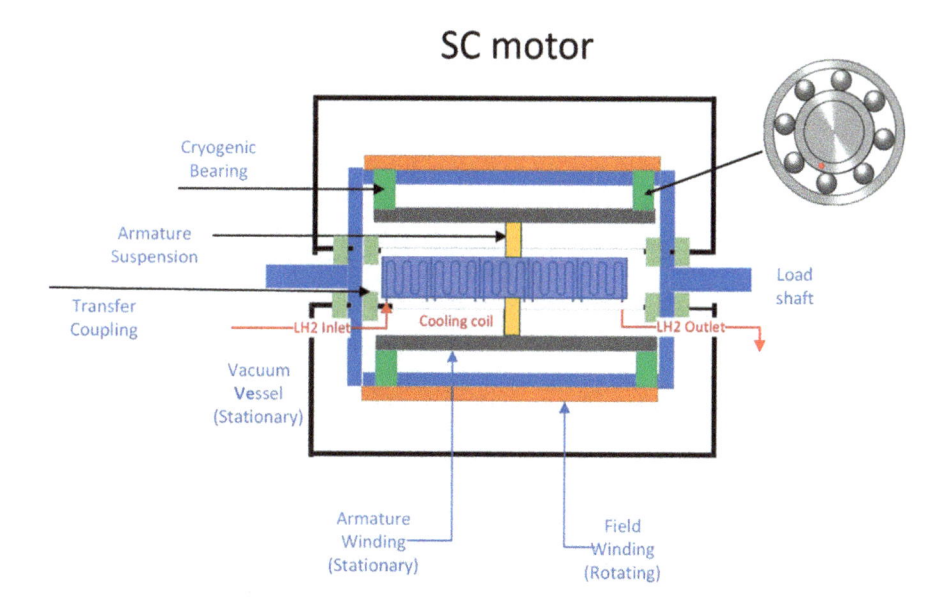

Fig. 10.6 Bearing position for a superconducting motor

bearing at both ends along the rotating shaft. When we consider the type of materials that are involved in the heat transfer between stator and rotor, we note that most likely stainless steel or ceramics cannot be used [13].

In the following, we discuss the simplified heat transfer between stator and rotor for some bearing ball material properties as shown in Fig. 10.7.

For a 25.4 mm radius sphere (Rs), made of SS316L with a compressive pressure of 1 bar, the deformation a_1 and thermal resistance (KW) R_{c1} of:

Set of Eq. (10.1):

$$a_1 = 1.11 \cdot \left(F_a \cdot \frac{R_s}{E_{SS316}} \right)^{1/3} \quad a_1 = 3.317 \times 10^{-4}\,m$$

$$R_{c1} = \frac{1}{(2k_{SS316} \cdot a_1)} - \frac{\ln(2)}{\pi \cdot k_{SS316} \cdot R_s} \quad R_{c1} = 749.45\,K/W$$

(10.1)

With stainless steel balls with a thermal conductivity of K_{SS316} @ 25 K = 2 W/mK [14] we find a thermal resistance of R_{c1} = 750 K/W. We therefore need to conclude that stainless steel balls cannot be used.

This first result now leads us to the design of an advanced cryogenic bearing, possibly also including some multifunctionality, depending on individual operating goals.

Step 1: Material Change

For a 25.4 mm radius sphere, made of Beryllium Copper C17200 with a compressive pressure of 1 bar, the deformation a_2 and thermal resistance R_{c2} of 17.3 K/W:

Set of Eq. (10.2):

Fig. 10.7 Conceptual view of bearing cooling

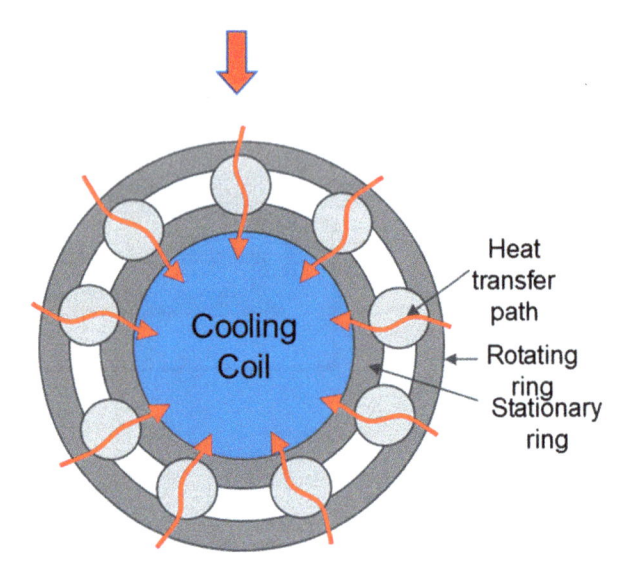

$$a_2 = 1.11 \cdot \left(F_a \cdot \frac{R_s}{E_{BeCu}} \right)^{1/3} \qquad a_2 = 3.833 \times 10^{-4} \, m$$

$$R_{c2} = \frac{1}{(2k_{BeCu} \cdot a_2)} - \frac{\ln(2)}{\pi \cdot k_{BeCu} \cdot R_s} \qquad R_{c2} = 17.276 \cdot K/W$$

(10.2)

A simple material change gives us a reduction in contact resistance by a factor of 37 when using Berylco copper instead of stainless steel. The thermal conductivity of Berylco 25 @ 25 K is approx. 75 W/mK [14]. Other high-strength CuNi materials may offer similar benefits.

When using regular high-purity copper, wear may become an issue to consider. Copper balls then need to be reinforced with either high-strength coating or other reinforcement means.

Wyatt [15] reports 5 K/W @ 60 K for copper, and 700 K/W @ 60 K for steel, a factor of 140, when comparing steel with pure copper. Obviously, as we know, contact resistances increase with decreasing temperature.

Wyatt has also shown that faster coldmass cooldown is possible as depicted in Fig. 10.8 if pure copper bearings are chosen [15] and describes design and test performance of a special bearing that was used to provide rotation of a circular variable interference filter.

Although appropriate for many applications Wyatt warns that unlike large steel bearings that are immune to damage at moderate g-levels, copper balls and races deteriorate very rapidly if chatter or hammering of the races against the balls occurs. However, Wyatt proves that in some instances this can be avoided by floating the lower copper race on springs. When the system is then subjected to vibration, the lower bearing race moves with the hub. This requires that the lower race be heat sunk to the base through flexible copper straps. According to the test results the bearing was subjected to a 10-min vibration test along the longitudinal axis at a 50 g level with no measurable ill effects detected [15].

Copper straps should be avoided due to the high thermal resistance below 30 K. This leads to the next step of a further bearing advancement in which we will redesign the bearing housing.

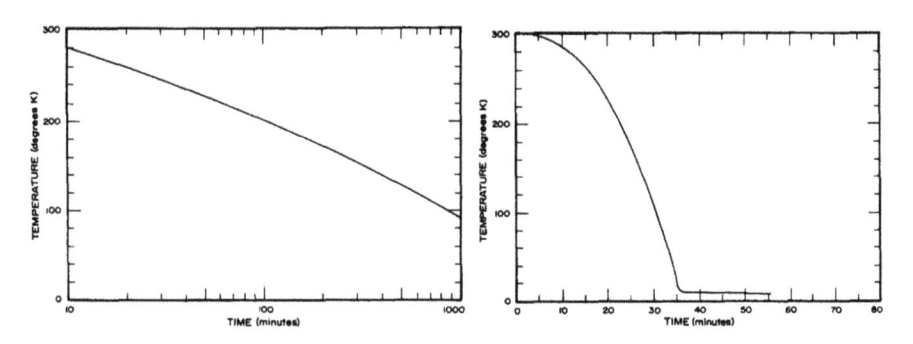

Fig. 10.8 (a) Low temperature bearing with SS balls, (b) with copper balls

Step 2: Bearing Housing Redesign

Several modifications are possible as shown in Fig. 10.9. For this we will keep the housing of strong stainless steel but change the balls to a high-conductivity material. We expand the bearing housing to add copper meshes around the circumference on the inner and outer race and confine them with the bearing balls in touch with a highly conducting support structure. For the inner and outer races, we add fill ports for high-pressure helium gas and crimp off the fill ends.

In summary, we create a copper beryllium (CuBe) ball bearing in a high-strength CuBe or thin stainless steel housing embedded in a rotating heat exchanger in the races with high helium fill pressure for forced flow convective heat transfer to the copper interface structure for field and armature coils.

In the next step, we advance this design further and add even higher heat transfer capability coupled with strong structural spokes as shown in Fig. 10.10 [16]. The added benefit here is that any heat transfer capability can be adapted for bigger gaps between field and stator coils.

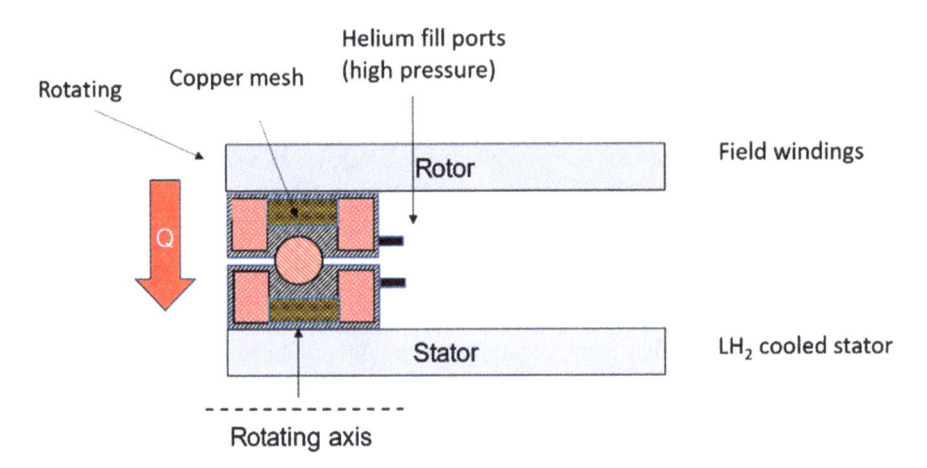

Fig. 10.9 Thermally conducting bearing between rotor/stator

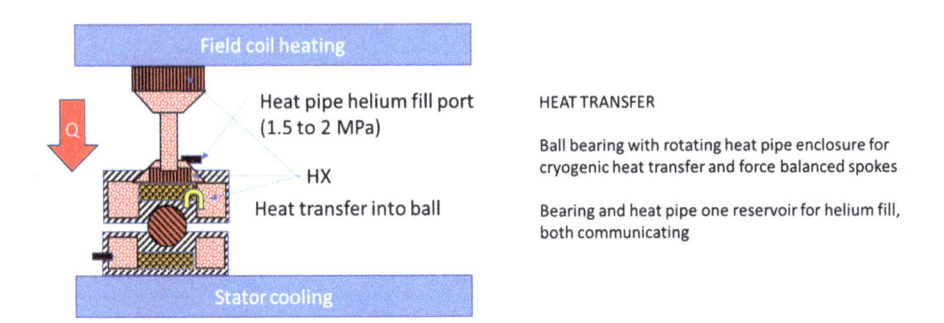

Fig. 10.10 Modified bearing with heat pipe between field coils and stator

Fig. 10.11 Rotating heatpipe

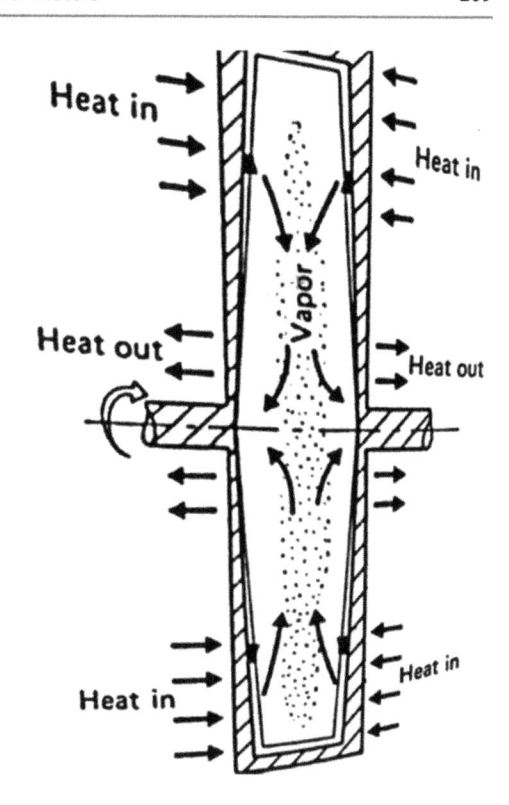

Step 3: Spoke-Type Ball Bearing Integration (with Heat Pipe)

Attached to those bearing chambers are gaseous helium-filled heat pipe spokes transferring heat from balls to chambers and to the top of the heat pipe heat exchanger. Rotating heat pipes are efficiently being used in different designs for motor rotors for room temperature applications [17] as shown in Fig. 10.11. Although cryogenic-type rotating heat pipes are rather unusual there is good reason to focus on this design for high heat transfer rates in the 30 W range, with up to 5000 rpms. A simulation model based on CFD calculations for heat transfer in rotating tubes needs to cover possible rotational angles under internal pressure. With that we take heat out from the outer diameter of the field coils transferring it to the stator shaft.

Pitfalls

Besides mechanical aspects, thermal contraction challenges may occur with improper material choice. Table 10.1 gives an overview on ball bearing and housing materials. Both should be matched since otherwise shrinkage mitigation can lead to a need for further design efforts. Below 40 K Berylco and stainless steel show nearly the same CTE and are therefore suitable materials [18, 19].

Table 10.1 Ball-bearing materials and their shrinkage

Material (typical values)	ΔL/l at 4 K [%]	ΔL/l at 40 K [%]	ΔL/l at 77 K [%]	ΔL/l at 100 K [%]	ΔL/l at 150 K [%]	ΔL/l at 200 K [%]	ΔL/l at 250 K [%]
Cu-2%Be-0.3%Co (Beryllium copper, Berylco 25)	0.316	0.315	0.298	0.277	0.219	0.151	0.074
Fe-9% Ni	0.195	0.193	0.188	0.180	0.146	0.100	0.049
Hastelloy C	0.218	0.216	0.204	0.193	0.150	0.105	0.047
Inconel 718	0.238	0.236	0.224	0.211	0.167	0.114	0.055
Invar (Fe-36%Ni)	–	0.040	0.038	0.036	0.025	0.016	0.009
50%Pb-50%Sn solder	0.514	0.510	0.480	0.447	0.343	0.229	0.108
AISI 304 steel	0.296	0.296	0.281	0.261	0.206	0.139	0.066
AISI 310 steel	–	–	–	0.237	0.187	0.127	0.061
AISI 316 steel	0.297	0.296	0.279	0.259	0.201	0.136	0.065
Ti6%Al4%V	0.173	0.171	0.163	0.154	0.118	0.078	0.036

10.2 Conclusion

- Rotating heat transfer between 2 parallel surfaces in vacuum is a very challenging task.
- Our preliminary analysis indicates the possibility that heat transfer with a modified ball-bearing design may be feasible.
- The ball bearing heat pipe spoke design allows the designer to use bearings as a structural component.
- The absence of a cryogenic exchange gas or liquid in the vacuum space simplifies the superconducting motor cooling.
- More research work is needed, combined with CFD modeling to further detail the respective design options.

10.3 Other Alternatives

- The spoke-like suspension elements shown in Fig. 10.3 can be replaced using stainless steel capillary tubes. Experiments need to prove that pulsating heat pipes could act as a preferred dual component, as a support means, and a heat transfer component.
- As shown in Fig. 10.12 a magnetic fluid seal is used to seal the helium environment. If this is acceptable for some drivetrains the vacuum environment can be replaced allowing a very low-pressure helium atmosphere to exchange heat between rotating surfaces. See also residual gas heat conduction in vacuum. Figure 10.12 depicts a superconducting magnet-bearing housing in indirect contact with a wall at liquid helium temperature.

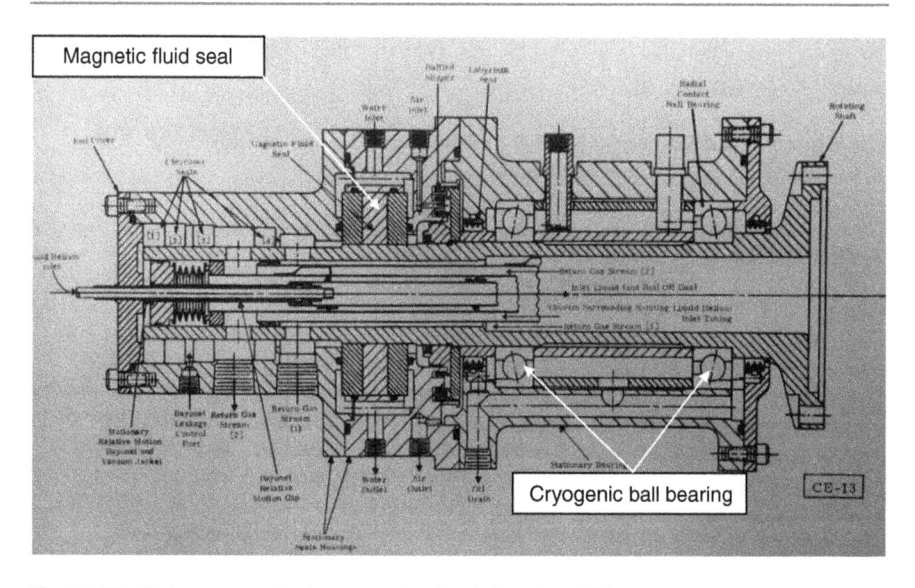

Fig. 10.12 Early superconducting motor bearing in housing (GE)

References

1. Morris W D **Heat transfer and fluid flow in rotating coolant channels** Research Studies Press Wiley 1981
2. Hofmann A **Self-regulating transfer modes of liquid helium to the rotor of a superconducting generator** Primärbericht 080103P07A Kernforschungszentrum Karlsruhe 1980
3. Schnapper C **Wärmetransport und Wärmeübergang in rotierenden Kanälen** KfK 2654 1978
4. Stautner W **Cryocoolers for Superconducting Generators** Chapter 6 in Atrey M (ed.) **Cryocoolers – Theory and Applications** International Cryogenics Monograph Series Springer Publishing 2020
5. Caughley A J *et al* **Cooling method for the rotor of a superconducting motor** 2024 *IOP Conf. Ser.: Mater. Sci. Eng.* 1301012008
6. Xu M Cryogenic refrigerator and rotary joint patent US 10,393,410 B2 2019
7. Feldman J Balachandran T Xiao J Stautner W Miljkovic N Haran K **Design of a Fully Superconducting Aircraft Propulsion Motor** EATS conference and other 2022
8. Lumsden G Ludbrook B Rehn N Fernandez FS Davies M, Chamritski V Signamneni S Badcock R **Additive manufacturing materials for structural optimisation and cooling enhancement of superconducting motors in cryo-electric aircraft** Supercond. Sci. Technol. 36 2023 105014 https://doi.org/10.1088/1361-6668/acf1d4
9. Kim Y Ki T Kim H Jeong S Kim J Jung J **High Temperature Superconducting Motor Cooled by On-Board Cryocooler** IEEE TRANSACTIONS ON APPLIED SUPERCONDUCTIVITY vol 21 no 3 2011 2217
10. Xiao J Balachandran T Samarakoon AJ Haran KA Spoke-Supported **Superconducting Rotor With Rotating Cryocooler** IEEE TRANSACTIONS ON MAGNETICS, VOL. 58, NO. 8, AUGUST 2022 9000405
11. Eitel L **Electrical power and signals through a slip ring: Where do they go?** 2018 https://www.motioncontroltips.com/electrical-power-and-signals-through-a-slip-ring-where-do-they-go/

12. Wikipedia   https://en.wikipedia.org/wiki/Slip_ring#/media/File:Electric_Motor_with_Slip_Rings.svg
13. Mikesell R P Scott R B **Heat Conduction Through Insulating Supports In Very Low Temperature Equipment** Journal of Research of the National Bureau of Standards vol 57 No 6 1956 Research Paper 2726
14. Simon N J Drexler E S Reed R P **Properties of Copper and Copper Alloys at Cryogenic Temperatures** NIST Monograph 177 1992
15. Wyatt C L Haycock R H **High thermal conductivity bearing for rotating devices at liquid helium Temperatures** Review of Scientific Instruments 45 pp 434-437 1974
16. Stautner et al. High heat transfer cryogenic bearing US Pat. Appl. 2024/0322638 A1
17. Dunn P D Reay D A **Heat pipes** 4th Edition Pergamon 1994
18. Ekin J W **Experimental techniques for low-temperature measurements** Oxford University Press 2006
19. Ekin J W Appendix data tables from **Experimental techniques for low-temperature measurements** Oxford University Press 2011

Airport Infrastructure Requirements for Liquid Hydrogen Supply and Distribution

11.1 Introduction

When considering the application of cryogenic LH$_2$ systems for aircraft, considerable effort and investment will be needed to develop infrastructure compatible with this energy carrier across an entire supply chain. These infrastructure developments include those associated with energy production, transportation, liquefaction, on-site storage, and dispensing. Given the size and complexity of these changes, the transition to LH$_2$ will require long-term commitment and substantial investment. Nevertheless, it represents a promising solution for a future zero-carbon aviation ecosystem.

11.2 Current Aviation Fuel and Energy Infrastructure

Before establishing the changes required in the energy infrastructure, a basic overview of the fossil-derived jet fuel development pathway will be provided. Much like most fossil fuels, the production of jet fuel begins with the extraction of crude oil from underground reservoirs. After extraction, the crude oil is transported to a refinery. Given the large volume of oil transported, the use of pipelines, railroad, or ships is typically preferred over shipping by heavy-duty truck.

After arrival at the refinery, the crude oil is subject to one of several refining processes. Fractional distillation is commonly applied first to separate the oil into hydrocarbon components of similar carbon-number range. After this point, catalytic cracking and hydrotreating can be utilized to further break down complex hydrocarbon molecules and remove undesirable impurities (e.g., sulfur, nitrogen, and oxygen). Jet fuels are kerosene-type mixtures of various hydrocarbons extracted with carbon number distributions typically between 8 and 16. This extracted range of

---

Lead Author: Phillip J. Ansell

© The Author(s), under exclusive license to Springer Nature Switzerland AG 2024
E. W. Stautner et al., *Aircraft Cryogenics*, International Cryogenics Monograph Series, https://doi.org/10.1007/978-3-031-71408-5_11

hydrocarbons is then subject to further blending to bring the fuel properties in line with defined standards, such as ASTM D1655. These standards define a variety of tests required to confirm physical properties, performance characteristics, and safety of a given fuel blend.

Following refinement, multiproduct pipelines are commonly used to transport jet fuel to airports. Some large, international airports feature pipelines that connect directly to an off-airport terminal, while some require the use of rail, ship, or truck delivery to either cover additional short-distance transportation from a fuel terminal or serve remote locations. Tanker trucks are then commonly used to transport jet fuel from off-site storage to dedicated on-airport storage areas. From the on-airport storage, aircraft are typically provided with fuel either through a refueler truck or a hydrant system with a dispenser truck. For small airports, mobile fuel trailers are also used. The jet fuel is pumped into the aircraft storage tanks for final use. An overview of the terminal servicing equipment currently used with conventional jet fuel is shown in Fig. 11.1 [1].

In addition to the physical properties of jet fuel defined by standards documents, additional best practices and standards exist for fuel handling and servicing. The standards for refueling are typically defined in ANSI and NFPA publications, which are defined to maintain safety regulations and quality control measures.

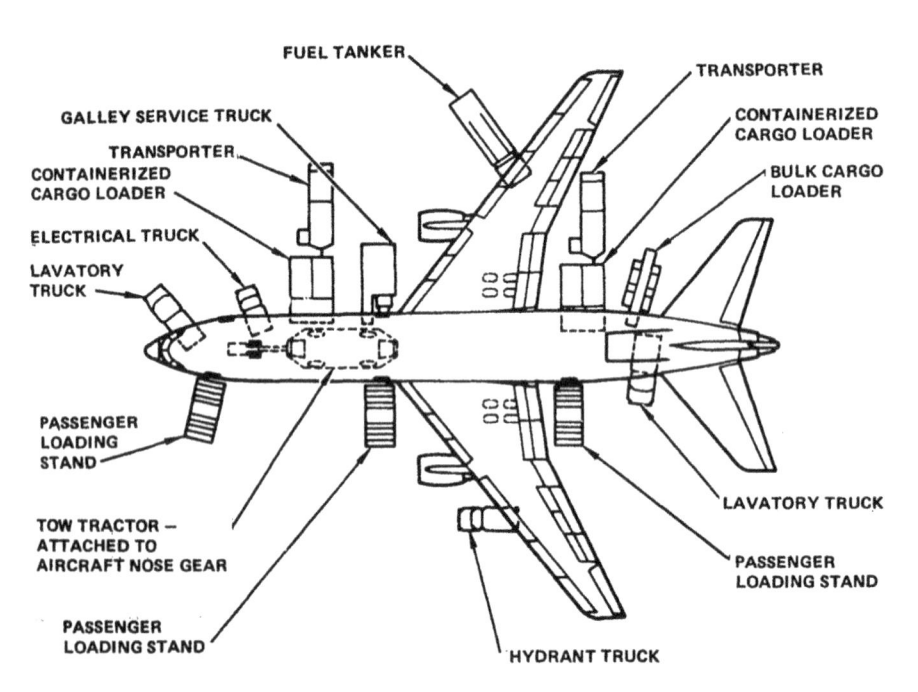

Fig. 11.1 Terminal servicing equipment used for commercial aircraft with conventional jet fuel [1]

11.3 Hydrogen Production and Distribution Networks

11.3.1 Hydrogen Production Methods

Being a fundamental building block of life and the most plentiful chemical substance in the universe, there are many ways that pure hydrogen can be produced for energy applications. Several examples of hydrogen production methods, often termed the "colors" of hydrogen, are shown in Table 11.1, alongside the net greenhouse gas impacts produced by direct and indirect emissions. In this table, the direct emissions result during the processing of the feedstock, whereas the indirect emissions are produced during the supply of the feedstock. As such, indirect emissions can result from transmission, electricity generation, or other recurring emissions required to supply the feedstock.

Globally, demand for hydrogen peaked at 95 million tons per year across 2022 [3]. However, across this same year, the vast majority of hydrogen was produced using fossil fuel resources (gray and black hydrogen), with less than 1% of production classified as "low emission" hydrogen production (blue and green hydrogen). Even most of the small production capacity of electrolytic hydrogen utilized grid electricity (yellow hydrogen), which features emission impacts that are highly variable and dependent on the grid mix that source the electricity.

As is evident from Table 11.1, the overall sustainability of hydrogen applications is incumbent upon development of renewable energy pathways for green hydrogen production. However, the long-term viability of hydrogen production is also dependent upon availability of other resources, like land and water. The reaction stoichiometry of water electrolysis indicates that 9 kg of H_2O is required in order to produce 1 kg of H_2. While the size of the water input requirements for electrolytic

Table 11.1 Colors of hydrogen production across various feedstocks, production technologies, and greenhouse gas emission impacts

Color	Feedstock	Production technology	Net GHG (kg CO_2e/kg H_2)
Gray	Natural gas	Steam methane reforming	9.5–15
Brown	Lignite	Gasification	19–27
Black	Coal	Gasification	19–27
Blue	Natural gas	Steam methane reforming with CCS	1–11
Turquoise	Natural gas	Pyrolysis	0.5–5
Green	Renewable electricity, water	Electrolysis	>0
Yellow	Grid electricity, water	Electrolysis	<1–30
Pink	Nuclear electricity, water	Electrolysis	>0
Red	Nuclear heat, water	Thermolysis	>0
Purple	Nuclear electricity and heat, water	Thermolysis and electrolysis	>0
Orange	Solar irradiance, water	Photolysis	>0

Based on Alvik et al. [2]

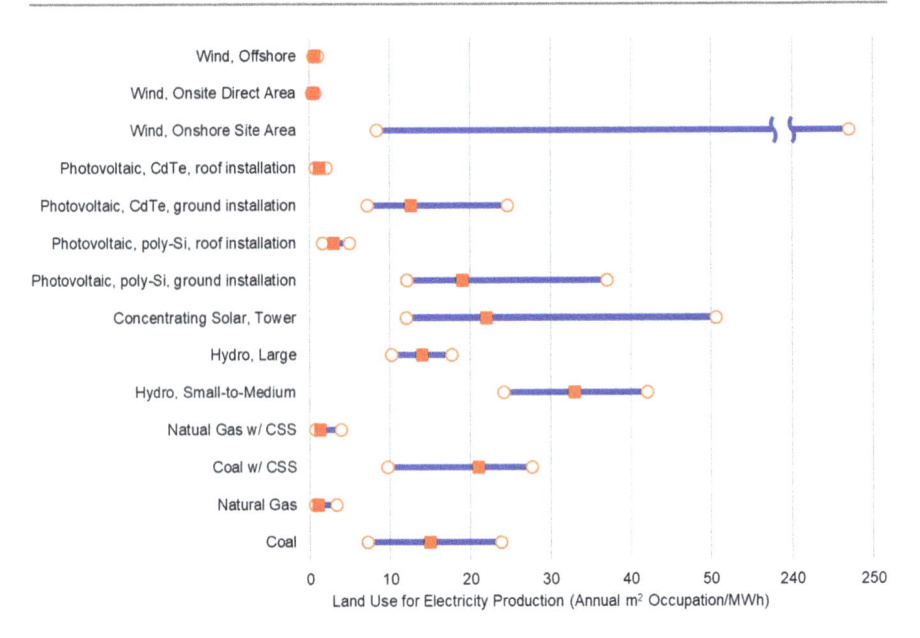

Fig. 11.2 Land occupation area required per year for 1 MWh of electricity production

hydrogen may seem concerning, meeting the long-term consumption demand with green hydrogen production is only approximately 2/3 that of the water consumed in 2014 for fossil fuel energy production and power generation [4].

The land area required for hydrogen production is also highly variable on the production pathway and mix of energy generation methods. In general, renewable energy production requires a greater footprint of land area required across the electricity production life cycle, as compared to fossil fuel power generation. Example land area occupation required, per MWh of electricity generated, is provided in Fig. 11.2, based on data from [5, 6]. Even while photovoltaic solar and wind energy sources require large land occupation areas, the ability to share land area with other residential, commercial, or agricultural uses can help to mitigate some of these challenges. Example approaches to reduce land area required include the use of offshore installations for wind energy and roof-mounted installations for photovoltaic solar systems. Of all power systems provided, nuclear energy provides the most compact source for electricity generation, which could be used for low-emission hydrogen production at large scales.

In addition to environmental sustainability and feasible resource requirements, the production of hydrogen for future aviation applications must also be economically viable. For this reason, the availability of low-cost renewable energy becomes an enabler for market adoption. Over the last decade, the levelized cost of electricity sourced by wind and solar renewable generation has decreased to levels below conventional fossil fuel sources. Further reductions in cost and increases in availability will be necessary for practical use in a variety of industries, though continued decarbonization of electrical grids and development of large-scale electrolyzer systems provide promise for maturation in hydrogen as a future energy carrier for aviation.

Active use of hydrogen in aviation also requires growth in distribution networks and careful attention to materials in contact with hydrogen volumes. Known material compatibility challenges exist in the form of hydrogen embrittlement, which can promote crack formation and fractures, as well as hydrogen permeation, which can permit transmission of hydrogen through materials to form leaks. Research into new, low-cost materials that mitigate these challenges is an active area of development. In the interim, hydrogen is largely incompatible with existing fuel distribution networks. While hydrogen can be blended with natural gas in concentrations of approximately 15% by volume, new hydrogen-dedicated pipelines will be needed for large-volume transportation for greater energy uses. Other methods for distribution in the interim include liquid cryogenic transportation or with the use of chemical carriers that are compatible with existing transportation infrastructure.

11.4 Hydrogen Infrastructure at Airports

The development of hydrogen systems for aviation will require significant changes in airport infrastructure to support this transition. An overview of compatibility of airport infrastructure for hydrogen needs into future aviation scenarios is provided in Table 11.2, based on forecasts provided by ACI and ATI [7]. From this overview, it is clear that significant development in refueling and other ground support equipment, as well as fuel handling safety and training, is required. Additional development in hydrogen storage and liquefaction also requires near-term attention, alongside potential infrastructure that can be utilized for on-site production.

Table 11.2 Compatibility of airport infrastructure for hydrogen aviation needs: × = further development needed, ? = development needs are uncertain, ✓ = no new developments are needed

		Timeline		
	Challenge	2023–2025	2025–2035	2035–2050
Airport land use plan	Storage	?	×	×
	Liquefaction	?	×	×
	Hydrogen production	✓	✓	?
Airside development plan	Terminals	✓	?	?
	Refueling equipment	×	×	×
	Ground support equipment	×	×	×
Utility infrastructure development plan	Electricity and water supply	✓	✓	?
Safety, security, and training	Aircraft Refueling	×	×	×
	Airport rescue and firefighting	×	×	×
	Employee training	×	×	×
	Policy and regulations	×	×	×

11.4.1 Airport Hydrogen Supply Chain

When considering airport infrastructure requirements for LH_2 use in aviation, a variety of supply chain configurations can be utilized. While conventional aviation fuel utilizes an external process for petroleum extraction, refining, and transportation to the airport premises, hydrogen has the advantage of on-site production potential, though it also requires on-site or off-site liquefaction to be suitable for aircraft applications. As highlighted in Sect. 11.3.1, there are various production pathways that can be leveraged to source the hydrogen for use in aviation, as further indicated in Fig. 11.3. Each production pathway imposes variable requirements in terms of resources used, electricity demand, land area required, and processing infrastructure, as well as varying implications of greenhouse gas emissions produced in the energy life cycle.

For the specific case of electrolytic hydrogen, Fig. 11.4 demonstrates the three primary supply chain pathways for LH_2 use at an airport. Since hydrogen can be produced on-site, it is possible to stage large-scale electrolyzers and liquefiers at the premises of an airport, with the only inputs needed being water and renewable electricity. This option avoids the need for dedicated hydrogen transport infrastructure to the airport and carries the potential for sourcing renewable energy on-site at airports as well. This approach is, however, also associated with the greatest land footprint and capital investment in infrastructure required by the airport. As such, developing LH_2 capacities fully on-site may be infeasible for some airports, due to limited space availability in areas surrounding the airport or funds available to support investment in new facilities.

Another option to make LH_2 available for dispensing at airports is to produce GH_2 off-site, such as through water electrolysis, and transport the hydrogen to the airport via pipeline, trucks, or trains. In particular, the use of pipelines is a rapid and efficient way to transport hydrogen with low operating costs, though requiring high

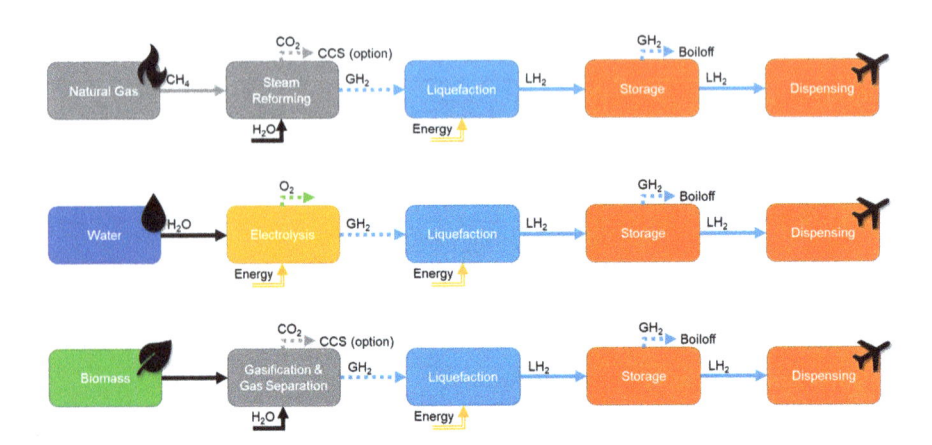

Fig. 11.3 Liquid hydrogen production for use at airports steam methane reformation, water electrolysis, and biomass gasification

11.4 Hydrogen Infrastructure at Airports

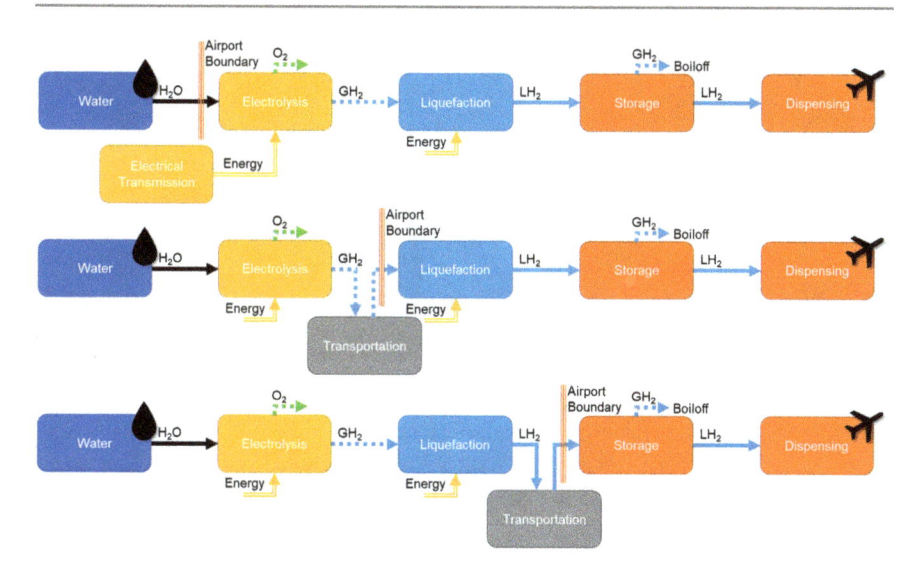

Fig. 11.4 Electrolytic hydrogen supply at airports through on-site electrolysis and liquefaction, off-site electrolysis and on-site liquefaction, and off-site electrolysis and liquefaction

infrastructure costs and development timelines. This approach would still require liquefaction on-site within the airport boundary, though doing so alleviates the need to maintain the cryogenic state of the LH$_2$ during the transportation process. Transportation scenarios that leverage on-site liquefaction are attractive for integrating the aviation energy system into energy transitions that are anticipated across a broad range of industries, where centralized, large-scale electrolyzers are leveraged to produce hydrogen for a variety of applications. If the development of new hydrogen-compatible pipelines is a feasible component of a future hydrogen economy, then an on-site liquefaction strategy becomes a very attractive option.

The final option is to produce and liquefy hydrogen off-site, then transport LH$_2$ to the airport via heavy-duty trucks or other cryogenic transportation approaches. This approach is, however, extremely limited in overall feasibility for large airports due to the significant volume of LH$_2$ needed to meet the energy demand. Given that LH$_2$ has an energy density approximately 25% that of conventional jet fuel, significantly more heavy-duty trucks or railcars would be needed to transport an equivalent energy of LH$_2$ for airport use, as compared to Jet A sources. While LH$_2$ pipelines are technically possible to develop, the great expense and complexity of this network makes it unlikely to be selected as a viable means for direct LH$_2$ transportation to airports.

11.4.2 Airport Local Distribution

Storage of LH$_2$ in an airport will require the construction of a dedicated tank farm, with vessels capable of meeting the cryogenic conditions and material compatibility requirements for long-term operations. Leveraging the decades of development in

Fig. 11.5 4700 m³ LH₂ spherical tank developed for NASA Kennedy Space Center: cross-section of tank design (left), construction of tank (right). (From Fesmire et al. [8])

LH₂ storage from launch vehicle and space exploration research, full-scale deployment of large-scale LH₂ spherical vessels have been constructed at locations such as NASA Kennedy Space Center. The most recent 4700 m³ tank produced at this site marks the largest LH₂ storage vessel ever produced. A cross-sectional view of this spherical tank is shown in Fig. 11.5, after Fesmire et al. [8]. See also Appendix 4. These large-scale storage tanks typically feature a vacuum insulation region between the inner and outer spheres, which are further filled with insulation materials. Historically, perlite powder has been used to fill this evacuated region, though the most recent large-scale LH₂ spheres have used glass bubbles due to improved thermal insulation performance. Additionally, current large-scale LH₂ spheres can be developed with internal heat exchangers within the inner tank, which allows these systems to provide long-term storage capabilities with zero boil off. In future infrastructure developments, multiple large-scale storage facilities will be needed to store >10,000 m³ of LH₂ to satisfy the energy demand of large international airports. Additionally, the tank farm must provide sufficient backup capacity in the event of a temporary outage in the liquefier system or GH₂ supply to the airport. While the specific storage needs of an airport will vary depending on the number of aircraft serviced, three dedicated LH₂ spherical tanks similar in design to that of Fig. 11.5 would meet the needs for several thousands of daily aircraft movements.

With LH₂ present in a bulk storage terminal at the airport facility, there also exists several approaches for distribution local to the airport for aircraft refueling. Approaches that have been considered include the use of truck fleets for LH₂ transportation to gates for refueling, pipelines and hydrant systems for refueling at gates, and designated remote-site refueling stations nearby or supplied by the bulk terminal. Each approach has different tradeoffs and the approach most suitable for a given airport will depend on airport area availability, the number of flights handled per year, total energy required for aircraft fleets, and the capital costs tolerable for new fuel handling infrastructure. While the use of remote refueling stations has manageable capital infrastructure cost, this strategy tends to be viewed as infeasible due to the greater extent of land needed for fuel infrastructure, high recurring costs, and long turnaround times required for aircraft.

11.4 Hydrogen Infrastructure at Airports

Local transportation of LH$_2$ by way of pipelines or trucks for refueling at gate both pose feasible options, though for the case of particularly large airports, the acquisition and management of large fleets of trucks likely poses an unappealing option in comparison to the development of a dedicated pipeline and hydrant system. These conclusions are reinforced by Hoelzen et al. [9], who indicated a crossover point in the lowest-cost distribution scenario when 125 kt$_{LH2}$/year is reached. For airports that distribute less than 125 kt$_{LH2}$/year, it is more economical to utilize LH$_2$ refueling trucks, while for airports that distribute more than this quantity, it is more economical to develop pipeline and hydrant infrastructure. These approaches for on-site hydrogen liquefaction and local distribution via vacuum-jacketed pipeline were also recommended by G.D. Brewer [1] in a detailed study of airport hydrogen infrastructure.

When developing networks of vacuum jacketed LH$_2$ pipelines, the fuel farm used for on-site bulk storage of LH$_2$ should ideally be positioned in close proximity (within approximately 1 mile) from the airport terminal. LH$_2$ pipelines are then installed to provide access at gates, maintenance areas, ground support hubs, cargo areas, and other locations across the airport. Inclusion of two independent supply loops ensures that, if closure of one pipeline is necessary due to leaks or other maintenance, supply to airport terminals is maintained. Example layouts of hydrogen pipeline systems are shown in Fig. 11.6 for SFO and ORD airports based on studies performed by Brewer [1] and Hanks et al. [10], respectively. In addition to the LH$_2$ supply lines, both layouts assume vented boil off hydrogen is captured and

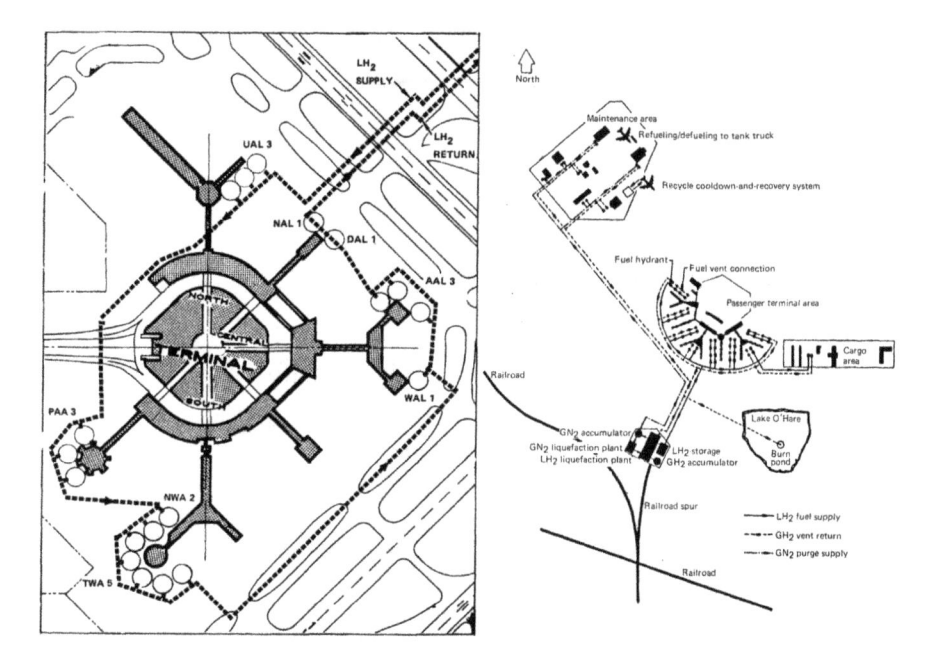

Fig. 11.6 Hydrogen pipeline network concepts for SFO airport, after Brewer [1] (left) and ORD airport, after Hanks et al. [10] (right)

transported to the liquefaction plant by way of a GH_2 return line. Additionally, supply lines for inert gas are necessary for hydrogen system purging at servicing stations or gates. While installation of these additional lines introduces additional cost for the infrastructure development and operation, the savings enabled by capturing the vented hydrogen far outweigh those of the infrastructure in cost trades. Also present in the plans for ORD in Fig. 11.6 is a burn pond, which serves as a backup system for safely disposing of vented GH_2 in the event where it cannot be properly transported and recovered at the liquefaction facility. Other concepts also have the hydrogen boil off used for power generation to support airport electrical loads.

When configuring pipeline systems for LH_2 delivery, lines can be installed above ground, directly buried, in a pipe gallery, or in a dedicated tunnel, as shown in Fig. 11.7. However, above-ground installations are typically undesirable from a safety perspective, and the required avoidance of these structures by ground vehicles could disrupt ground traffic and operational efficiency. Similarly, direct burial of hydrogen lines inhibits the ability to maintain, service, or expand the distribution

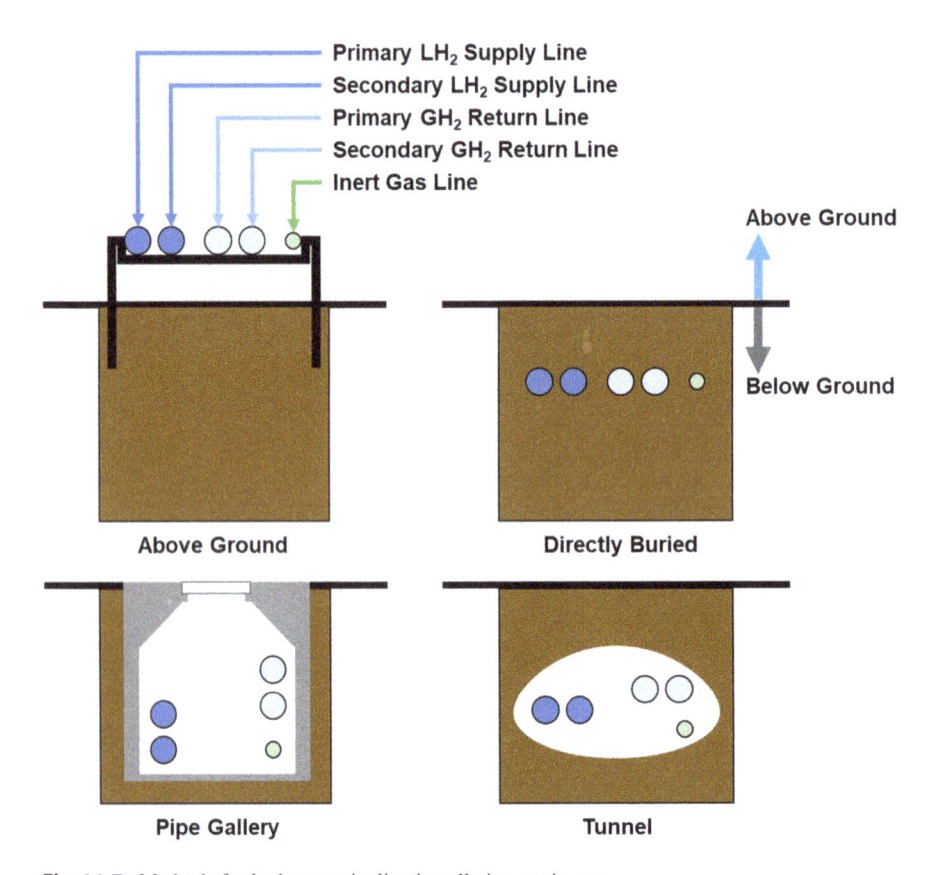

Fig. 11.7 Methods for hydrogen pipeline installation at airports

11.4 Hydrogen Infrastructure at Airports

system. Direct burial approaches also open the possibility of accumulation of hydrogen-air mixtures and makes removal of hydrogen from faulty pipelines difficult. For these reasons, installation of hydrogen pipeline infrastructure is currently envisaged in underground pipe galleries or tunnels. A pipe gallery can be developed by open-cut trenching to place the pipeline infrastructure, which is then covered by removable grates to permit venting and maintain servicing access. Tunnels, on the other hand, can be bored without the need for direct excavation from surface level. Air intake pits, exhaust pits, and tunnel access ports must also be periodically placed along the distribution network to similarly meet safety and servicing needs. Placing remotely controlled service isolation valves periodically along the distribution network also allows supply lines to be securely closed when leaks are identified in certain segments, or prior to performing pipeline maintenance.

The hydrogen distribution pipelines are then routed to a network of hydrant pits at the airport gates, such as that shown in Fig. 11.8. These hydrant pits include hose connections for both LH_2 supply and GH_2 ventilation return. While the LH_2 is supplied to the aircraft, the GH_2 return lines are used to capture gas volumes that are displaced by the addition of liquid content, while also recapturing new boil off produced as LH_2 filling commences. Valves on each line are used to open or close flow of hydrogen during the refueling process. The hydrant pit also provides access to an inert gas line for purging tasks. While the need for a full system warm-up is not

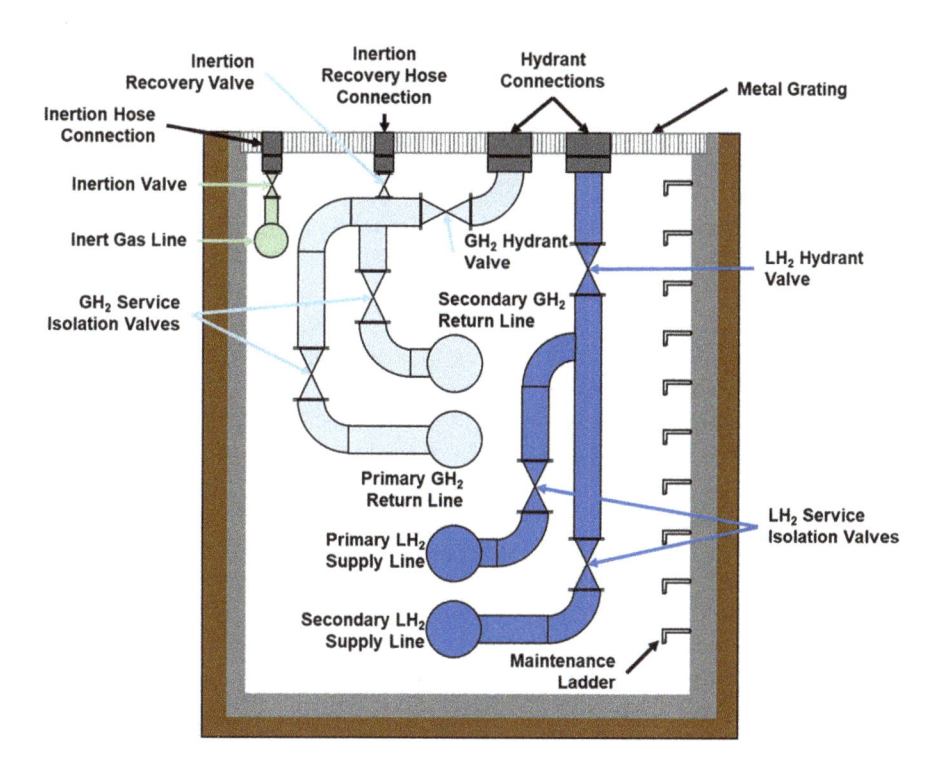

Fig. 11.8 Cross-sectional view of notional hydrogen hydrant pit for installation at airport gates

anticipated at the gates, purging of transfer lines may serve as an important safety feature of standard operating procedures during refueling. However, hose connection points for inert gas lines will be necessary at maintenance areas when complete system warmup and servicing tasks are completed. GHe serves as an ideal inert gas for this process, as it remains gaseous at LH_2 temperatures, whereas N_2 freezes into a solid state. GN_2 could be utilized for inerting GH_2 transfer lines, for example, but GHe is recommended for LH_2 transfer lines. However, as Helium is more expensive than other inert gas options, recovery of gases obtained during the purging process is envisioned. Since the purging process will produce a mixed flow of GH_2 and GHe, an inertion gas recovery hose is also envisioned that is exhausted into the GH_2 return line. A gas separation process can then be used to appropriately ensure the GHe is removed from the return line before the bulk GH_2 volume reaches the hydrogen liquefaction station.

11.4.3 Refueling with LH_2

Given the difference in physical properties between Jet A fuel and LH_2, different infrastructure and standard procedures for refueling will be necessary with the adoption of hydrogen. However, many of the refueling methods envisioned utilize similar ground support equipment and infrastructure, though with pumps, safety equipment, and materials intended for compatibility with hydrogen. As highlighted in Sect. 11.4.2, gate refueling can be performed using hydrant fuelers or dedicated tanker trucks. Examples of ground support equipment, hydrant hose connections, and supporting safety equipment envisioned for hydrant fueling from several studies are shown in Fig. 11.9. Since LH_2 storage tanks are not anticipated to reside within wing box volumes, the fuel access ports on the aircraft may not be within immediate reach to ground service personnel. For this reason, the use of a refueling boom can help to ensure fill ports can be reached, while also keeping ground service personnel clear of frostbite or ignition hazards that may otherwise be present if performing manual hose connections for refueling. The use of dedicated boom systems also ensures that rigid transfer lines with adequate thermal support can be utilized during the fueling process.

A similar set of concepts for refueling at gates or other remote areas with the use of tanker trucks is shown in Fig. 11.10. Even if a hydrant system is installed at an airport, refueling capability by way of tanker trucks will still be necessary for aircraft parked at remote maintenance or cargo locations. Furthermore, these trucks can also be utilized to offload LH_2 volume from an aircraft in an unlikely event where an aircraft becomes disabled, prior to return of the aircraft to maintenance areas. Another concept for gate refueling is also presented in Fig. 11.11, consisting of a boom fueling system that is installed directly at the gate. The concept for operating this refueling system is conceptually similar to that of a hydrant fueler vehicle, just with the hydrant access routed directly through the boom system at the gate. This concept was preferred by Hanks et al. [10] due to the potential of reducing

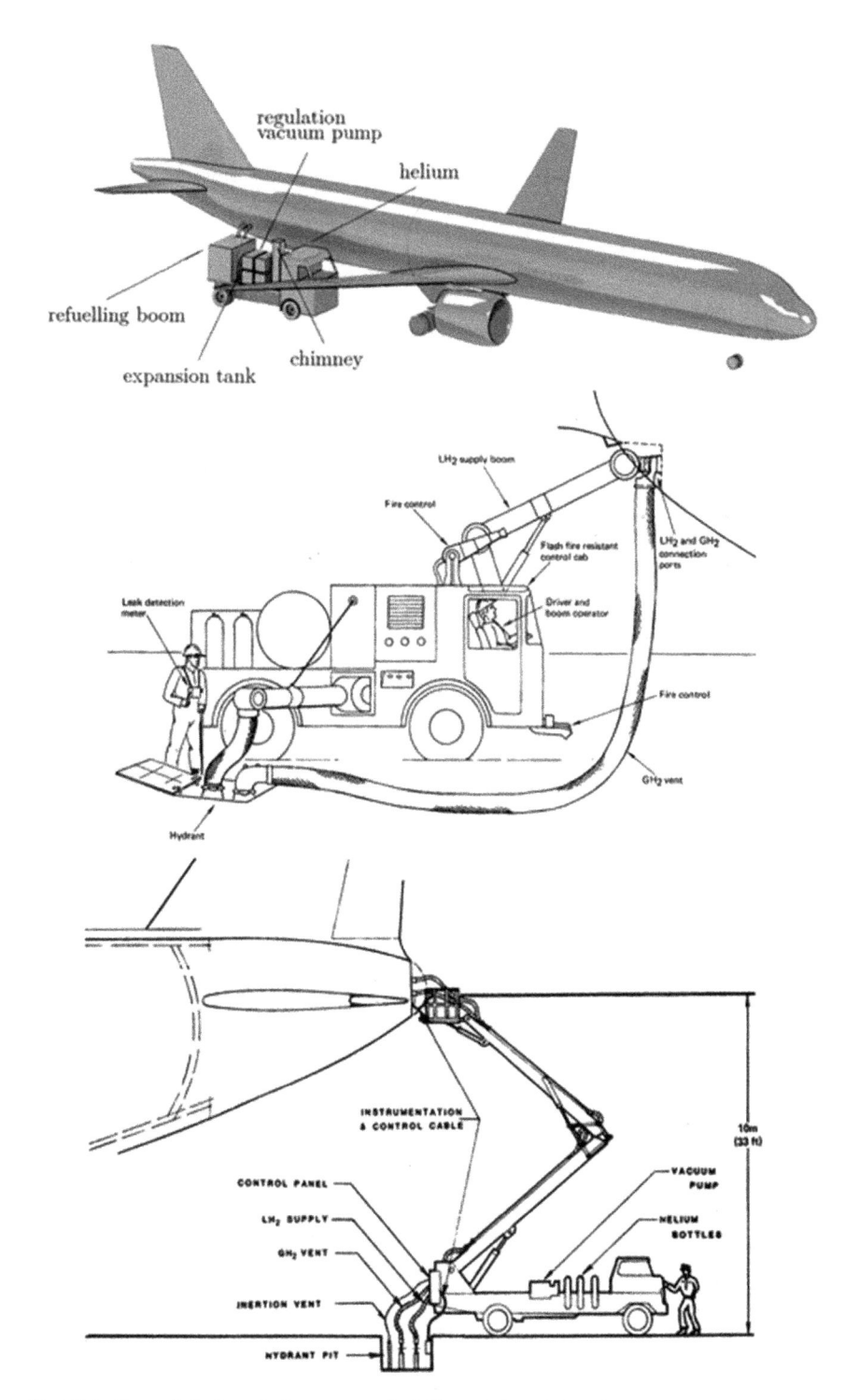

Fig. 11.9 Example hydrant fueler concepts for LH$_2$ servicing of aircraft: dispenser sized for single-aisle aircraft, after Mangold [11] (top); hydrant fueler with dispenser boom and safety equipment, after Hanks et al. [10] (middle); hydrant fueler with boom for aft-tank refueling, after Brewer [1] (bottom)

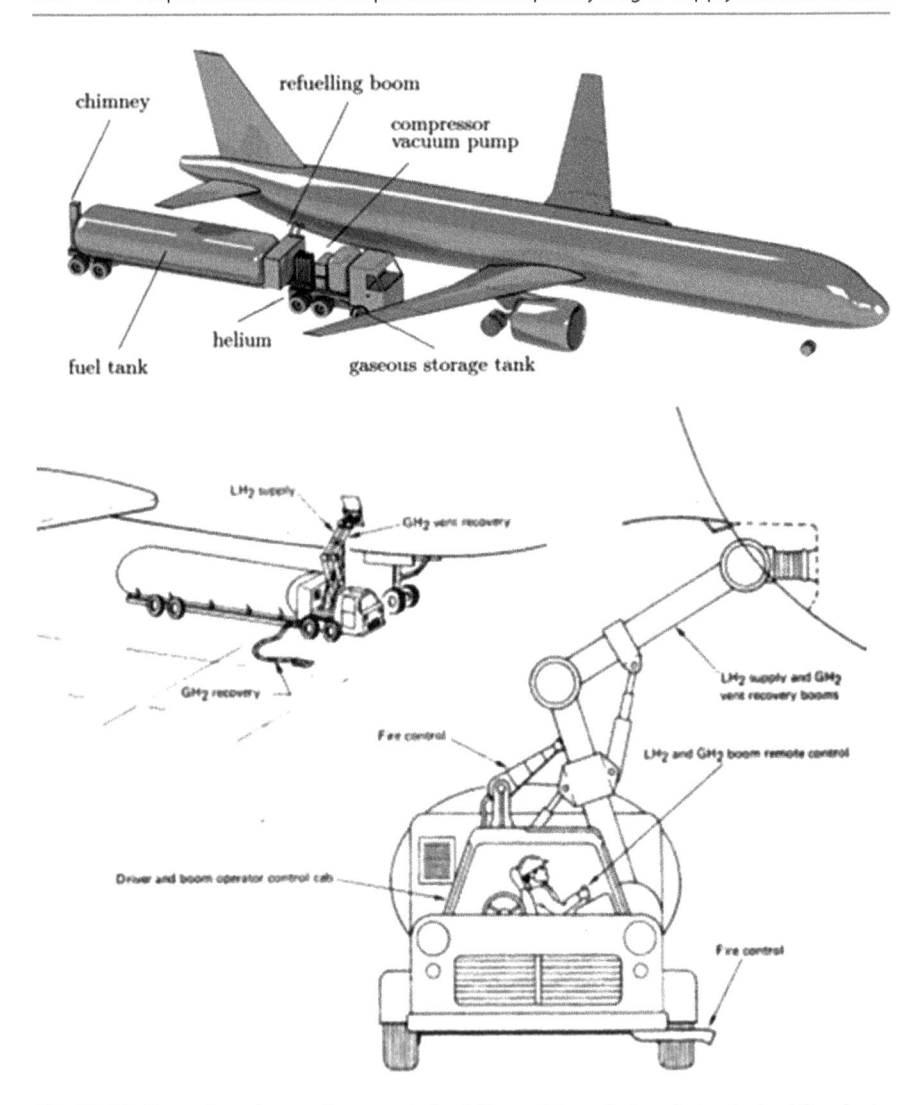

Fig. 11.10 Example tanker truck concepts for LH$_2$ servicing of aircraft: truck sized for single-aisle aircraft, after Mangold [12] (top); tanker truck with dispenser boom and safety equipment, after Hanks et al. [10] (bottom)

congestion of ground support equipment in ramp areas and reduced potential for damage to the hose connection interfaces during transit of hydrant fueler equipment.

The refueling process utilizes several steps, including positioning of fuel handling components, hose connection and securing, GH$_2$ and LH$_2$ transfer line purging, hose chill down, LH$_2$ fueling with GH$_2$ recovery, transfer line purging, hose disconnection, and removal of fueling equipment. In instances where clean break disconnect systems are used, where the hydrant hose connection, aircraft fill port, and both ends of the transfer line can all be closed prior to removal, it may not be necessary to purge

11.4 Hydrogen Infrastructure at Airports

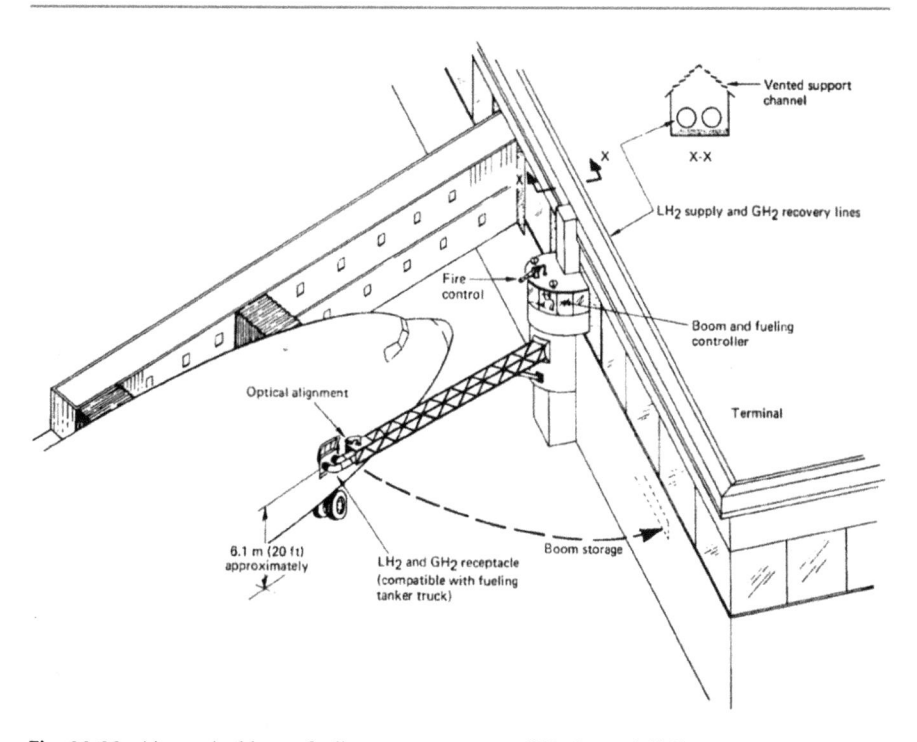

Fig. 11.11 Air terminal boom fueling system concept of Hanks et al. [10]

transfer lines before and after each refueling procedure. Eliminating this purging step is highly desirable, as it would streamline ground operations and reduce capital and recurring costs for hydrogen aviation infrastructure.

Estimates of the turnaround time for fueling LH_2 aircraft significantly vary due to differing assumptions around the equipment and standard operating procedures taken during the aircraft turnaround process. The most prominent of these assumptions is the flow rate of LH_2 through fuel delivery hoses. Some studies have assumed volumetric flow rates of LH_2 to be the same as those of current Jet A fueling systems, resulting in a significant increase in overall refueling time needed. This fueling time increase is due to the lower energy density of LH_2 as compared to Jet A fuel, resulting in fueling time increases 2–3 the length of current systems, depending on the fuel volume required. However, most studies on LH_2 airport infrastructure instead assume an energy flow rate, rather than volume flow rate, to be approximately consistent with that of current Jet A systems. Assumed volume flow rates closer to 11,000–17,000 L/min, or mass flow rates of 13–20 kg/s have been utilized, which have resulted in fueling-stage time requirements comparable to those of Jet A. These higher flow rates are well justified by existing systems for LH_2 fueling utilized on the US Space Shuttle program and other space exploration applications, as well as fuel flow systems with rated limits above those conventionally used in existing markets. Additionally, multiple filling hoses can be utilized, or

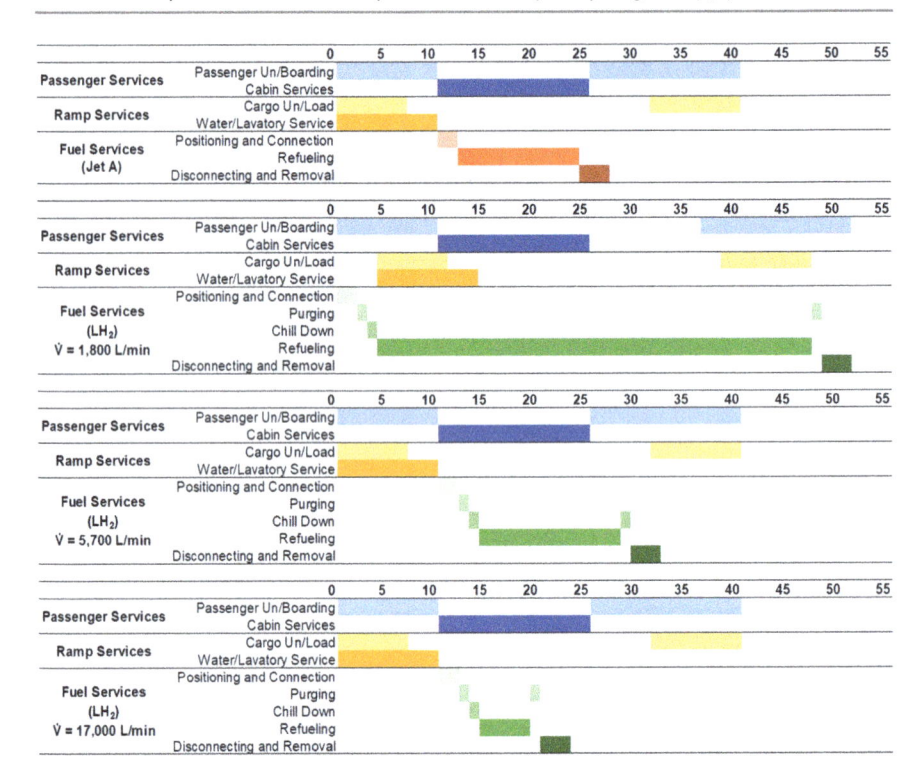

Fig. 11.12 Representative turnaround time for single-aisle, 200-pax aircraft being fueled for a 2500 nmi mission, with schedule of gate services, including Jet A refueling considerations and LH$_2$ refueling with different flow rates assumed

hoses with larger diameter, so long as the velocity of the fuel within the line is limited to that below that which would produce a hazard of static charge buildup.

A series of representative turnaround times for aircraft including either Jet A or LH$_2$ refueling are shown in Fig. 11.12, assuming a single-aisle, 200-passenger configuration with sufficient energy for a 2500 nmi flight. The top chart shows the anticipated turnaround time, including time required for Jet A fueling, followed by the anticipated turnaround time for LH$_2$ fueling scenarios with different volumetric flow rates. For aircraft of this class, the turnaround time is typically not dictated by the refueling time required, as passenger unboarding, cabin services, and subsequent passenger boarding processes serve as the limiting factor for the duration of the gate phase. If fueling volumetric flow rates of existing aircraft systems are maintained, assuming $\dot{V} = 1800$ L/min, the anticipated turnaround time is extended by more than 25%. Since the time required for the full refueling process dictates the duration of the gate phase, it is assumed to begin immediately upon arrival at the gate, and ramp services are not initiated until the initial connection, purging, and chilldown stages are completed. Doing so limits the ground support vehicle traffic in the vicinity of the gate. If fuel services do not commence until after ramp services are completed, this would further delay the departure of the aircraft.

11.4 Hydrogen Infrastructure at Airports

If the flow rate of the LH_2 fuel system is increased to the more practical value of V = 5700 L/min, then the refueling time becomes far more comparable to that of the baseline Jet A system, and a typical standard order of the gate services can be utilized. It is notable that this volume flow rate also coincides with the initial slow fill LH_2 refueling rate used the Space Shuttle program [12]. For the Space Shuttle Program, following this slow fill stage the flow rate was then further increased to 5 times this slow fill value, suggesting that fill rates at or above 5700 L/min are quite reasonable. The final chart shows the turnaround time assuming an LH_2 flow rate of V = 17,000 L/min, which is the maximum rate typically suggested in related studies for hydrogen airport infrastructure. At these aggressive flow rates, the time required for refueling is significantly reduced, relative to conventional Jet A systems. Most analysis on the turnaround time for LH_2 aircraft place the net duration of the gate stage to be within approximately ±10% of the time required for conventional Jet A aircraft, suggesting that excessive time spent for gate services can be mitigated with proper infrastructure.

11.4.4 Safety and Standards

At the time of developing this section, no fueling standards have been developed specific to hydrogen applications in aviation. Previous efforts to standardize hydrogen safety for aviation are represented in ISO/PAS 15594:2004, which was withdrawn shortly after initial publication in 2004. Current standards for aircraft fuel servicing with Jet A or Jet A-1 are established in NFPA 407. While safety of GH_2 and LH_2 are incorporated into other NFPA codes, including NFPA 2 and NFPA 55, these documents do not include aviation-specific considerations. Other standards for hydrogen refueling used in other industries, such as ISO 19880, will serve as useful starting points for developing standards for hydrogen use in aviation. Currently, multiple standardization organizations, trade organizations, and government entities are actively working to establish safety protocols for hydrogen storage, distribution, and fueling at airports. See also Appendix 3.

The reason for developing safety protocols specific to hydrogen is linked to the difference in physical properties of LH_2 as compared to Jet A. Several of the different material properties are shown in Table 11.3. Current safety challenges are related to the extremely low temperatures of LH_2, overpressurization of storage vessels when subject to excessive boil off, wide range of hydrogen flammability limits, low energy input required for ignition, rapid vaporization and high burn rates, and high flame temperature. However, the high flame velocity results in a consolidated flame structure, with low thermal radiation of hydrogen combustion, which is viewed by some as actually providing benefits in material safety over conventional hydrocarbon fuels, alongside other potential benefits.

In a detailed hazard analysis of LH_2 application at airports, Benson et al. [13] concluded that the current unknown overhauls required in firefighting procedures represented the highest risk for airport operations. For example, the extreme buoyancy of GH_2 is advantageous for having flammable vapors clear from ground sources, though advection from wind also introduces complications by introducing

Table 11.3 Physical properties of hydrogen and Jet A

	Hydrogen	Jet A
Molar mass (g/mol)	2.016	168
Specific energy, LHV (MJ/kg)	120	43.1
Energy density (MJ/L)	8.49	35.3
Liquid density, at boiling point (g/cm^3)	0.071	0.811
Specific heat (J/gK)	9.69	1.98
Boiling point (°C)	−253	269
Melting point (°C)	−259	−51
Auto-ignition temperature (°C)	585	440
Minimum ignition energy in air (mJ)	0.02	0.25
Flammability limit in air (% vol)	4.0–75.0	0.8–6.0
Detonation limit in air (% vol)	13.0–65.0	1.1–3.3
Flame temperature (°C)	2022	2158
Laminar flame velocity in air (cm/s)	265	43
Thermal energy radiation (%)	17–25	33–43
Buoyancy in air (m/s)	1.2–9.0	N/A
Diffusion in air (cm^2/s)	0.61	0.05

Fig. 11.13 Visualization of spark-free zone for refueling on single-aisle aircraft, based on recommendation of Brewer [1]

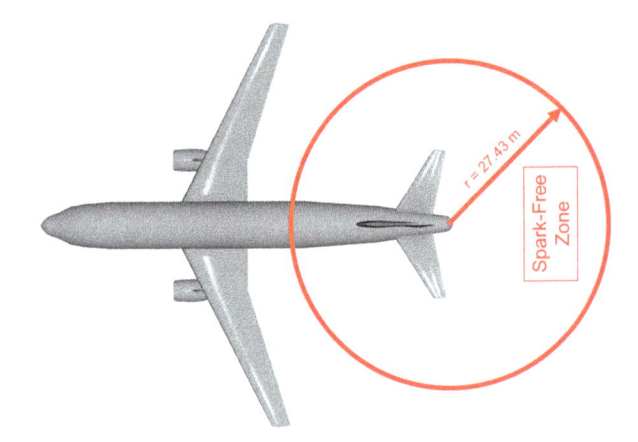

uncertainty in the size and position of potentially flammable GH$_2$ cloud regions. Hydrogen flames are also nearly invisible and, as such, are difficult to detect and extinguish. With these examples in mind, the unconventional nature of these considerations requires very careful development of firefighting practices.

Given the low input energy required for ignition, the designation of expanded spark-free zones around the aircraft will be necessary. Brewer [1] suggests a radius of 27.43 m from the fueling location would require restrictions to vehicles with open ignition systems or those not properly bonded or grounded. This study assumed the refueling port configured at the tail of the aircraft, resulting in a visualization of this spark-free radius shown in Fig. 11.13 for a representative single-aisle aircraft. Such a safety designation may require additional apron area at airport gates, which could constrict taxiways at existing airports.

References

1. Brewer, G.D., "LH2 Airport Requirements Study," NASA CR-2700, 1976.
2. Alvik, S., Ozgun, O., et al., "Hydrogen Forecast to 2050: Energy Transition Outlook 2022," DNV, 2022.
3. "Global Hydrogen Review 2023," International Energy Agency, 2023.
4. Beswick, R.R., Oliveira, A.M., Yan, Y., 2021. Does the green hydrogen economy have a water problem? ACS Energy Letters 6, 3167–3169.
5. "Carbon Neutrality in the UNECE Region: Integrated Life-cycle Assessment of Electricity Sources," United Nations Economic Commission for Europe, 2022.
6. Ritchie, H., "How does the land use of different electricity sources compare?," OurWorldInData.org, 2022.
7. ACI and ATI, "Integration of Hydrogen Aircraft into the Air Transport System," 2021.
8. Fesmire, J.E., Swanger, A.M., Jacobson, J.A., and Notardonato, W.U., "Energy Efficient Large-Scale Storage of Liquid Hydrogen," Advances in Cryogenic Engineering: Proceedings of the Cryogenic Engineering Conference (CEC) 2021, 19-23 July 2021.
9. Hoelzen, J., Flohr, M., Silberhorn, D., Mangold, J., Bensmann, A., and Hanke-Rauschenbach, R., "H2-powered aviation at airports – Design and economics of LH2 refueling systems," Energy Conversion and Management: X, Vol. 14, 2022.
10. Hanks, G.W. et al., "An exploratory study to determine the integrated technological air transportation system ground requirements of liquid-hydrogen-fueled subsonic, long-haul civil air transports," NASA CR-2699, 1976.
11. Mangold, J., "Economical Assessment of Hydrogen Short-Range Aircraft with the Focus on the Turnaround Procedure," MSc Thesis, Institute of Aircraft Design, University of Stuttgart, 2021.
12. Osipov, V.V., Daigle, M.J., Muratov, C.B., Foygel, M., Smelyanskiy, V.N., and Watson, M.D., "Dynamic Model of Rocket Propellant Loading with Liquid Hydrogen," Journal of Spacecraft and Rockets, Vol. 48, No. 6, 2011, pp. 987–998.
13. Benson, C.M., Holborn, P.G., Rolt, A.M., Ingram, J.M., and Alexander, E., "Combined Hazard Analyses to Explore the Impact of Liquid Hydrogen Fuel on the Civil Aviation Industry," Proceedings of ASME Turbo Expo 2020 Turbomachinery Technical Conference and Exposition, GT2020-14977, 2020.

Technology Gaps, Cryogenic Power Electronics, Cryogenic Current Leads, Transfer Lines, Liquid Hydrogen Pumps

12.1 Summary and Outlook

The effort to bring cryogenics and superconductivity "On board" an aircraft, while progressing, still faces several, considerable gaps in key technology areas. In this review, we analyzed lightweight liquid hydrogen tanks and pumps as key components developing future maturation. However, a holistic view must include a complete cryogenic flow distribution scheme of an all-electric aircraft with all its power lines, cooling, and fuel cell feed lines.

As indicated, the TRL level of a composite tank is still very low, despite some recent, encouraging progress. To close the gap, the following needs to happen:

First, creating and selecting suitable composite fibers, or, specific high-strength carbon fibers, for reinforced structural tanks, with or without a liner, need to be developed. Second, an enhanced epoxy matrix mixture with properties that work well at cryogenic temperatures needs to be found. Most of the resins we use derive from compositions and mixes that were created 40 years ago. There is considerable research going on regarding a specific increase in certain material properties, e.g., thermal resin conductivity and resin strength, but those need to be validated until found suitable for future aircraft tanks. A new impregnation material for example is TELENE®, an organic olefin-based thermosetting dicyclopentadiene resin [1].

To tag a significant meaning of the Weibull statistics on the reliable tank design, feasibility studies need to be run on many composite designs with a low-pressure rating (e.g., tanks made from PEEK, fatigue tests on liners, etc.) to achieve minimum tank mass weight fraction that results in a high payload. This will retire by building a series of test tanks.

The cryogenic airframe infrastructure for the cryoaircraft has yet to be optimized. It needs to be refined step by step, with respect to liquid and gaseous hydrogen transfer lines (sizing and routing) to provide dedicated cooling temperatures and mass flows, with appropriate heat sinking capability for all components, including built-in contingencies, and for complying with all hydrogen safety features.

Transfer lines, originally only meant to transfer cryogenic liquids, have a long history of use and have been modified for use with superconducting cables turning into cable cryostats with a high TRL level, most likely composite based. Powerful batteries are still required for initial take off, but those still need to be developed for light weight and efficiency.

Pumps are at the heart of the cryoaircraft. Present-day LH_2 pumps are inadequate for use on an aircraft. Specific pumps need to be developed with respect to weight and reliability and need to be safe against cavitation. Pumping specs, e.g., required flow rates, power rating, discharge pressure, etc., need to be defined for any flow circuit designed, so that liquid withdrawal from hydrogen tanks at inflight pitch angles can be achieved with a minimum liquid depletion level, ensuring tank sizing is adequate for the complete mission profile.

Finally, a technology roadmap needs to be developed urgently to enable the successful development of an aircraft-qualified liquid hydrogen pump that is maintenance-free, follows safety standards and protocols and meets all consolidated specs for motor cooling and fuel cell feeds. Those stumbling blocks can only be removed with a significant investment in future technologies fully integrated into a future green hydrogen economy.

12.2 Technology Gaps

Table 12.1 summarizes the technology requirements where Research and Industry needs to be actively working on aircraft cryogenics. Gaps are classified with respect to development need and urgency with increasing priority. Classes 4 and 5 for example indicate a technology gap, but prior technology exists that may have to be modified for airworthiness. Robust, reliable fuel cell technology, paired with light weight battery stacks have been omitted (non-cryogenic) but are key factors for the success of any all-electric aircraft.

12.2.1 Cryogenic Power Modules: Knowledge Gap

Cryogenic power electronics will be embedded into the cryogenic aircraft infrastructure. The storage of cryogenic liquids in an aircraft provides a unique opportunity to tap into the cold environment for increasing the efficiency of those elements by operating them below 80 K.

As electrical resistivity of circuits reduces at low temperatures, GaN devices up to 650 V and higher did not show a carrier freezeout at 20 K. This technology is still in its infancy state and cryocooled power electronics will prove to be highly efficient in an aircraft [3–5]. It is expected that unknown phenomena will be detected as this technological area is further researched.

12.2 Technology Gaps

Table 12.1 Cryogenic technology gaps

Class	Aircraft component/module
1	Lightweight, composite transfer lines, composite bayonets
1	Transfer lines that collect and transfer vapor bubbles back to the cryotank
1	Cryolines without vapor lock in various operating conditions, like climb and descent
1	Quick connect vacuum jacketed lines (presently available for LNG only)
1	Electric breakdown insulation of lines for use with gaseous and liquid hydrogen
2	Liquid hydrogen pump without servicing need and built-in cavitation-free elements
2	Lightweight, efficient phase separator for liquid hydrogen
2	Lightweight, efficient compressors or after coolers, for liquid hydrogen
3[a]	Lightweight dual-walled composite liquid hydrogen tank with 100,000 liquid fill/removals resilient to many initial chilldowns without creating stress concentration spots
3	Novel suspension systems with built-in fatigue indicators and built-in warning sensors
3	Low liquid hydrogen boil-off loss during coasting or tarmac parking
3	Cryogenic, service-free fuel tank vibration dampers, designed for all flight conditions
3	New sloshing baffles to reduce mechanical stresses on tank shell/boil-off
3	Validating current tank skin and piping against hydrogen, lightning ignition
4	Hydrogen detection sensors in the ppb range for cryogenic and room temperature, with automatic self-diagnostics/calibration and leak-free wire routing
4	Sensor technology with automatic resetting and built-in self-calibration features at negligible pressure drop
4	Dedicated instrumentation and safety organs that fulfill electric aircraft specifications
4	Self-calibrating level and flow meters and other system health indicators
4	Safety valves and other process flow components with no leak rate
4	Strain-stress sensors deeply embedded in the composite matrix to indicate tank lifetime degradation, for example, polymer-based piezoelectric sensors, fiber Bragg-based elements, etc.
5	Liquid hydrogen heat exchangers
5	Heat exchangers for all cryomodules used for power electronics, etc.
5	Heat exchangers for feeding warm vapor into fuel cells
6	Superconducting, high power rated/low weight electric motor without iron shielding and low AC losses
6	Robust miniature high-power cryocoolers for 20 K, Stirling-based
6	High power HTS current leads cooled with liquid hydrogen, warm end cooling
6	Rotating cooling means with hydrogen
6	Leak-free rotary couplings
7	Cryogenic power module aircraft infrastructure
8	Development of an extensive aviation fuel and energy infrastructure

[a]Understanding of the long-term effects of hydrogen on the mechanical properties of reinforced composite materials. Understanding how rapid fills and chill downs will interact with the composite matrix, e.g., hairline cracks over time

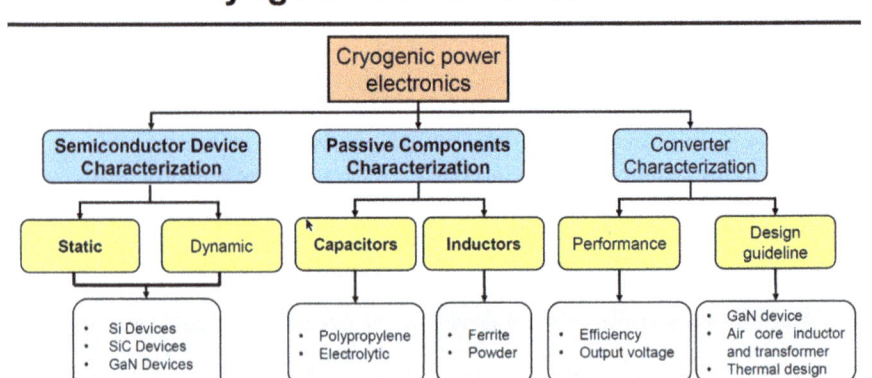

Cryogenic Power Electronics

```
                    Cryogenic power
                     electronics
          ┌───────────────┼───────────────┐
Semiconductor Device   Passive Components    Converter
 Characterization       Characterization   Characterization
    ┌─────┴─────┐      ┌─────┴─────┐      ┌─────┴─────┐
  Static     Dynamic  Capacitors  Inductors Performance  Design
                                                        guideline
    │                    │          │         │            │
• Si Devices       • Polypropylene • Ferrite • Efficiency • GaN device
• SiC Devices      • Electrolytic  • Powder  • Output      • Air core inductor
• GaN Devices                                  voltage       and transformer
                                                           • Thermal design
```

12.2.2 Cryogenic Current Leads: Knowledge Gap

Cryogenic, superconducting current leads are a "must have" in an aircraft. This is the only way to reduce the weight of the power lines to increase the payload. To achieve that, several activities are currently carried out in light of the CHEETA Phase II program. Although superconducting current leads are known for many superconducting applications, the particular use in an aircraft in different inclination angles and power consumption rates to room temperature is challenging.

12.2.3 Flight Verification

Airborne liquid hydrogen and superconducting power drivetrains need to be tested under flight conditions. Component verification needs to be tested in real operating conditions which may lead to some surprises. Worldwide efforts just recently started in the United States, Europe, Australia and China to test airworthiness of all new superconducting and cryogenic drivetrain components, including fuel tanks and cells.

12.2.3.1 Further Technologies We Can Tap into to Accelerate Breakthrough Toward Higher TRL

In general, most superconducting applications by far provide the greatest synergy when designing composite tanks. Many reinforced epoxy mixtures with different fibers work for superconducting magnet applications. An up-to-date review of current resins for magnet impregnation is given by Feldman in [2].

Superconducting industry has a long history in nuclear fusion, particle detectors, and medical applications (MRI) that routinely handle composites.

Other neighboring modalities are shown in Appendix 4.

Last but not least we need to create special aircraft standards for use with liquid and gaseous hydrogen, and in particular with respect to electric breakdown in cables and transfer lines, together with NHA, universities, government agencies and industry.

References

1. E. Barzi et al, Performance Improvement of LTS Undulators for Synchrotron Light Sources, 2LOr1E-01, ASC 2024
2. J. Feldman, W. Stautner, C. Kovacs, N. Miljkovic, K. S. Haran "Review of materials for HTS magnet Impregnation" Supercond. Sci. Technol. 37 033001 2024
3. A. Mantooth, M. Hossain, Y. Wei Cryogenic Power Electronics Component Static Characterization, CHEETA, 2022
4. AM Hossain, A. U. Rashid, R. Sweeting, Y. Wei, H.r Mhiesan, A. Mantooth, D. Woldegiorgis. "Cryogenic Characterization and Modeling of Silicon Superjunction MOSFET for Power Loss Estimation," AIAA 2020-3660. AIAA Propulsion and Energy 2020 Forum 2020
5. H. Mhiesan, M. M. Hossain, A. U. Rashid, Y. Wei and A. Mantooth, "Survey of Cryogenic Power Electronics for Hybrid Electric Aircraft Applications," 2020 IEEE Aerospace Conference, , pp. 1-7, https://doi.org/10.1109/AERO47225.2020.9172807 2020

Further General Reading[1]

W. Stautner, P. Ansell, K. Haran, "CHEETA: All-electric Aircraft takes Cryogenics and Superconductivity "On board", *IEEE Electrification Magazine*, June 2022

K. S. Haran, S. Kalsi, T. Arndt, H. Karmaker, R. Badcock, B. Buckley, T. Haugan, M. Izumi, D. Loder, J. Bray, P. Masson, E. W. Stautner, Topical Review "High power density superconducting rotating machines – development status and technology roadmap", *Supercond. Sci. Technol.*, vol. 30, no 12, pp. 1–41, 2017. https://doi.org/10.1088/1361-6668/aa833e.

P. Wheeler, K. Haran, "Electric/Hybrid-Electric Aircraft Propulsion System", *Proceedings of the IEEE*, vol. 109, No. 6, 2021.

T. Balachandran, K. Haran, "A fully superconducting air-core machine for aircraft propulsion", IOP Conf. Series: Materials Science and Engineering 756 (2020) 012030, IOP Publishing, https://doi.org/10.1088/1757-899X/756/1/012030

W. Stautner, P. Ansell, K. Haran, D. Mariappan, "Liquid Hydrogen Tank Design for Medium and Long Range All-electric Airplanes", Invited plenary talk, CEC-ICMC 21, 2021

Further Detailed Reading

Lauer, M. G., and Ansell, P. J., "A Parametrization Framework for Multi-Element Airfoil Systems Using Bézier Curves," 2022 AIAA Aviation Forum, 2022. https://doi.org/10.2514/6.2022-3525

Lauer, M. G., and Ansell, P. J., "Aerodynamic Shape Optimization of a Transonic, Propulsion-Airframe-Integrated Airfoil System," 2022 AIAA Aviation Forum, 2022. https://doi.org/10.2514/6.2022-3662

---

[1] For further general Reading on Cryogenics, please see Appendix A6.

White, A.S., Waddington, E., Merret, J.M., Greitzer, E.M., Ansell, P.J., and Hall, D.K., "System-level Utilization of Low-grade, MW-scale Thermal Loads for Electric Aircraft," AIAA Paper 2022-3291, 2022.

Waddington, E., Merret, J.M., and Ansell, P.J., "Impact of LH2 Fuel Cell-Electric Propulsion on Aircraft Configuration and Integration," AIAA Paper 2021-2409, 2021.

Ansell, P.J. and Haran, K.S., "Electrified Airplanes—A Path to Zero-Emission Air Travel," IEEE Electrification Magazine, June 2020.

Ansell, P.J., "Hydrogen-Electric Aircraft Technologies and Integration," IEEE Electrification Magazine, June 2022.

Jois, H. and Ansell, P.J., "Analytical Framework for Design of Aero-Propulsive Geometries with Powered Wakes," AIAA SciTech 2023.

White, A.S., Waddington, E., Merret, J.M., Greitzer, E.M., Ansell, P.J., and Hall, D.K., "Trade-Space Assessment of Liquid Hydrogen Propulsion Systems for Electrified Aircraft," AIAA Paper 2023-4345, 2023.

Waddington, E., Jois, H., Lauer, M.G., Patel, Y., and Ansell, P.J., "Hybridization Impact on Emissions for Hydrogen Fuel-Cell/Turbo-Electric Aircraft," AIAA Paper 2023-3873, 2023.

Ranjan, P., Das, G.K., Waddington, E., Lauer, M.G., Ansell, P.J., and James, K., "Preliminary Design and Optimization of a Hydrogen-Electric Aircraft Concept," AIAA Paper 2023-4344, 2023

Ansell, P.J., "Required Developments for Integration of Sustainable Hydrogen in Aviation," AIAA Paper 2023-4476, 2023.

Mahesh, K., Shah, A., and Ansell, P.J., "Multi-fidelity Analysis and Reduced-Order Modeling of Wing-Body Aerodynamics," AIAA Paper 2023-4346, 2023.

Waddington, E.G., Merret, J.M., and Ansell, P.J., "Impact of Liquid-Hydrogen Fuel-Cell Electric Propulsion on Aircraft Configuration and Integration," Journal of Aircraft, 2023.

M. Podlaski, L. Vanfretti, H. Nademi and H. Chang, "UAV Dynamic System Modeling and Visualization using Modelica and FMI," The Vertical Flight Society's 76th Annual Forum & Technology Display, Virginia Beach Convention Center on Oct. 6–8, 2020.

M. Podlaski, L. Vanfretti, A. Khare, H. Nademi, P. Ansell, K. Haran and T. Balachandran, "Initial Steps in Modeling of CHEETA Hybrid Propulsion Aircraft Vehicle Power Systems using Modelica," AIAA/IEEE Electric Aircraft Technologies Symposium (EATS), 26–28 August 2020, Virtual Event. https://doi.org/10.2514/6.2020-3580

M. Podlaski, L. Vanfretti, T. Bogodorova, T. Rabuzin and M. Baudette, "RaPId – A Parameter Estimation Toolbox for Modelica/FMI-Based Models Exploiting Global Optimization Methods," SYSID 2021 – 19th IFAC Symposium on System Identification – Learning Models for Decision and Control, Padova, Italy, 14–16 July 2021.

M. Podlaski, R. Niemiec, L. Vanfretti, and F. Gandhi, "Multi-Domain Electric Drivetrain Modeling for UAM-Scale eVTOL Aircraft" The Vertical Flight Society's 77th Annual Forum & Technology Display, May 11–13, 2021.

M. Podlaski, A. Khare, L. Vanfretti, M. Sumption and P. Ansell, "Multi-Domain Modeling for High Temperature Superconducting Components for the CHEETA Hybrid Propulsion System," 2021 AIAA/IEEE Electric Aircraft Technologies Symposium, 11–13 August 2021, Denver, Colorado.

E. Segerstrom, M. Podlaski, A. Khare and L. Vanfretti, "Parameter Estimation and Model Validation of Quanser AERO using Modelica and RaPId," 2021 AIAA Propulsion & Energy Forum, August 9–11, 2021, Denver, Colorado.

C. Canham, M. Podlaski and L. Vanfretti, "Guidance, Navigation, and Control enabling Retrograde Landing of a First Stage Rocket," 14th International Modelica Conference, Linköping, September 20–24, 2021.

M. Podlaski, L. Vanfretti, J. Monteneiri, A. Khare, J. Lewin and E. Segall, "Towards VR-based Early Design Interaction for Electric Vertical Take-off & Landing (eVTOL) Cyber-Physical Models," NAFEMS World Congress 2021, Salzburg, Austria, October 25–29, 2021.

M. Podlaski, L. Vanfretti, R. Niemiec, F. Gandhi, "Extending a Multicopter Analysis Tool using Modelica and FMI for Integrated eVTOL Aerodynamic and Electrical Drivetrain Design," American Modelica Conference 2022, October 26–28, 2022, Dallas, Texas, US.

References

Abhijit Khare, Meaghan Podlaski, Wolfgang Stautner, Joshua Feldman, Luigi Vanfretti, Kiruba Haran, Phillip Ansell, "Design and Analysis of Cryogenic Cooling System for Superconducting Motor," 2021 joint 23rd Cryogenic Engineering Conference and International Cryogenic Materials Conference, July 19–23, 2021, virtua

Meaghan Podlaski, Luigi Vanfretti, Abhijit Khare, Mike Sumption, Phillip Ansell, "Electro-Thermal Modeling of HTS Power Lines for Cryogenically-Cooled Electric Aircraft Design," 2021 joint 23rd Cryogenic Engineering Conference and International Cryogenic Materials Conference, July 19–23, 2021, virtual conference.

M. Podlaski, L. Vanfretti, A. Khare, M. Sumption, and P. Ansell, "Electro-Thermal Modeling of HTS Power Lines for Cryogenically-Cooled Electric Aircraft Design," IEEE CSC & ESAS SUPERCONDUCTIVITY NEWS FORUM (global edition), No. 50, October 2021.

M. Podlaski, L. Vanfretti, A. Khare, M. Sumption, and P. Ansell, "Multi-Domain Modeling and Simulation of High Temperature Superconducting Transmission Lines under Short Circuit Fault Conditions," IEEE Transactions on Transportation Electrification, vol. 8, no. 3, pp. 3859–3869, Sept. 2022, https://doi.org/10.1109/TTE.2021.3131271.

M. Podlaski, A. Khare, L. Vanfretti, "Electro-Thermal Modeling of Fuel Cells and Batteries for CHEETA Aircraft," AIAA Paper 2023–3990, 2023.

Y. Wei, M. Hossain, D. Woldegiorgis, X. Du and H. A. Mantooth, "Power Relay Based Multiple Device Cryogenic Characterization Method and Results," IEEE Open Journal of Industry Applications, vol. 3, pp. 211–223, 2022.

Y. Wei, M. M. Hossain and A. Mantooth, "Dynamic Characterizations of 650 V, 900 V and 1200 V SiC MOSFETs under Low Temperatures," 2022 IEEE Aerospace Conference (AERO), 2022, pp. 1–8.

Y. Wei, M. M. Hossain, A. Stratta and H. A. Mantooth, "Cryogenic Four-switch Buck-Boost Converter Design for All Electric Aircraft," 2022 IEEE Transportation Electrification Conference & Expo (ITEC), 2022, pp. 337–344.

Y. Wei, M. M. Hossain and H. A. Mantooth, "Static and Dynamic Cryogenic Characterizations of Commercial High Performance GaN HEMTs for More Electric Aircraft," 2022 International Power Electronics Conference (IPEC-Himeji 2022- ECCE Asia), 2022, pp. 2300–2306.

Y. Wei, M. M. Hossain and H. A. Mantooth, "Low Temperature Evaluation of Silicon Carbide (SiC) based Converter," 2022 IEEE Applied Power Electronics Conference and Exposition (APEC), 2022, pp. 230–236.

Y. Wei, M. M. Hossain and H. A. Mantooth, "Overcurrent Test for GaN HEMT with Cryogenic Cooling", 2022 IEEE Transportation Electrification Conference and Expo, Asia-Pacific (ITEC Asia-Pacific 2022). (Accepted)

Y. Wei, M. M. Hossain, R. Sweeting and A. Mantooth, "Functionality and Performance Evaluation of Gate Drivers under Cryogenic Temperature," 2021 IEEE Aerospace Conference (50100), 2021, pp. 1–9, https://doi.org/10.1109/AERO50100.2021.9438136.

M. M. Hossain, A. U. Rashid, Y. Wei, R. Sweeting and H. A. Mantooth, "Cryogenic Characterization and Modeling of Silicon IGBT for Hybrid Aircraft Application," 2021 IEEE Aerospace Conference (50100), 2021, pp. 1–8, https://doi.org/10.1109/AERO50100.2021.9438422.

M. M. Hossain, Y. Wei, A. Ur Rashid, R. Sweeting and H. A. Mantooth, "Cryogenic Evaluation and Modeling of a 900 V Cascode GaN HEMT," 2021 IEEE 12th Energy Conversion Congress & Exposition – Asia (ECCE-Asia), 2021, pp. 7–12, https://doi.org/10.1109/ECCE-Asia49820.2021.9479307.

Y. Wei, M. M. Hossain, R. Sweeting and H. A. Mantooth, "Cryogenic Temperature Characterizations of State-of-the-art Cascode 900 V GaN FET," 2021 AIAA/IEEE Electric Aircraft Technologies Symposium (EATS), 2021, pp. 1–11, https://doi.org/10.23919/EATS52162.2021.9703452.

M. M. Hossain, A. Ur Rashid, Y. Wei, R. Sweeting and H. A. Mantooth, "Static Characterization and Modeling of GaN HEMT at Cryogenic Temperature in Saber," 2021 AIAA/IEEE Electric Aircraft Technologies Symposium (EATS), 2021, pp. 1–10, https://doi.org/10.23919/EATS52162.2021.9704829.

Y. Wei, M. M. Hossain and A. Mantooth, "Evaluation of the High Performance 650 V Cascode GaN FET under Low Temperature," 2021 IEEE eighth Workshop on Wide Bandgap

Power Devices and Applications (WiPDA), 2021, pp. 236–241, https://doi.org/10.1109/WiPDA49284.2021.9645109.

Y. Wei, M. M. Hossain and A. Mantooth, "Comprehensive Cryogenic Characterizations of a Commercial 650 V GaN HEMT," 2021 IEEE International Future Energy Electronics Conference (IFEEC), 2021, pp. 1–6, https://doi.org/10.1109/IFEEC53238.2021.9661762.

Md Maksudul Hossain et al, "Electrical characterization of a 1200 V GaN HEMT at cryogenic temperatures," 2022 IOP Conf. Ser.: Mater. Sci. Eng. 1241 012041

M. M. Hossain, Y. Wei, and H. A. Mantooth, "LTspice Modeling for GaN-GIT HEMT Including Cryogenic Temperature," 2022 IEEE Aerospace Conference.(Accepted)

Y. Wei, M. M. Hossain, and H. A. Mantooth, "GaN HEMT and Air Core Magnetics Based Power Converters Evaluations at Cryogenic Temperature," 2022 IEEE Aerospace Conference. (Accepted)

Y. Wei, M. M. Hossain, and H. A. Mantooth, "Evaluation and Modeling of SiC based Power Converter for Low Temperature Operation," IEEE Transactions on Industry Applications, 2023.

Ranjan, P., Zheng, W., and James, K.A., "Mission-Adaptive Lifting System Design using Integrated Multidisciplinary Topology Optimization," AIAA Paper 2020–3143, AIAA Aviation Forum, 2020. https://arc.aiaa.org/doi/abs/10.2514/6.2020-3143

Ranjan, P., Das, G.K. and James, K.A., 2021, August. A Large-Scale Aero-Structural Optimization Framework for Electric Aircraft Design. In 2021 AIAA/IEEE Electric Aircraft Technologies Symposium (EATS) (pp. 1–29). IEEE.

Das, G.K., Ranjan, P. and James, K.A., 2022. 3D Topology Optimization of Aircraft Wings with Conventional and Non-conventional Layouts: A Comparative Study. In AIAA AVIATION 2022 Forum (p. 3725).

Mustafeez-ul-Hassan, Z. Yuan, H. Peng, A. I. Emon, Y. Chen and F. Luo, "Model Based Optimization of Propulsion Inverter for More-Electric Aircraft Applications Using Double Fourier Integral Analysis," 2020 AIAA/IEEE Electric Aircraft Technologies Symposium (EATS), 2020, pp. 1–7.

M. ul-Hassan, A. I. Emon, H. Peng, C. Kushan and F. Luo, "Investigation on Conducted EMI for Single and Parallel Connected Inverters," 2021 IEEE 12th International Symposium on Power Electronics for Distributed Generation Systems (PEDG), 2021, pp. 1–6.

M ul-Hassan, Z. Yuan, A. I. Emon and F. Luo, "A Framework for High Density Converter Electrical-Thermal-Mechanical Co-design and Co-optimization for MEA Application," 2021 IEEE Energy Conversion Congress and Exposition (ECCE), 2021, pp. 3120–3125.

Ul Hassan, M., Emon, A. I., Azadeh, Y., & Luo, F. (2022). Comparative evaluation of different DC-AC converter topologies for cryogenic applications utilizing superconducting materials. IOP Conference Series: Materials Science and Engineering, 1241(1), 012043.

Luo, F., Emon, A. I., Azadeh, Y., & Ul Hassan, M. (2022). Review of Power Electronics Converters and associated components/systems at cryogenic temperatures. International Journal of Powertrains, 11(2/3), 1.

M ul-Hassan, Y. Wu, V. Solovyov and F. Luo, "Liquid Nitrogen Immersed and Noise Tolerant Gate Driver for Cryogenically Cooled Power Electronics Applications," 2022 IEEE Applied Power Electronics Conference and Exposition (APEC), 2022, pp. 555–56.

Mustafeez-ul-Hassan, A. I. Emon, Z. Yuan, H. Peng and F. Luo, "Performance Comparison and Modelling of Instantaneous Current Sharing Amongst GaN HEMT Switch Configurations for Current Source Inverters," 2022 IEEE Applied Power Electronics Conference and Exposition (APEC), 2022, pp. 2014–2020.

Mustafeez-ul-Hassan, A. I. Emon, F. Luo and V. Solovyov, "Design and Validation of a 20 kVA, Fully Cryogenic, 2-Level GaN Based Current Source Inverter for Full Electric Aircrafts," in IEEE Transactions on Transportation Electrification

F. Luo et al, "Investigation About Operation and Performance of Gate Drivers for Power Electronics Converters for Cryogenic Temperatures", Accepted in ECCE-Europe-2022

F. Luo et al "Isolated Cryogenic Auxiliary Power Supply (CAPS) for GaN Based Converters", Accepted in ECCE USA 2022

References

F. Luo et al "Influence of Layout Parasitics and its Optimization in Two-Level Gallium-Nitride Based Current Source Inverter", Accepted in ECCE USA 2022

F. Luo et al "Analysis of a Switching Event and its Impact on Gate Drive in Gallium-Nitride Based Bi-Directional Switches", Accepted in WIPDA 2022

F. Luo et al "Electrical-Thermal-Mechanical Considerations for the Design of a Compact Two-Level Current Source Inverter using GaN Devices", Submitted in APEC 2023

F. Luo et al "GaN Based Active Clamp Flyback (ACF) Auxiliary Power Supply for Cryogenic Power Electronics Conversion", Submitted in APEC 2023

W. Stautner, CHEETA: An All-Electric Aircraft Takes Cryogenics and Superconductivity on Board" in IEEE Electrification magazine: "Hydrogen-Electric Aircraft Technologies", Vol 10, no. 2, June 2022

W. Stautner, Liquid Hydrogen Cryogenics for Long and Medium Range All-Electric Aircraft – General Guidelines and Outlook, GE Research, internal summary report, July 2022

W. Stautner, Liquid Hydrogen Tank Design for Medium and Long Range All-electric Airplanes, Invited talk, CEC/ICMC 21, July 2021

David K. Hall, Edward M. Greitzer and Choon S. Tan, 2022, "Mitigation of BLI Circumferential Distortion Using Non-Axisymmetric Fan Exit Guide Vanes", Journal of Turbomachinery

Durgesh Chandel, John D. Reband, David K. Hall, Thanatheepan Balachandran, Jianqiao Xiao, Kiruba S. Haran and Edward M. Greitzer, 2022, "Fan and Motor Co-optimization for a Distributed Electric Aircraft Propulsion System", IEEE Transactions on Transport Electrification

Chandel, D., Reband, J., Hall, D.K., Balachandran, T., Xiao, J., Haran, K.S., and Greitzer, E.M., "Conceptual Design of Distributed Electrified Boundary Layer Ingesting Propulsors for the CHEETA Aircraft Concept," AIAA Paper 2021-3287, 2021.

T. Balachandran, N. J. Salk, D. Lee, M. D. Sumption and K. S. Haran, "Methods of Estimating AC Losses in Superconducting MgB2 Armature Windings With Spatial and Time Harmonics," in IEEE Transactions on Applied Superconductivity, vol. 32, no. 6, pp. 1–7, Sept. 2022, Art no. 4702407, https://doi.org/10.1109/TASC.2022.3181535.

S. Sirimanna, T. Balachandran, N. Salk, J. Xiao, D. Lee and K. Haran, "Electric Propulsors for Zero-Emission Aircraft: Partially superconducting machines," in IEEE Electrification Magazine, vol. 10, no. 2, pp. 43–56, June 2022, https://doi.org/10.1109/MELE.2022.3165952.

J. Xiao, T. Balachandran, A. J. Samarakoon and K. S. Haran, "A Spoke-Supported Superconducting Rotor With Rotating Cryocooler," in IEEE Transactions on Magnetics, vol. 58, no. 8, pp. 1–5, Aug. 2022, Art no. 9000405, https://doi.org/10.1109/TMAG.2022.3150786.

T. Balachandran, J. David Reband, J. Xiao, S. Sirimmana, R. Dhilon and K. S. Haran, "Co-design of an Integrated Direct-drive Electric Motor and Ducted Propeller for Aircraft Propulsion," 2020 AIAA/IEEE Electric Aircraft Technologies Symposium (EATS), 2020, pp. 1–11.

Balachandran, T., Lee, D., Salk, N., and Haran, K. S., "A fully superconducting air-core machine for aircraft propulsion," In IOP Conference Series: Materials Science and Engineering, Vol. 756, No. 1, March 2020, p. 012030. https://iopscience.iop.org/article/10.1088/1757-899X/756/1/012030

T. Balachandran, Y. Zhao, S. Sirimanna, J. Xiao, and K.S. Haran, "Designing and Commissioning an Experimental Setup to Evaluate AC Losses in Superconductors Under Transverse Rotating Fields," IEEE Transactions on Applied Superconductivity, Vol. 33, No. 5, 2023.

T. Balachandran and K.S. Haran, "Instantaneous Loss Integration Method to Estimate AC Losses in Superconductors with Spatial and Time Harmonics," IEEE Transactions on Applied Superconductivity, 2023.

J. Xiao, Y. Zhao, R. Dutta, and K. Haran, "Rotating Cryocooler Performance for Superconducting Rotor," 2023 IEEE Power and Energy Conference at Illinois (PECI).

J. Feldman, J. Xiao, T. Balachandran, and K. Haran, "Design of a Fully Superconducting Aircraft Propulsion Motor," AIAA Paper 2023-4474, 2023.

Sebastian, M.P., Haugan, T.J., and Kovacs, C.J., "Design and Scaling Laws of a 40-MW-class Electric Power Distribution System for Liquid-H2 Fuel-Cell Propulsion," AIAA Paper 2021–3310, 2021.

Kovacs, C.J., Haugan, T.J., Sumption, M.D., Tomsic, M., and Rindfleisch, M., "Current sharing and stability in an extremely low AC-loss MgB2 conductor," IOP Conference Series: Materials Science and Engineering, Vol. 1241, 2022.

Kovacs, C.J., Haugan, T.J., and Sumption, M.D., "Metal composite T-junction terminals for MW-class aerospace electric power distribution," Journal of Physics: Conference Series, Vol. 1975, 2021.

Sumption, M., "AC Loss of Superconducting Strands at High frequencies, and the impact for Harmonic Contributions," AIAA Paper 2023–4473, 2023.

Appendices

A1: Material Data and Selection

Abstract This Appendix gives an overview on material selection for tanks, cryogenic components, like heat exchangers or liners, for composite tanks.

Since a hydrogen-powered aircraft will not "fly" with heavy base material, the focus here is on material embrittlement and material selection, with supply of rare data made available for engineers and researchers.

Keywords Hydrogen environment embrittlement; HEE tables; Material selection; Embrittlement tables; Cryogenic aircraft alloys; Liner alloys

Cryogenic Material Properties Database with Emphasis on Aircraft Aluminum Alloys

Material Selection, Hydrogen Environment Embrittlement (HEE) Tables [1].

Table A1.1 HEE Indexes for selected metals tested at 24 °C under high hydrogen pressure for selected steels.

Table A1.2 HEE (ctd.)

Further materials, like superalloys, please see [1] (Tables A1.3, A1.4, A1.5, A1.6, and A1.7).

For further details on material compatibility with hydrogen, please see also Sandia Report SAND2012-7321 of 2012 [4].

Table A1.1 HEE Indexes for selected metals tested at 24 °C under high hydrogen pressure

Alloy system	Material (HEE Tested at 24 °C)	H pressure (MPa)	Qualitative rating for HEE	HEE index, (Ratio H/He) NTS	EL	RA	Smooth ductility (%), in helium or air EL	RA	Tensile strengths, in helium or air (MPa) NTS	YS	UTS
Nickel based	Nickel (electroformed)	68.9	Extreme	0.31		0.75		89	827		331
	Nickel 270	68.9	High	0.70	0.92		56		531	34	792
	Nickel 301	68.9	Extreme		0.35		34			482	
	K-Monel (precipitated)	68.9	Extreme	0.45					1729		958
	K-Monel (annealed)	68.9	High	0.73					992		689
Titanium based	Titanium (pure)	68.9	Small	0.95	0.96	1.00	32	61	868	365	434
	Ti-6Al-4V (annealed)	34.5	High	0.89	0.90	0.82	15	44	1426	1006	1040
	Ti-6Al-4V (annealed)	68.9	High	0.79	1.00	1.00	15	48	1674	909	1075
	Ti-6Al-4V (STA)	48.2	Severe	0.69							
	Ti-6Al-4V (STA)	68.9	Severe	0.58	0.85	0.95	13	48	1571	1082	1130
	Ti-5Al-2.5Sn (ELI)	68.9	High	0.81	0.90	0.86	20	45	1385	730	779
	Ti-11.5Mo-6Zr-4.5Sn (STA)	34.5	Extreme		0.18	0.20	22	63.4		551	785
	Alpha-2 TiAl alloy	13.8	Extreme		0.54	0.38	4.1	4		978	1068
	Gamma-TiAl alloy	13.8	Extreme		0.39		0.85			537	537
Copper based	Copper (OFHC)	68.9	Negligible	1.00	1.00	1.00	63	94	599	269	289
	Aluminum bronze	68.9	Negligible		1.02	1.05	48	67		220	599
	Be-Cu alloy 25	68.9	Small	0.93	1.00	0.98	22	72	1344	544	648
	GRCop-84 (Cu-8Cr-4Nb)	34.5	Negligible	1.03		1.20	20	42		241	413
	NARloy-Z (Cu-3Ag-0.5Zr)	40.0	Negligible	1.10		0.92		24		138	269
	70–30 Brass	68.9	Negligible		0.98	1.20	59	70		124	365

Aluminum based	1100-T0	68.9	Negligible	1.38	0.93	1.00	42	93	124	34	110
	2011	68.9	Negligible		1.01	0.94	57	18		227	296
	2024	68.9	Negligible		0.95	0.97	19	36		324	441
	5086	68.9	Negligible		1.05	1.03	20	55		193	303
	6061-T6	68.9	Negligible	1.07	1.00	1.08	19	61	496	227	269
	6063	68.9	Negligible		1.00	1.01	15	83		158	193
	7039	68.9	Negligible		1.00	1.01	14	85		152	179
	7075-T73	68.9	Negligible	0.98	0.80	0.94	15	37	799	372	455

NTS notched tensile strength, *EL* plastic elongation, *RA* reduction of area, *YS* yield strength, *UTS* ultimate tensile strength

Table A1.2 HEE Indexes for selected steels tested at 24 °C under high hydrogen pressure

Alloy system	Material (HEE Tested at 24 °C)	H pressure (MPa)	Qualitative rating for HEE	HEE index, (Ratio H/He) NTS	EL	RA	Smooth ductility (%), in helium or air EL	RA	Tensile strengths, in helium or air (MPa) NTS	YS	UTS
Austenitic steels	CG-27 (precip. Hardened)	68.9	Extreme		0.34		29			806	1164
	Tenelon	68.9	High		0.85		65			496	875
	A302B	68.9	High	0.79	0.85	0.50			1564		827
	A286 (sol treat @1640 °F)	68.9	Negligible	0.97	1.10	0.98	26	44	1605	847	1089
	216	68.9	Negligible		0.99		45			586	785
	304 L	68.9	High	0.87	0.92	0.91	86	78	703	165	531
	304 N	68.9	High	0.93	0.84		43			627	847
	304 LN	68.9	High		1.00	0.75	62	72		379	765
	305	68.9	High	0.89	1.03	0.96	63	78	1137	351	620
	308 L (304 L weld wire)	68.9	High	0.86	0.83	0.60	53	71	813	358	586
	309S	68.9	Small		0.96	0.97	85	76		241	558
	310	68.9	Small	0.93	1.00	0.96	56	64	799	220	531
	316	68.9	Negligible	1.00	0.95	1.04	59	72	1109	441	648
	321	34.5	High	0.88	0.83	0.90	77	66	779	200	579
	347	34.5	Small	0.92	1.00	1.00	38	70	1178	455	689
	18-2-12 (Nitronic 32)	68.9	Severe		0.64	0.47	75	78		482	861
	21-6-9 + 0.1 N (Nitronic 40)	68.9	High		0.89	0.80	65	74		434	744
	21-6-9 + 0.3 N (Nitronic 40)	68.9	High			0.85	56	78		462	799
	22-13-5 (Nitronic 50)	68.9	Negligible		1.00	1.00	51	67		586	937
	18-18 plus	68.9	Severe		0.67		63			517	909
	18-2-Mn	68.9	Severe		0.64		51			730	1006
	18-3-Mn	68.9	Small		0.92		50			531	785

Table A1.3 Susceptibility table of materials to embrittlement in hydrogen at 10,000 psi and 72 °F[a] [2]

Material	Strength ratio, H_2/He Notched	Strength ratio, H_2/He Unnotched	Unnotched ductility Elongation, % He	Unnotched ductility Elongation, % H_2	Unnotched ductility Reduction of area, % He	Unnotched ductility Reduction of area, % H_2
Extremely embrittled						
18Ni-250 maraging steel	0.12	0.68	8.2	0.2	55	2.5
410 stainless steel	0.22	0.70	15	1.3	60	12
I 042 steel (quenched and tempered)	0.22	–	–	–	–	–
I 7–7 pH stainless steel	0.23	0.92	17	I. 7	45	2.5
Fe-9Ni-4Co-0.20C	0.24	0.86	15	0.5	67	15
H II	0.25	0.57	8.8	0	30	0
Rene 41	0.27	0.84	21	4.3	29	II
Electro-formed nickel	0.31	–	–	–	–	–
4140	0.40	0.96	14	2.6	48	9
Inconel 718	0.46	0.93	17	1.5	26	I
440C	0.50	0.40			3.2	0
Severely embrittled						
Ti-6Al-4 V (STA)	0.58	–	–	–	–	–
430F	0.68	–	22	14	64	37
Nickel 270	0.70	–	56	52	89	67
A515	0.73	–	42	29	67	35
HY-100	0.73	–	20	18	76	63
A372 (class IV)	0.74	–	20	10	53	18
1042 (normalized)	0.75	–	–	–	59	27
A533-B	0.78	–	–	–	66	33
Ti-6Al-4 V (annealed)	0.79	–	–	–	–	–
AISI 1020	0.79	–	–	–	68	45
HY-50	0.80	–	–	–	70	60
Ti-5Al-2.5Sn (ELI)	0.81	–	–	–	45	39
Armco iron	0.86	–	–	–	83	50
Slightly embrittled						
304 ELC stainless steel	0.87	–	–	–	78	71
305 stainless steel	0.89	–	–	–	78	75
Be-cu alloy 25	0.93	–	–	–	72	71
Titanium	0.95	–	–	–	61	61
Negligibly embrittled						
310 stainless steel	0.93	–	–	–	64	62
A286	0.97	–	–	–	44	43
7075-T73 aluminum alloy	0.98	–	–	–	37	35
316 stainless steel	1.00	–	–	–	72	75
OFHC copper	1.00	–	–	–	94	94
NARloy-Z	1.10	–	–	–	24	22
6061-T6 aluminum	1.10	–	–	–	61	66
Alloy 1100 aluminum	1.40	–	–	–	93	93

[a]Chandler et al. [3]

Table A1.4 Ultimate tensile and yield strength of typical cryogenic aircraft alloys

Aluminum alloy type	Temperature (K)	Ultimate tensile strength (UTS) MPa	Yield strength MPa	Length change at UTS in %
7079 T6	300	521.89	460.09	9.0
	200	549.36	470.39	9.0
	77.4	642.06	576.83	4.0
	20	693.57	645.50	3.0
2119 T6	300	409.67	293.91	9.0
	200	430.56	307.35	9.5
	77.4	519.15	361.89	12.2
	20	604.98	294.59	16.5
7178 T6	300	644.12	602.24	7.5
	200	659.92	640.69	4.0
	77.4	748.50	716.23	1.2
	20	806.19	778.72	1.0
5032 H32	300	409.27	293.91	9.0
	200	430.66	307.35	9.5
	77.4	519.15	361.99	12.2
	20	584.38	294.59	16.5
6061 T6	300	296.75	254.08	10.0
	200	326.18	268.50	
	77.4	398.97	296.75	
	20	461.46	337.15	
2024 T4	300	446.36	412.02	18.0
	200	460.09	432.62	
	77.4	537.10	501.98	
	20	653.05	563.78	
7039 T6	300	371.50	326.87	
	200	394.85	343.35	
	77.4	446.36	384.55	
	20	508.16	412.02	
2014 T6	300	480.69	436.55	9.7
	200	504.72	459.11	9.5
	77.4	576.83	521.89	11.7
	20	662.67	545.93	13.6
2020 T6	300	543.87	515.71	8.0
	200	561.72	522.58	6.3
	77.4	651.68	603.61	4.0
	20	696.31	637.94	2.3
2219 T6	300	449.10	358.46	9.8
	200	477.26	379.16	9.3
	77.4	565.15	433.60	12.1
	20	661.98	543.96	15.3
5456 H343	300	394.85	309.02	8.7
	200	391.42	309.02	9.3
	77.4	508.16	363.95	13.0

(continued)

Table A1.4 (continued)

Aluminum alloy type	Temperature (K)	Ultimate tensile strength (UTS) MPa	Yield strength MPa	Length change at UTS in %
	20	597.43	393.87	8.7
7075 T6	300	545.93	506.69	9.2
	200	572.71	530.82	8.7
	77.4	644.12	603.61	5.2
	20	693.57	650.89	3.2
5086 H343	300	322.75	254.08	10.4
	200	329.62	260.95	12.0
	77.4	442.92	302.15	25.0
	20	583.70	326.18	20.2
7002 T6	300	459.40	393.38	16.7
	200	486.87	420.85	18.0
	77.4	573.69	478.73	19.80
	20	711.91	530.72	18.90

Redrawn and modified after Conte [6]

Table A1.5 Mechanical aluminum property data for Alloy 2219-T87 [8]

Alloy designation	2219-T87				
Specification	MIL-A-8920 A, ASTM B209				
Form	Sheet				
Thickness	0.1 to 0.319 cm				
Condition	T87				
Testing temperature, K		**297**	**195**	**144**	**77**	**20**
Tension, longitudinal						
TUS, MN/m² (MPa)	Avg	471	504	513	589	685
	Min	458	489	–	562	647
TYS, MN/m² (MPa)	Avg	388	410	427	474	501
	Min	372	379	–	421	469
Elong. %	Avg	9.5	9.8	12.5	11.5	14.8
	Min	6	6	–	7.0	8.5
E, GN/m²	Avg	73.8	73.77	–	80.7	85.5
	Min	70.3	69.6	–	73.8	75.2
Poisson's ratio		0.3	0.24	–	0.28	0.27
NTS, MN/m²	Avg	480	514	–	589	658
$K_t = 6.3$	Min	475	512	–	586	656
NTS, MN/m²	Avg	441	–	–	523	532
$K_t = 10$	Min	437	–	–	519	508
NTS, MN/m²	Avg	321	378	354	426	461
$K_t = 22$	Min	–	–	–	–	–

(continued)

Table A1.5 (continued)

Tension, transverse						
TUS, MN/m²	Avg	473	506	–	598	701
	Min	452	483	–	571	648
TYS, MN/m²	Avg	388	404	–	467	507
	Min	356	374	–	432	463
Elong. %	Avg	9.0	7.0	–	9.9	12.8
	Min	7.0	6.5	–	6.0	7.5
E, GN/m²	Avg	73.8	82.0	–	84.8	83.4
	Min	67.6	80.66	–	77.2	75.8
Work hardening coeff.						
Poisson's ratio		0.31	0.30	–	0.24	0.27
NTS, MN/m²	Avg	480	507	–	577	656
$K_t = 6.3$	Min	474	500	–	576	656
NTS, MN/m²	Avg	433	–	–	505	554
$K_t = 10$	Min	432	–	–	503	528
Testing temperature, K	297	77	20			
Fatigue, axial loading						
S_N at 10^5 cycles, MN/m²	207	221	358			
Ratio S_N/TUS at 10^5 cycles	0.45	0.38	0.53			
S_N at 10^6 cycles, MN/m²	152	117–172	276			
Ratio S_N/TUS at 10^6 cycles	0.32	0.22–0.29	0.4			
Fatigue, axial loading (notched specimens)						
S_N at 10^5 cycles, MN/m²	75.8	75.8	100			
S_N at 10^6 cycles, MN/m²	51.7	48.2	55.2			

Table A1.6 Aluminum property data for Alloy 2219-T81 [8]

Alloy designation	2219-T81
Specification	MIL-A-8920 A, ASTM B209
Form	Sheet
Thickness	0.1 to 0.319 cm
Condition	T81

Testing temperature, K		297	195	77	20
Tension, longitudinal					
TUS, MN/m²	Avg	447	481	566	667
	Min	438	460	553	600
TYS, MN/m²	Avg	348	374	423	476
	Min	339	362	413	439
Elong. %	Avg	8.8	9.1	10.7	13.4
	Min	4.3	6.5	6.8	7.0
E, GN/m²	Avg	68.5	72.4	80.0	82.0
	Min	64.1	66.9	73.1	76.5
Poisson's ratio		0.327	0.335	0.335	0.33
NTS, MN/m²	Avg	427	459	514	565
$K_t = 6.3$	Min	416	442	498	483
NTS, MN/m²	Avg	363	351	414	467
$K_t = 19$	Min	349	335	399	458

(continued)

Table A1.6 (continued)

Tension, transverse					
TUS, MN/m^2	Avg	450	488	563	673
	Min	436	478	555	625
TYS, MN/m^2	Avg	341	367	414	466
	Min	322	330	405	452
Elong. %	Avg	9.7	9.6	10.1	12.1
	Min	7.3	8.0	7.0	6.0
E. GN/m^2	Avg	69.0	73.8	79.3	81.4
	Min	66.2	70.3	73.8	74.5
Work hardening coeff.					
Poisson's ratio		0.325	0.325	0.335	0.34
NTS, MN/m^2	Avg	432	449	514	590
$K_t = 6.3$	Min	412	440	504	567
NTS, MN/m^2	Avg	363	322	392	423
$K_t = 19$	Min	353	318	354	392

Table A1.7 Cryogenic Fatigue Strength data for sheet and bar of 7075-T73 [10]

Alloy	7075-T73									
					Fatigue strength at cycles (kg/mm^2)			Ratio, fatigue/ultimate strength		
Form	Surface finish (rms)	Test (°F)	Temperature (°C)	Fatigue test type	10^5	10^6	10^7	10^5	10^6	10^7
0.050-inch sheet (1.27 mm)	16	70	21	Flexural	42 (29.5)	34 (23.9)	29 (20.4)	0.57	0.46	0.39
		−320	−196		54 (38.0)	49 (34.5)	45 (31.6)	0.57	0.52	0.47
		−423	−253		62 (43.6)	54 (38.0)	–	0.60	0.52	–
	140	70	21	Flexural	32 (22.5)	26 (18.3)	25 (17.6)	0.44	0.36	0.35
		−320	−196		43 (30.2)	38 (26.7)	36 (25.3)	0.47	0.41	0.39
		−423	−253		50 (35.2)	41 (28.8)	–	0.51	0.42	–
7/8-inch bar (2.89 mm)	32	70	21	Axial	52 (36.6)	44 (30.9)	42 (29.5)	0.69	0.59	0.56
		−320	−196		66 (46.4)	59 (41.5)	58 (40.8)	0.73	0.65	0.64
		−423	−253		78[a] (54.8)	66 (46.4)	–	0.74	0.62	–

[a]Maximum stress is equal or above yield strength temperature

Material Selection, Fracture Toughness

For selecting appropriate aircraft materials, a thorough knowledge of the mechanical as well as thermophysical properties is mandatory. This chapter presents an overview on cryogenic data rare to find and gives the engineer a first glimpse on what the design of a tank liner, a heat exchanger, cryogenic lines, or other components may look like. After any decision on selected materials, it is highly recommended to run material's testing to confirm those properties. Test houses down to 4 K are established at KIT (Germany) and at NIST (US) and at PNNL (US). Those houses also developed the appropriate cryogenic measuring techniques. Simple dunking a component into liquid nitrogen and testing it on a universal testing machine may give erroneous results. Fracture toughness and fatigue tests are highly recommended (Fig. A1.1).

Alloys 7075-T6, 7079-T6, 7118-T6, 7275-T6[7].

Yield and tensile strengths (ksi) and elongation (%) vs. temperature (F) for 7000 series aluminum alloy sheet [7] (Figs. A1.2 and A1.3).

With:

TUS: Tensile ultimate strength.
TYS: Tensile yield strength.
Elong.: Elongation.
E: Young's modulus.
NTS: Notched tensile strength.
Long.: Longitudinal orientation.
Trans.: Transverse orientation.
S_N: The greatest stress which can be sustained for a given number of cycles without fracture.
K_t: Stress concentration factor.
(corrected data)

Alloy 6061-T6 [9] (Figs. A1.4, A1.5, A1.6, and A1.7).
Alloy 7075-T6 [10]

Fig. A1.1 Fracture toughness of common cryogenic materials [5]

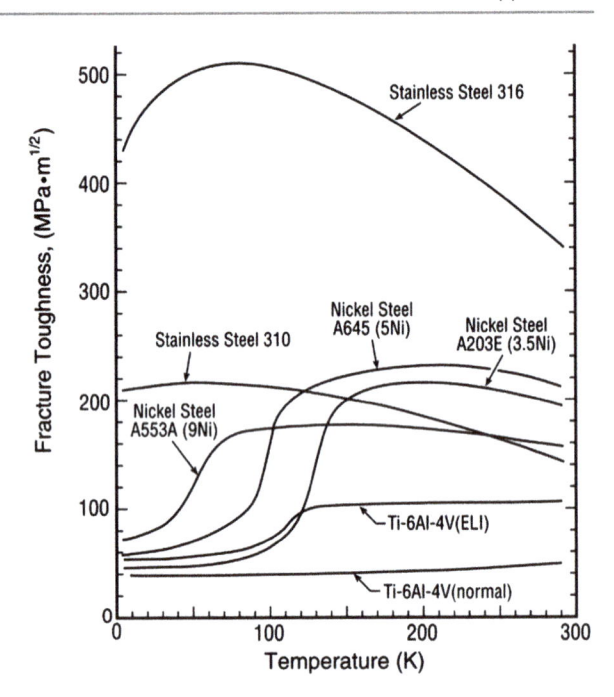

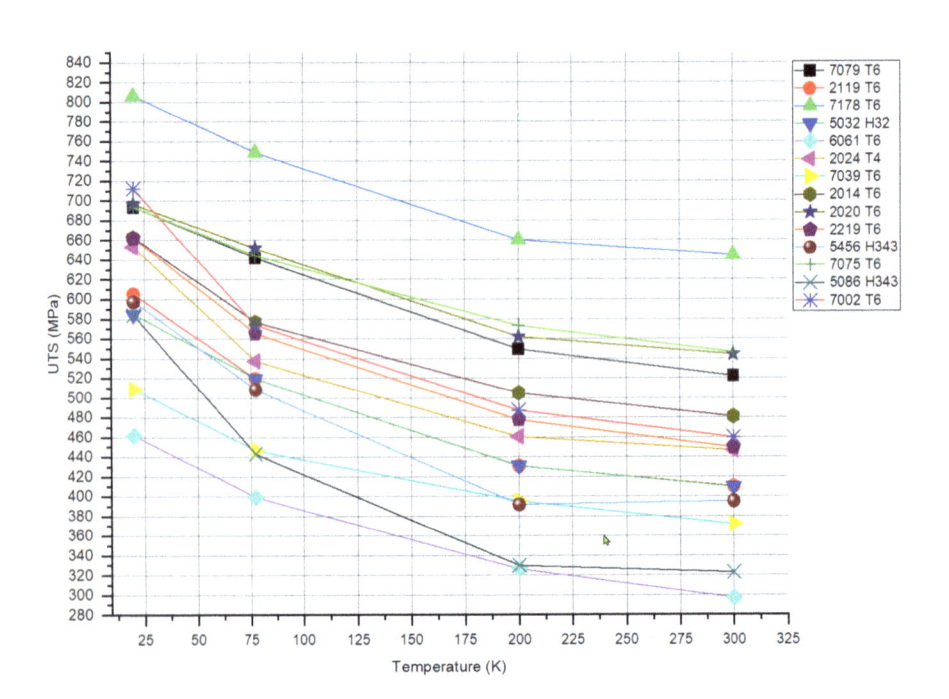

Fig. A1.2 UTS of aluminum aircraft alloys down to 20 K. (After Conte [6])

Appendices

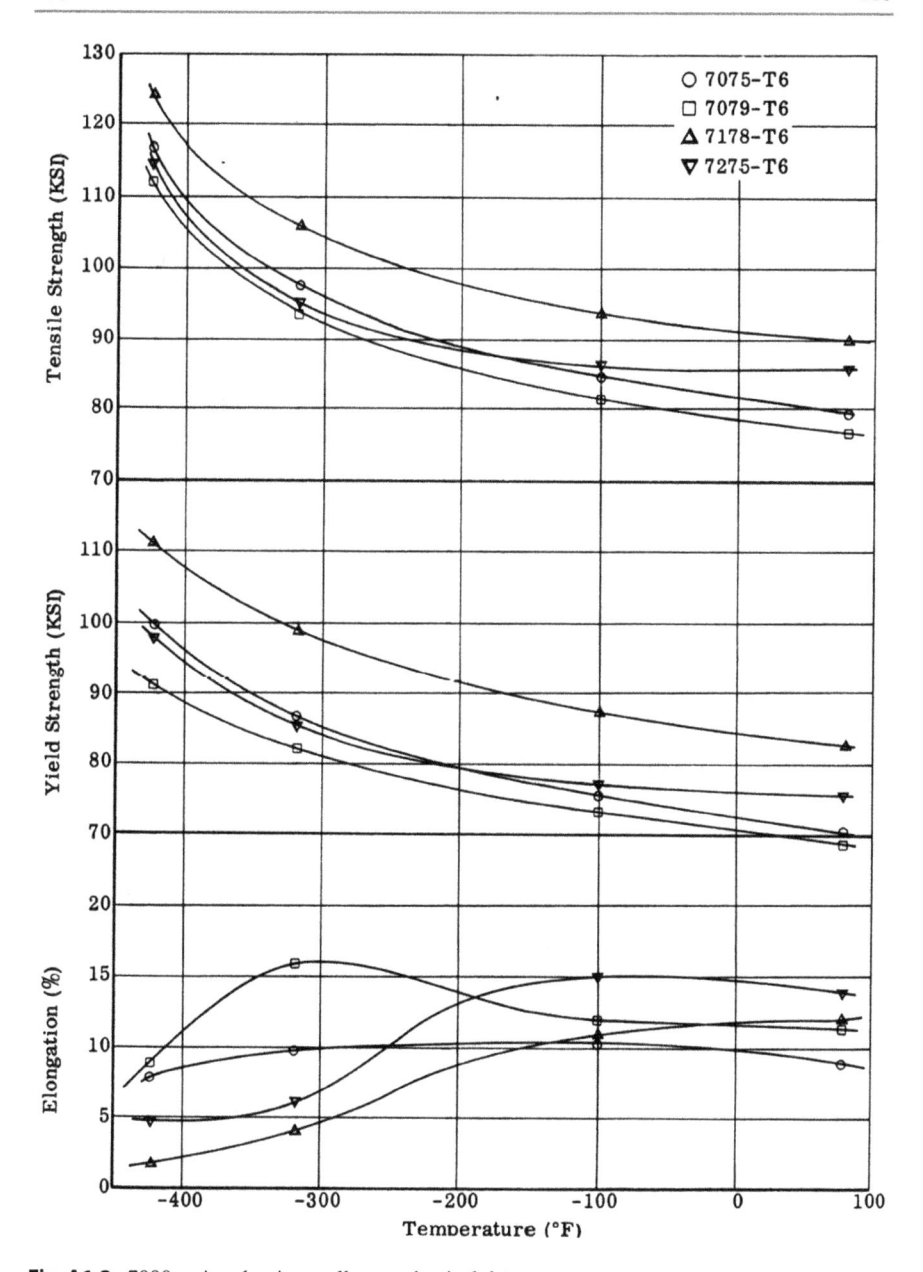

Fig. A1.3 7000 series aluminum alloy mechanical data

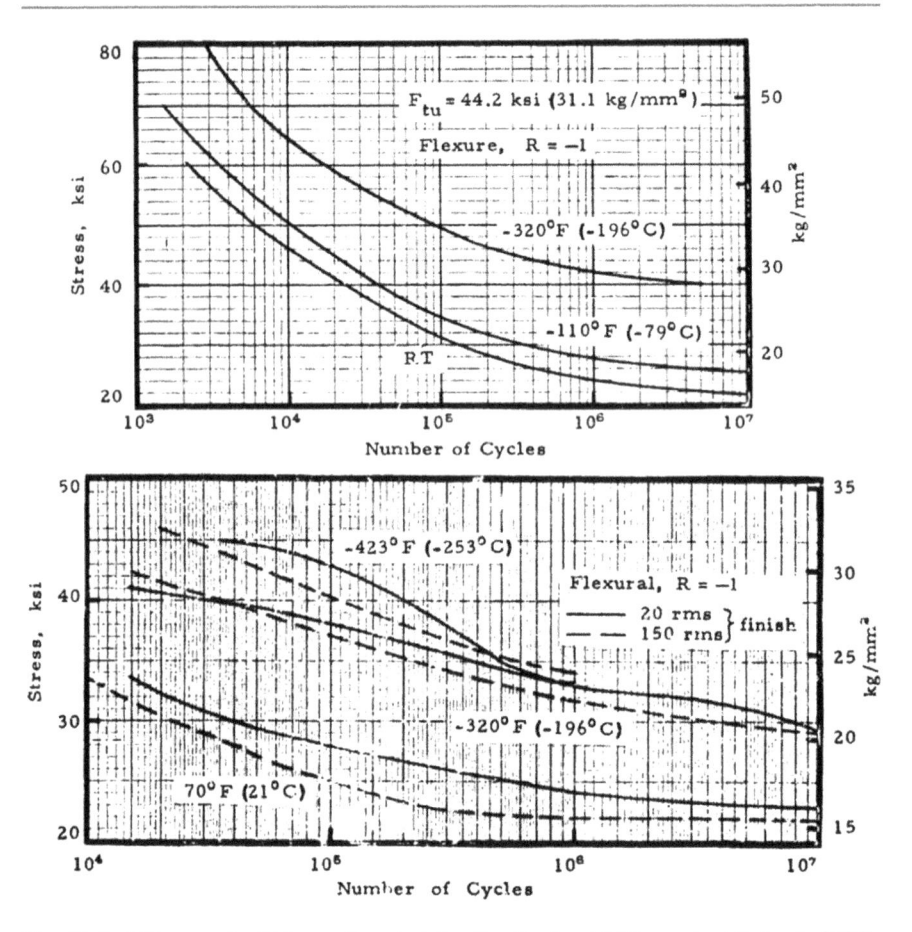

Fig. A1.4 S-N curves for 19.1 mm diameter tube alloy (top) and 1.27 mm sheet (below) of 6061-T6 at room and cryogenic temperatures [9]

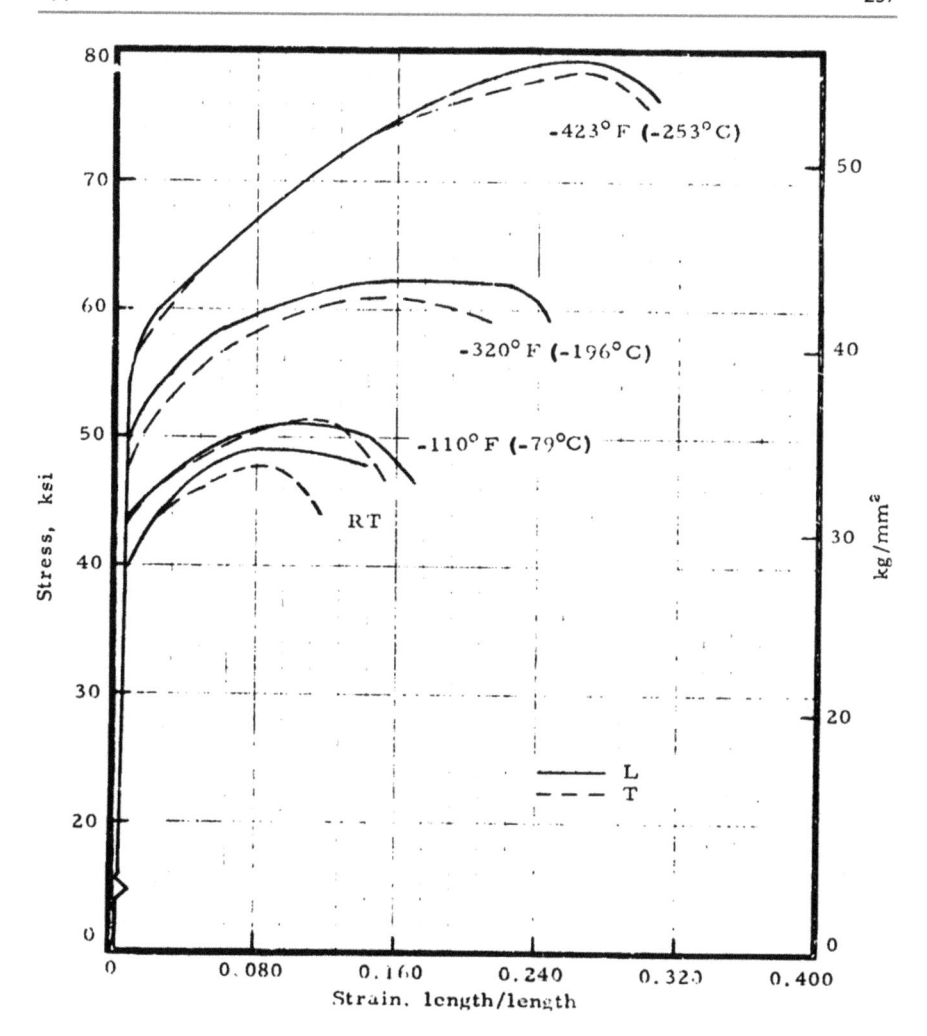

Fig. A1.5 Stress-strain diagram for 6061-T6 sheet (2.54 mm) at room and low temperatures [9]

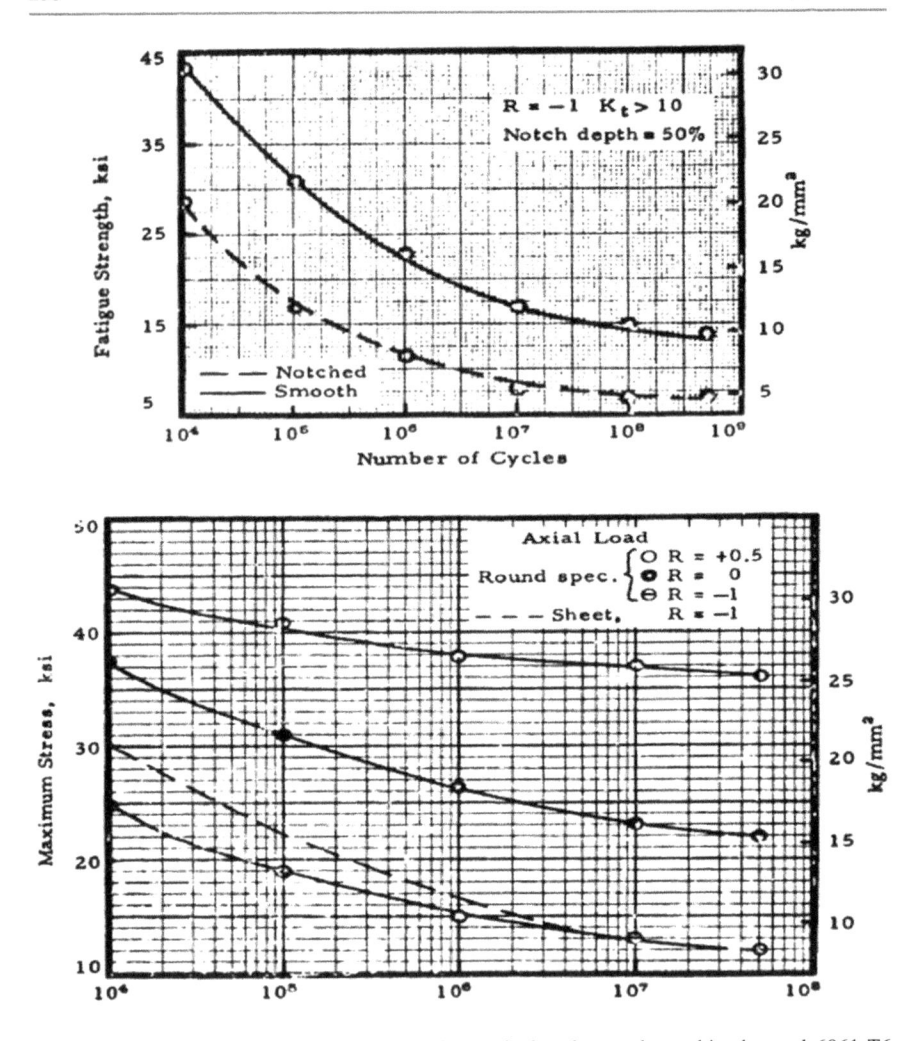

Fig. A1.6 (above) Fatigue strength of sharply notched and smooth machined round 6061-T6 specimens, (below) S-N curves at room temperature for 6061-T6 sheet (1.63 mm) and round specimen [9]

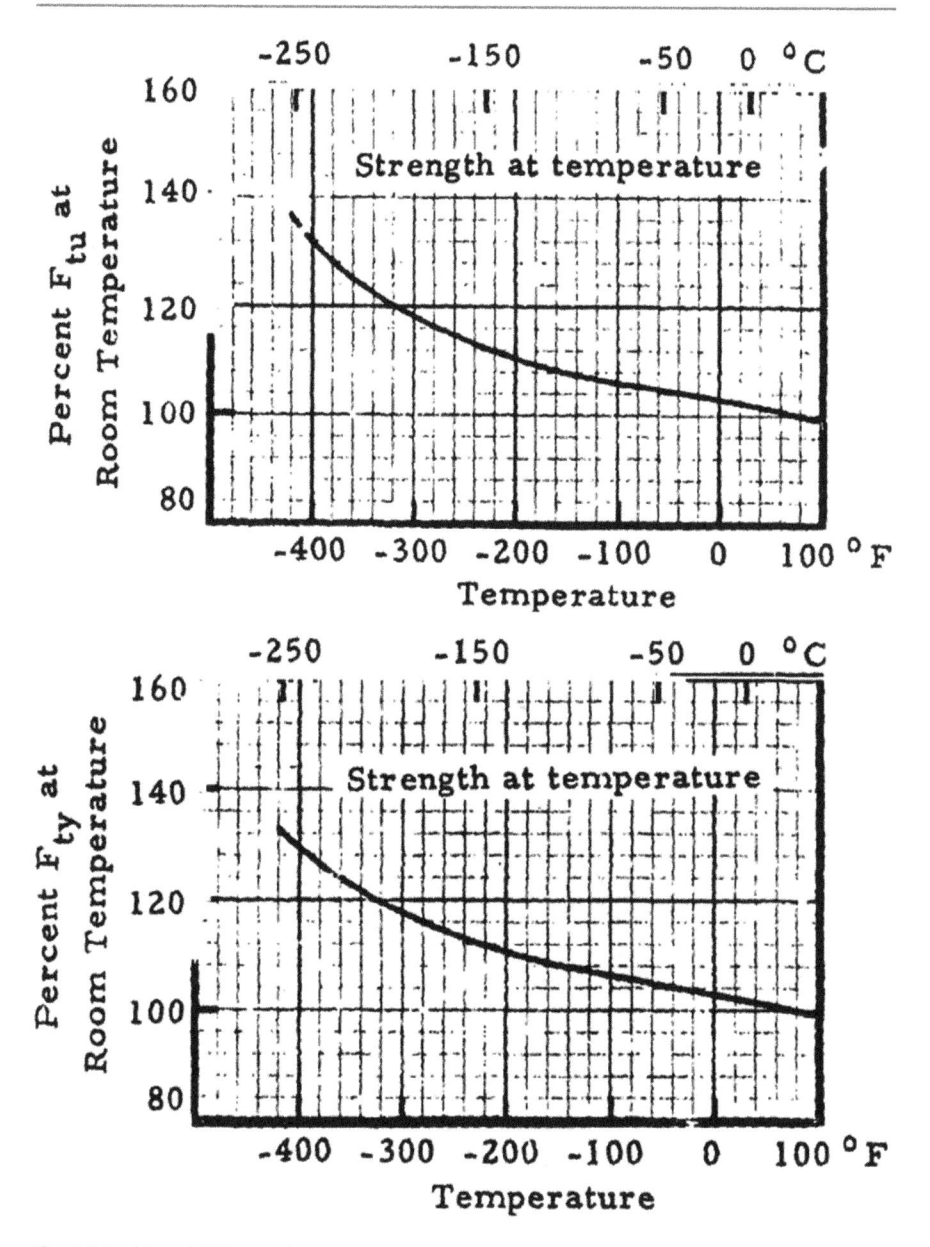

Fig. A1.7 (above) Effect of low temperature on the UTS of 7075-T6, (below) Effect of low temperature on the tensile strength of 7075-T6 [10]

A2: Physical and Engineering Properties of Hydrogen

Abstract This Appendix complements the material side with necessary hydrogen properties in all its states. Those rarely available data have been made available with approval by NBS/NIST. Most of the thermophysical data, including those for heat transfer, give the engineer the opportunity to arrive at novel solutions for accommodating liquid hydrogen in an aircraft.

Keywords Liquid hydrogen properties; Rare physical parameters; All-electric aircraft hydrogen properties

With appreciation to NASA and NBS/NIS T for permission to publish data (Figs. A2.1, A2.2, A2.3, A2.4, A2.5, A2.6, A2.7, A2.8, A2.9, A2.10, A2.11, A2.12, A2.13, A2.14, A2.15, A2.16, A2.17, A2.18, A2.19, A2.20, A2.21, A2.22, A2.23, A2.24, A2.25, A2.26, A2.27, A2.28, A2.29, A2.30, A2.31, A2.32, A2.33, A2.34, A2.35, A2.36, A2.37, A2.38, A2.39, A2.40, A2.41, A2.42, and A2.43, Tables A2.1 and A2.2).

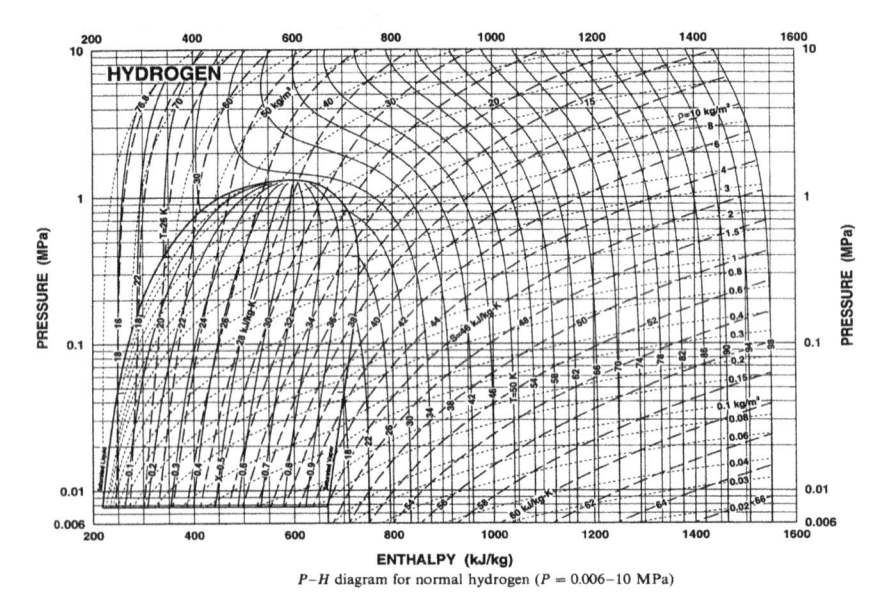

Fig. A2.1 P-H diagram for normal hydrogen up to 10 MPa [13]

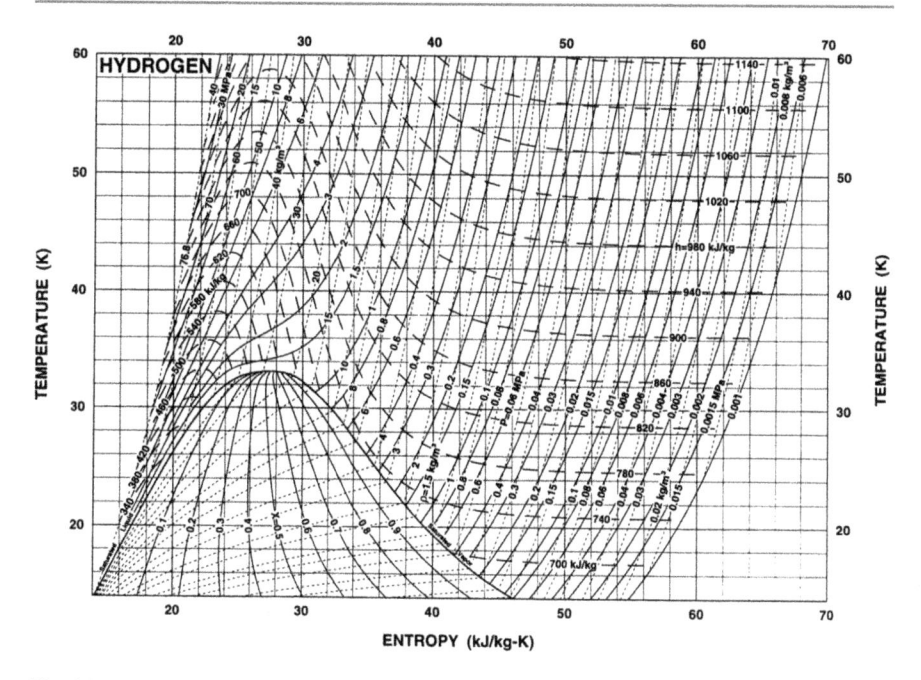

Fig. A2.2 T-S diagram for normal hydrogen [13]

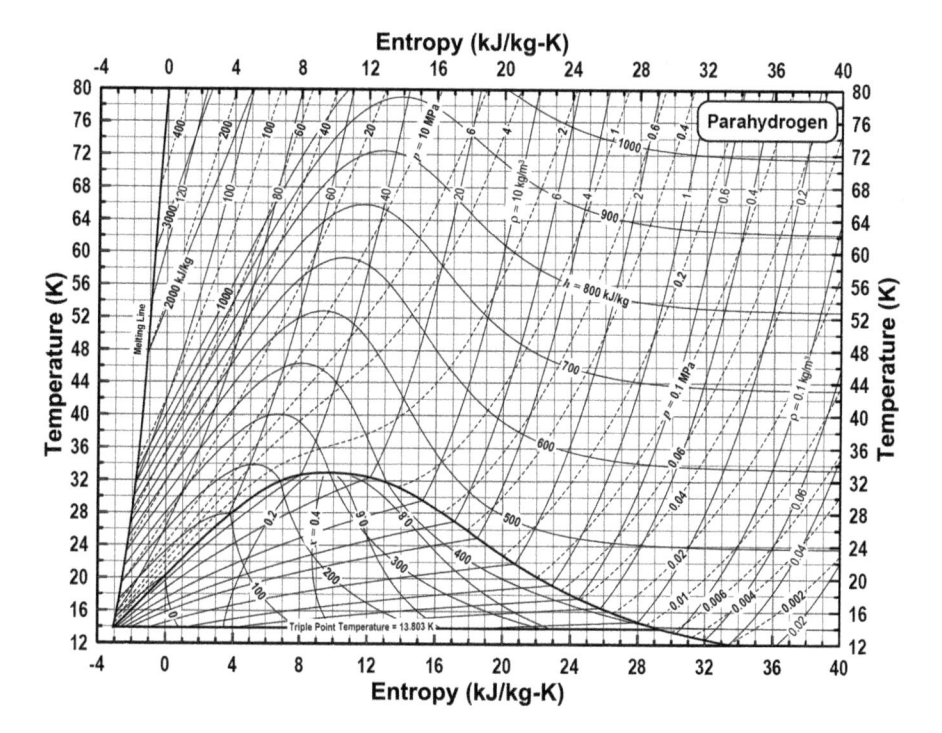

Fig. A2.3 T-S diagram for parahydrogen [14]

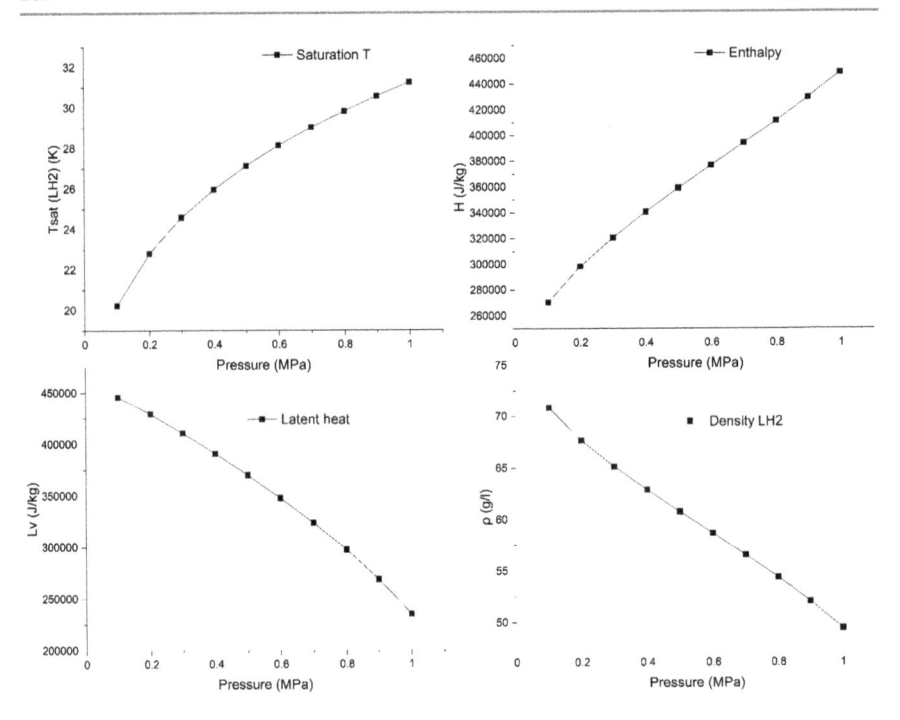

Fig. A2.4 T-S diagram for parahydrogen [15], created from Gaspak Ver. 3.35

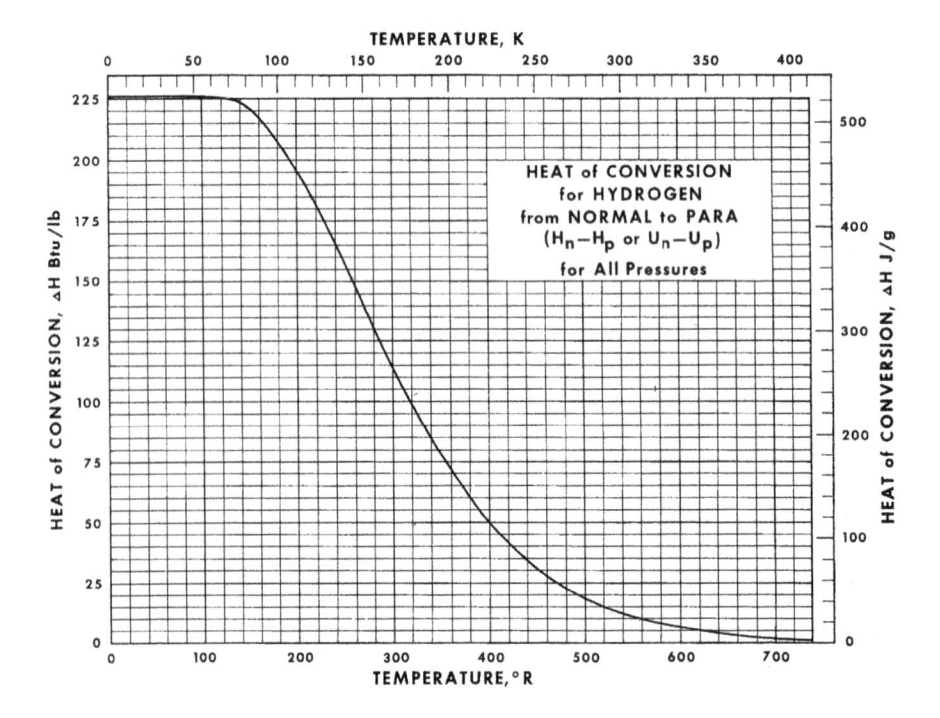

Fig. A2.5 Normal to para heat of conversion diagram [16]

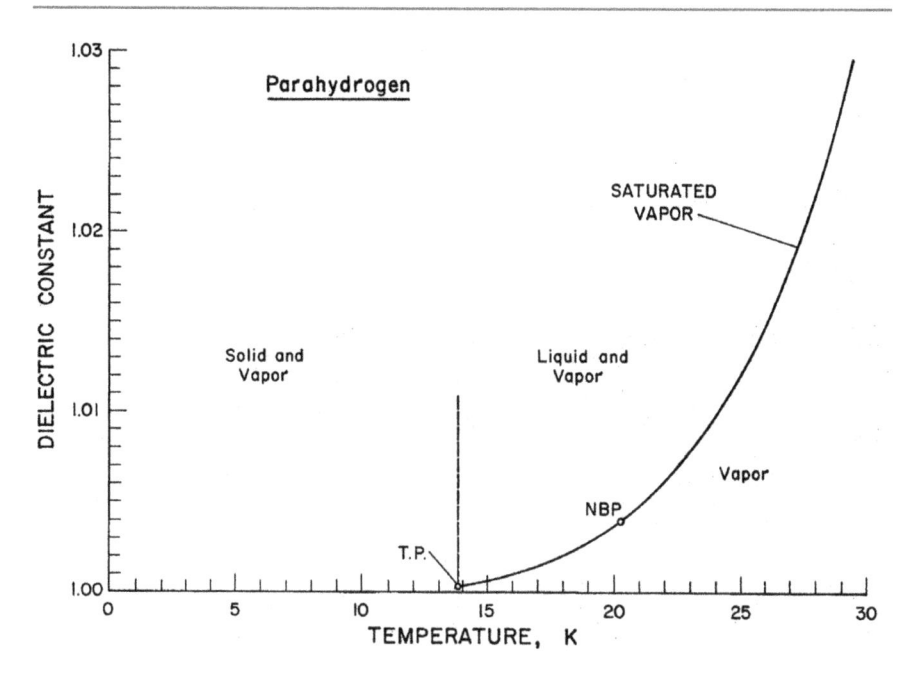

Fig. A2.6 Dielectric constant of saturated para hydrogen vapor [16]

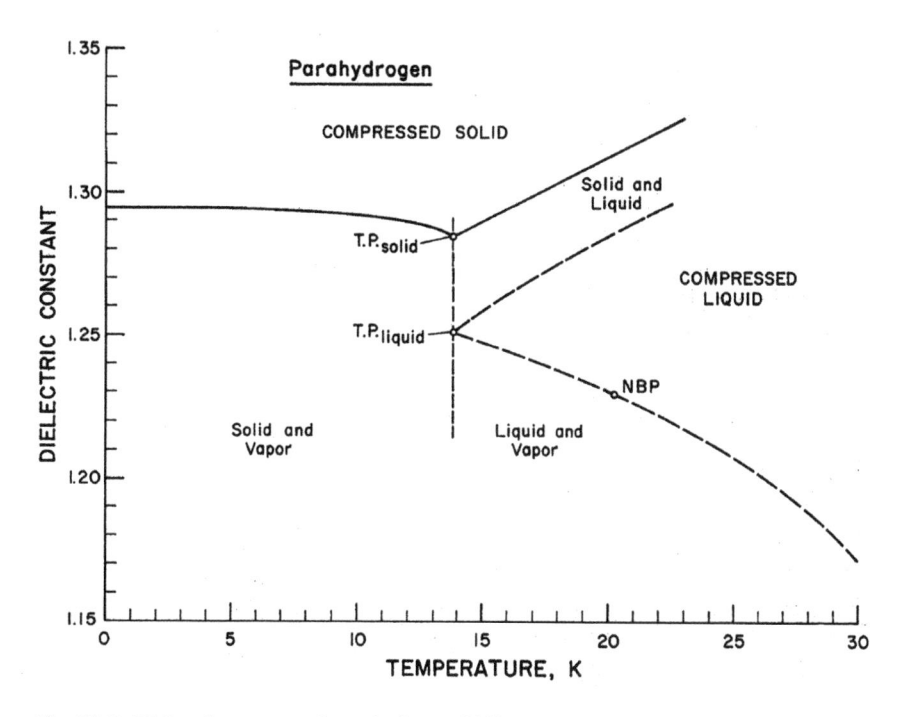

Fig. A2.7 Dielectric constant of para hydrogen [16]

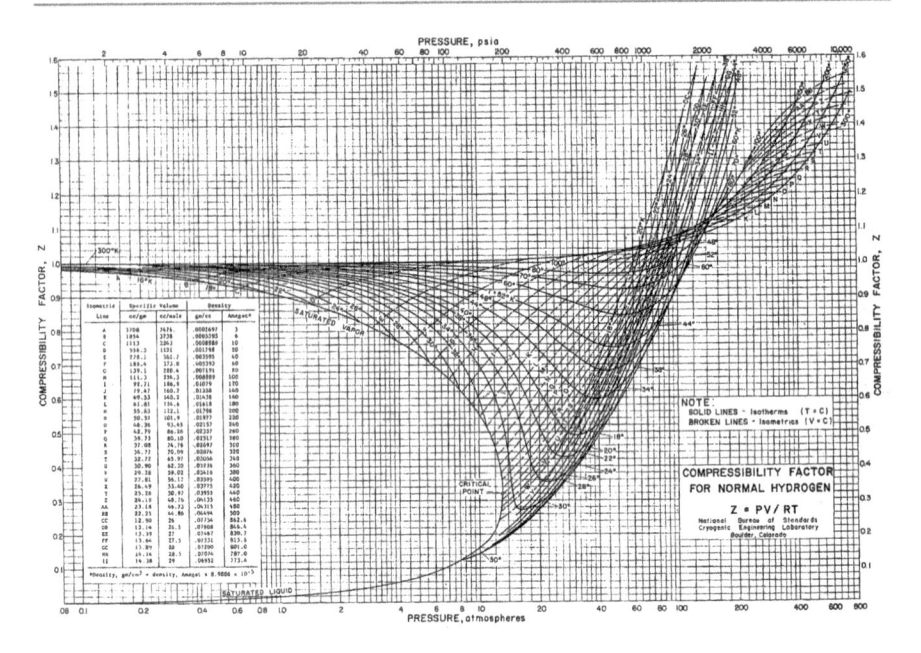

Fig. A2.8 Compressibility factor for normal hydrogen [16]

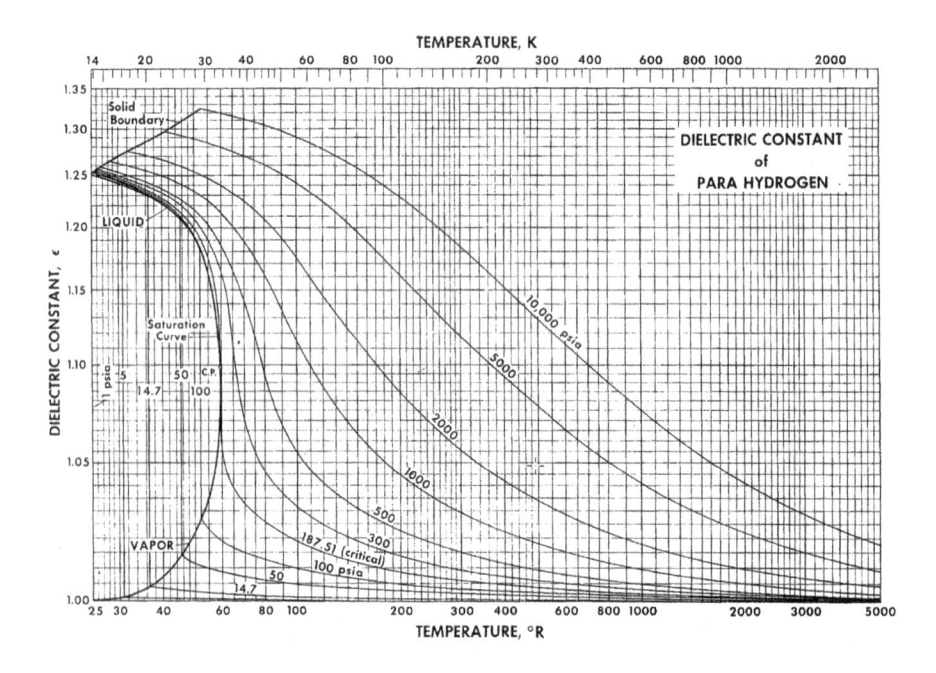

Fig. A2.9 Dielectric constant of para hydrogen [16]

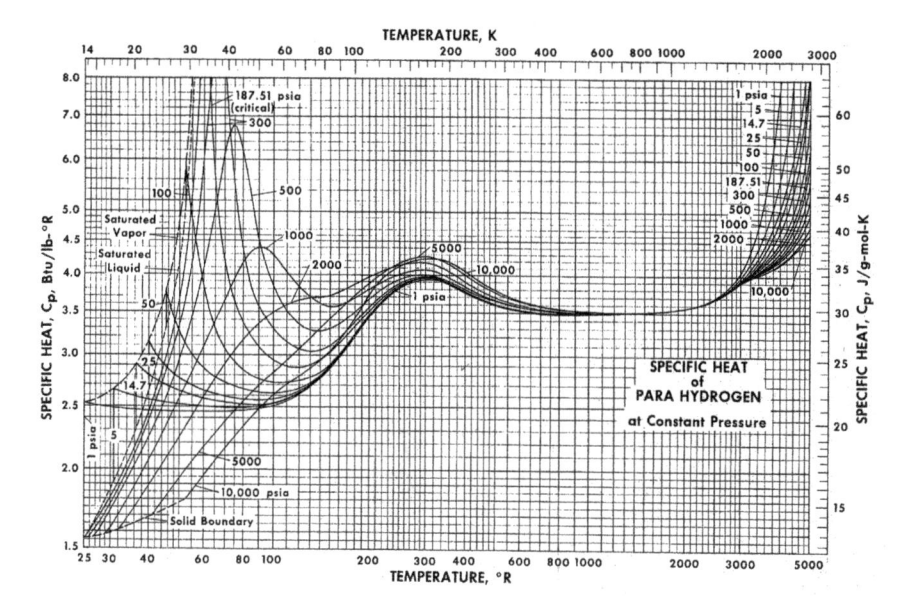

Fig. A2.10 Specific heat of para hydrogen at constant pressure [16]

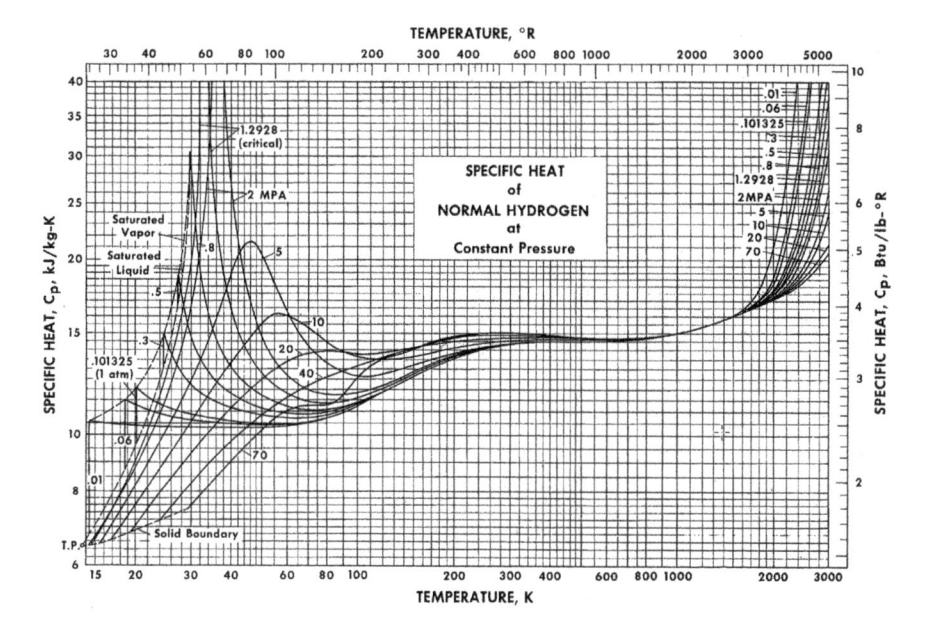

Fig. A2.11 Specific heat of normal hydrogen at constant pressure [16]

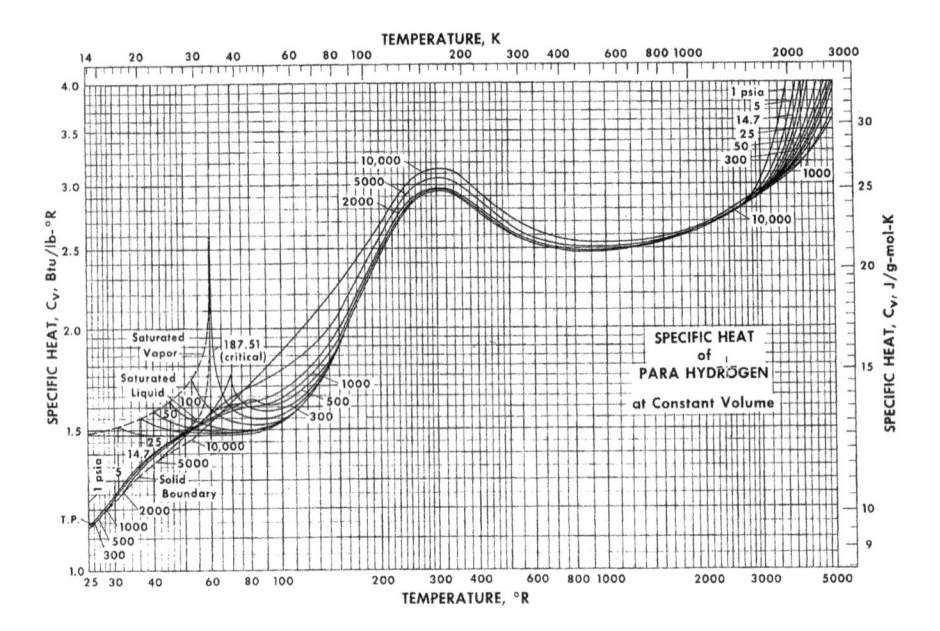

Fig. A2.12 Specific heat of para hydrogen at constant volume [16]

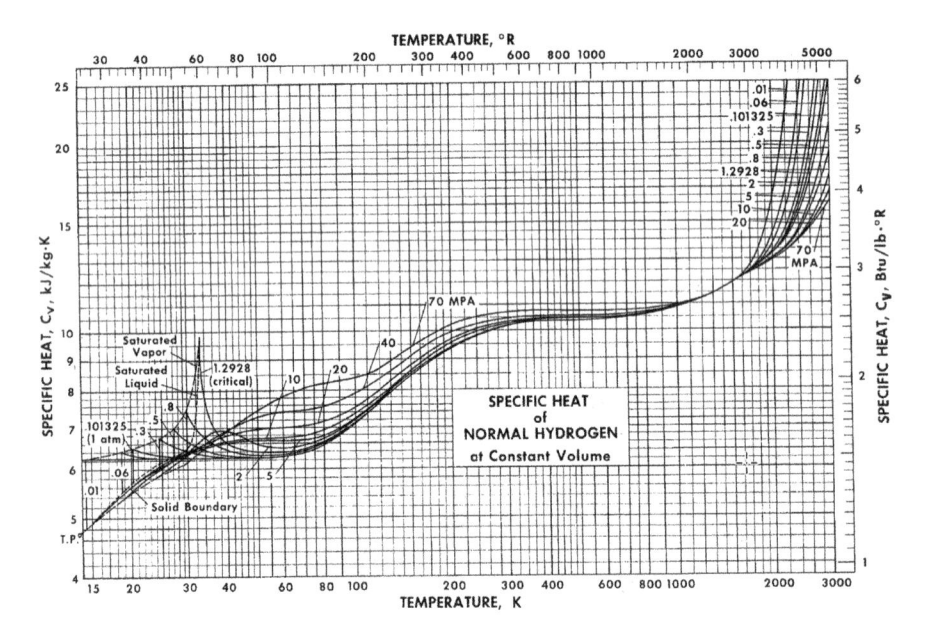

Fig. A2.13 Specific heat of normal hydrogen at constant volume [16]

Appendices

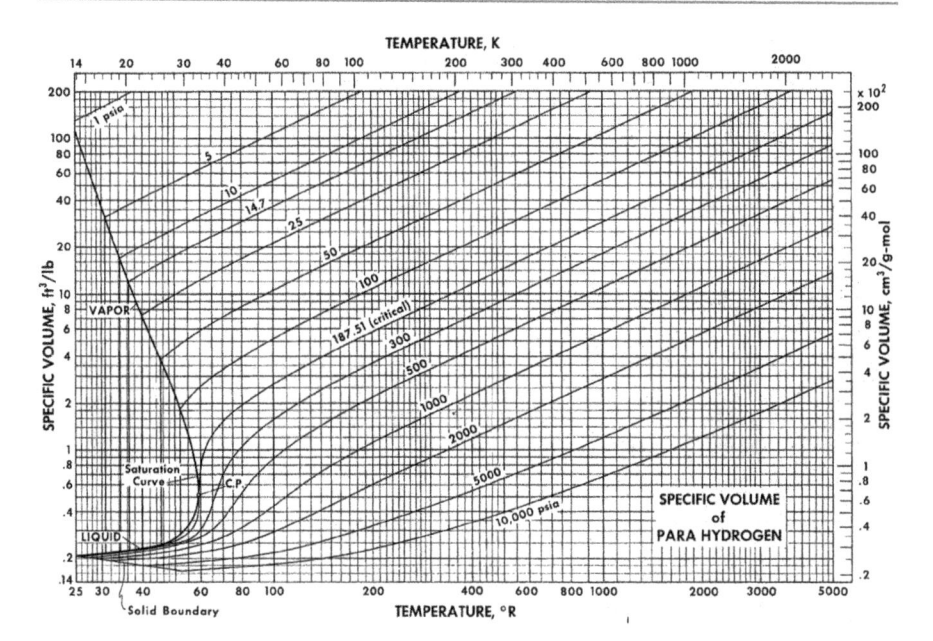

Fig. A2.14 Specific volume of para hydrogen [16]

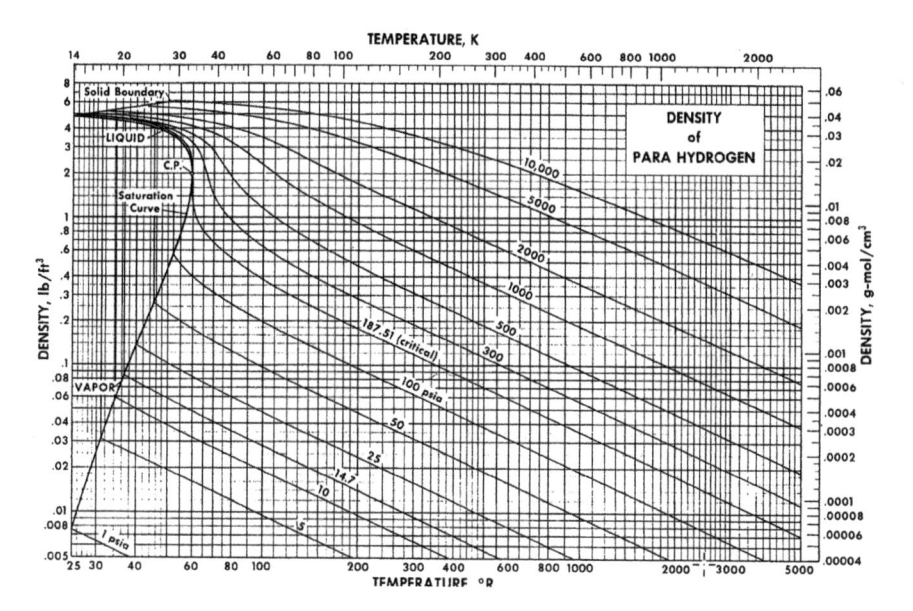

Fig. A2.15 Density of para hydrogen [16]

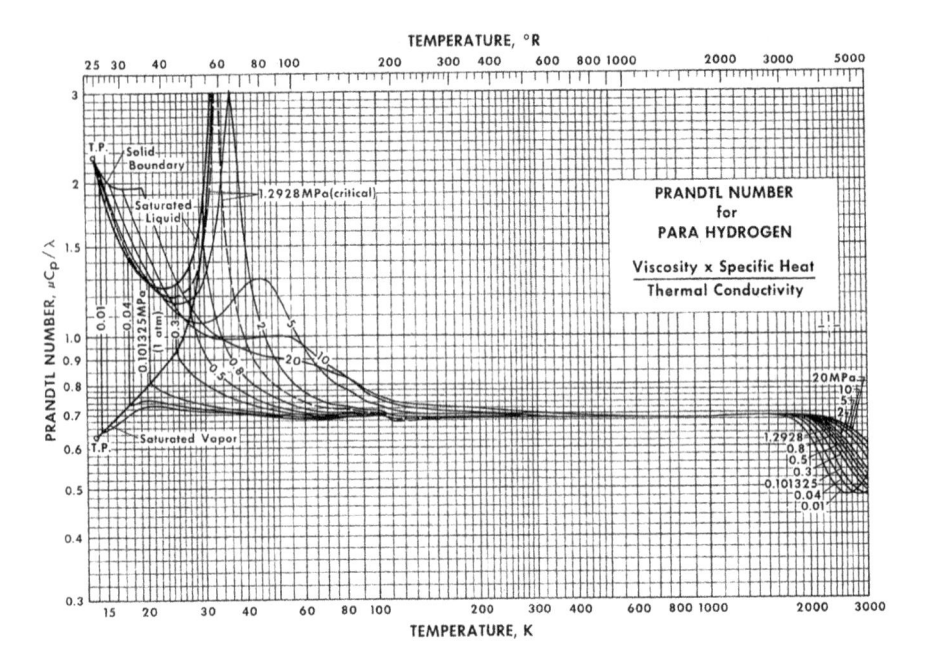

Fig. A2.16 Prandtl number for para hydrogen [16]

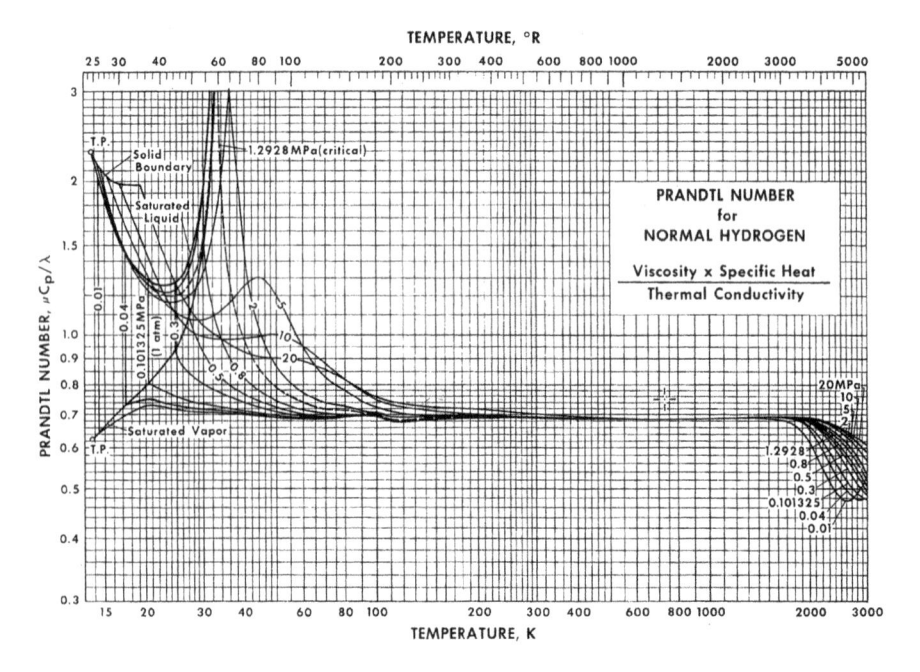

Fig. A2.17 Prandtl number for normal hydrogen [16]

Appendices

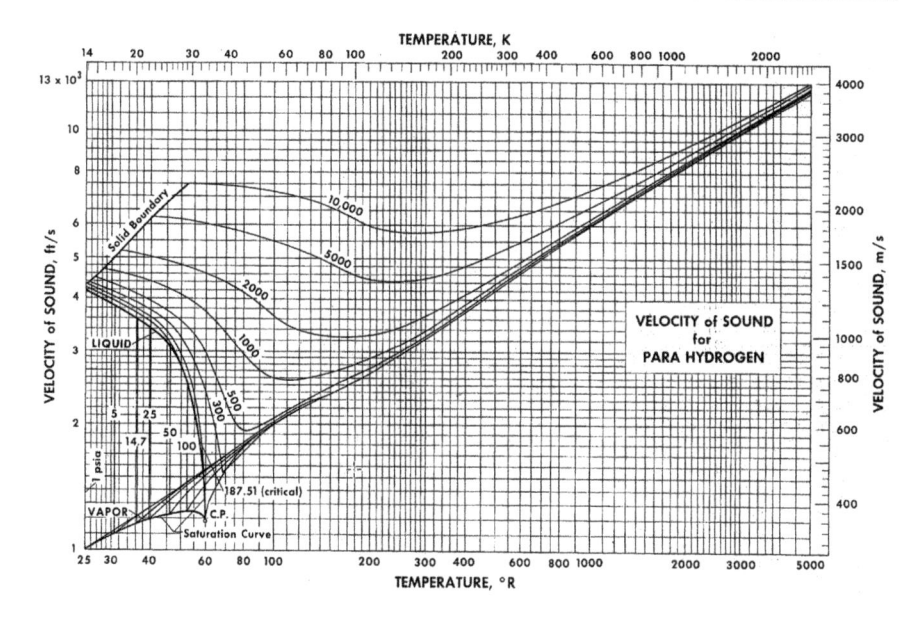

Fig. A2.18 Velocity of sound for para hydrogen [16]

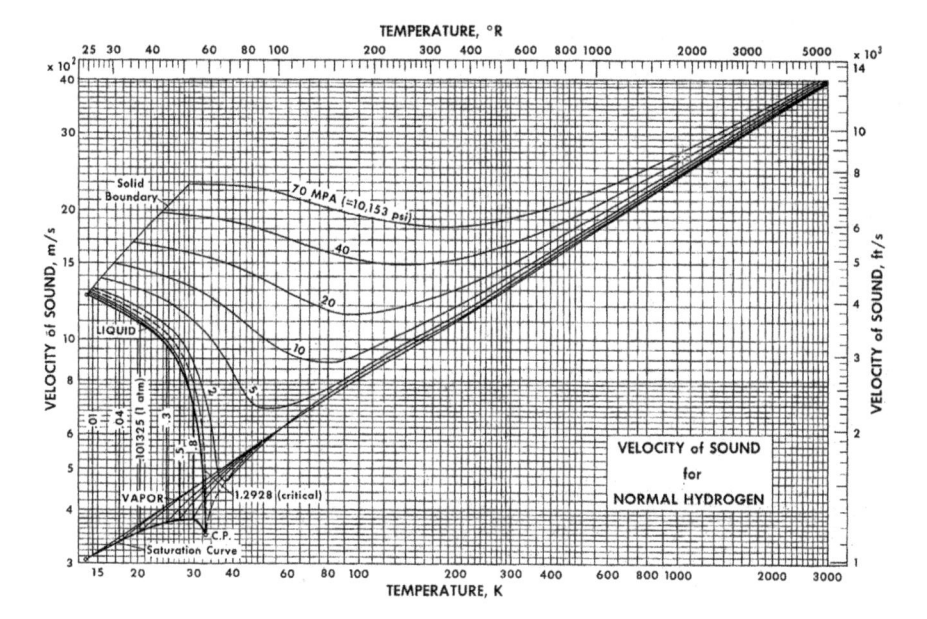

Fig. A2.19 Velocity of sound for normal hydrogen [16]

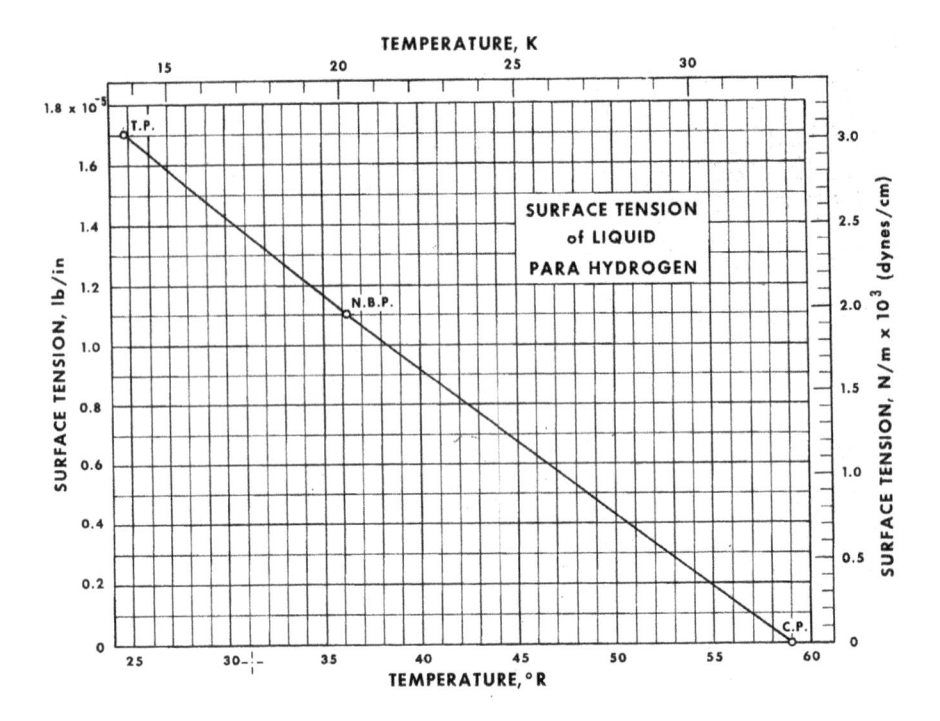

Fig. A2.20 Surface tension of liquid para hydrogen [16]

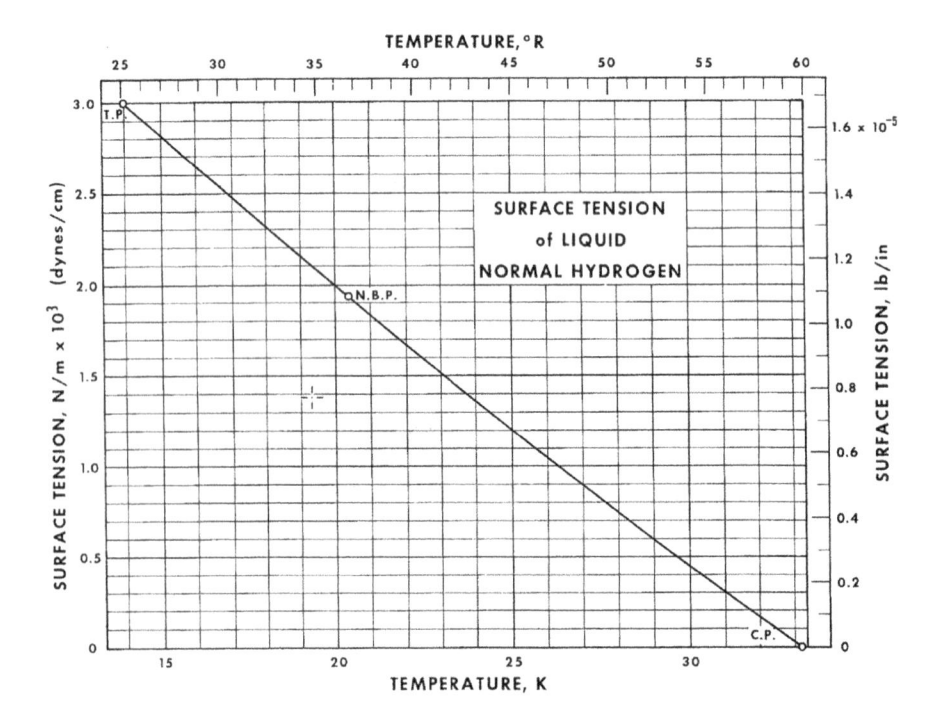

Fig. A2.21 Surface tension of liquid normal hydrogen [16]

Appendices

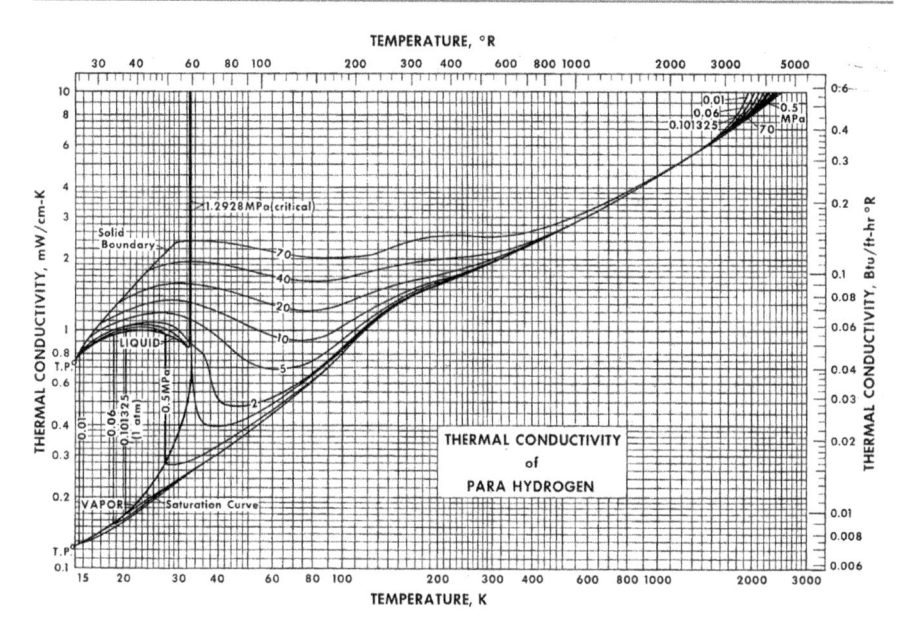

Fig. A2.22 Thermal conductivity of para hydrogen [16]

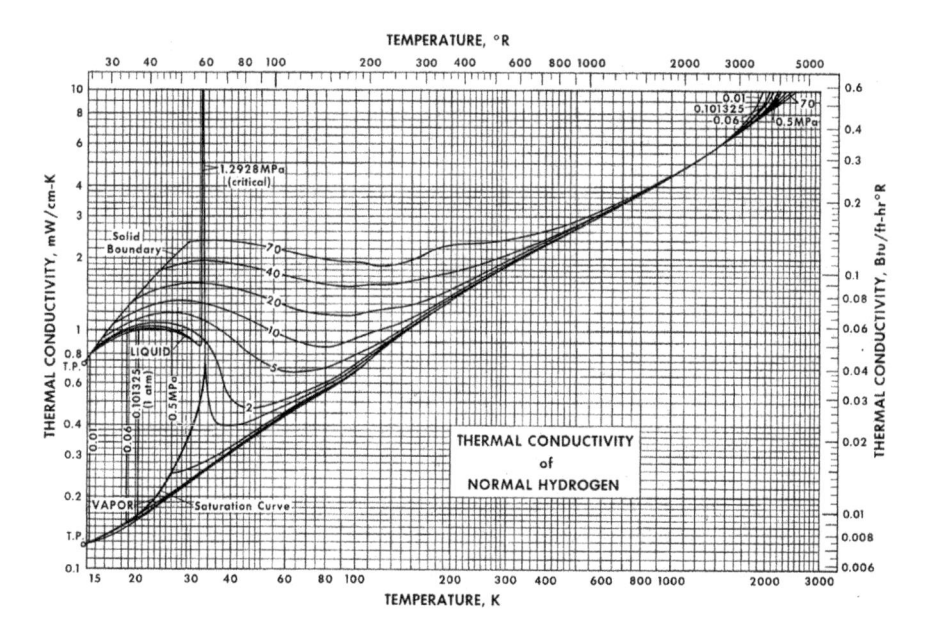

Fig. A2.23 Thermal conductivity of normal hydrogen [16]

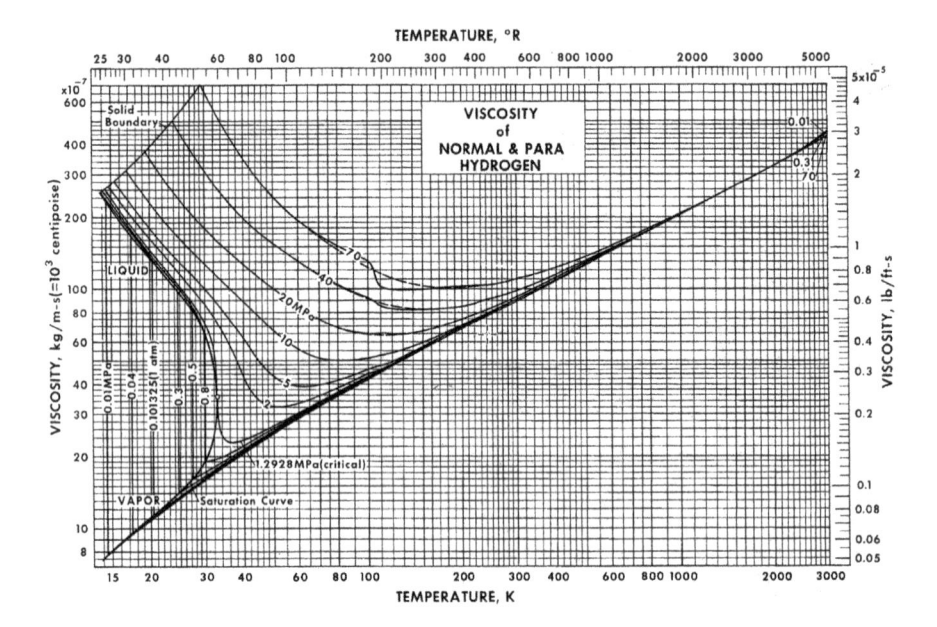

Fig. A2.24 Viscosity of normal and para hydrogen [16]

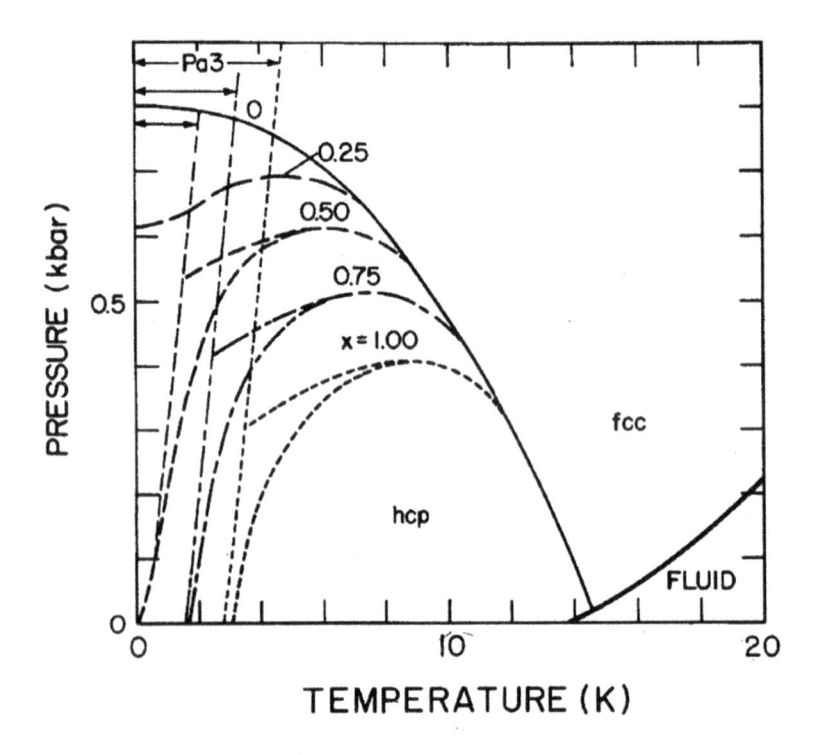

Shown on a common temperature scale at various ortho mole fractions x. Solid line, x = 0; long-dash line, x = 0.25; short-dash line, x = 0.50; dash-dot line, x = 0.75; dotted line, x = 1. Alternate branches are shown for hcp-fcc transition. Heavy solid line is melting curve.

Fig. A2.25 Proposed phase diagram for solid hydrogen [16]

Fig. A2.26 Temperature dependencies of the mechanical properties of normal and para hydrogen [16]. 1—Young's modulus (E), 2—shear modulus (G), 3—tensile strength (σt), 4—nominal yield stress (σV), ε—relative elongation

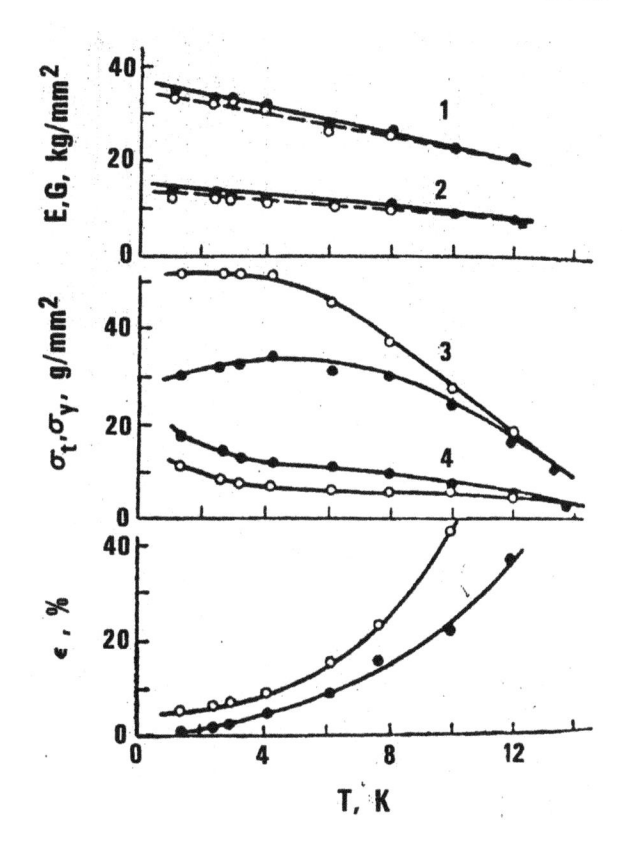

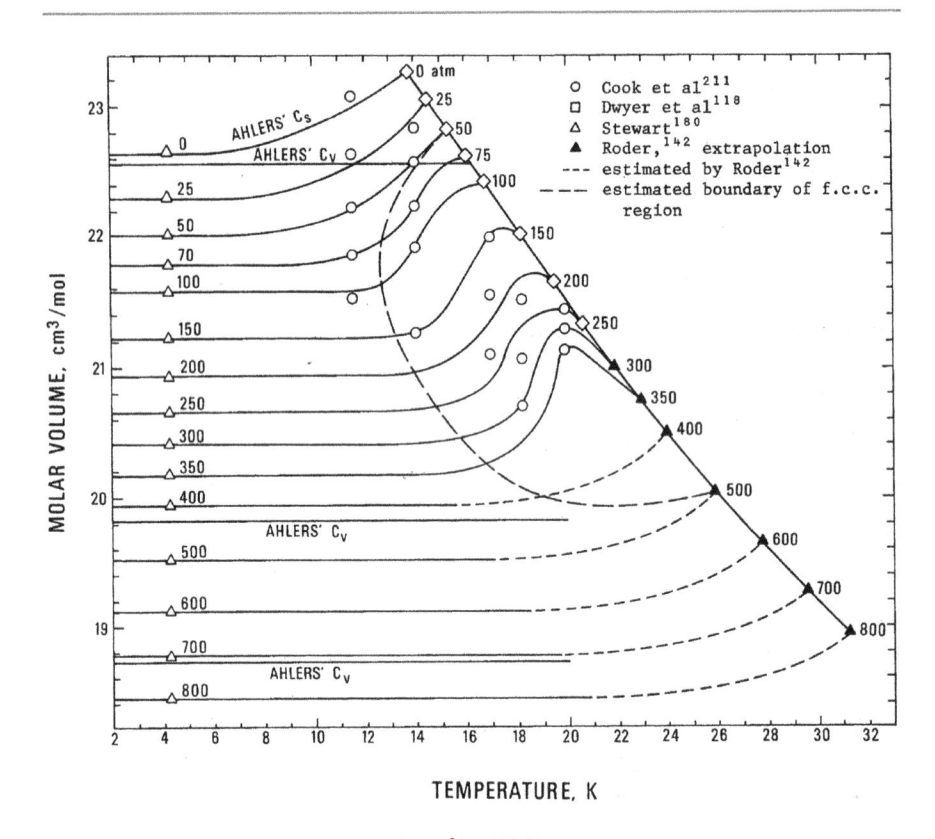

Fig. A2.27 Molar volume of hydrogen in cm³/mol [16]

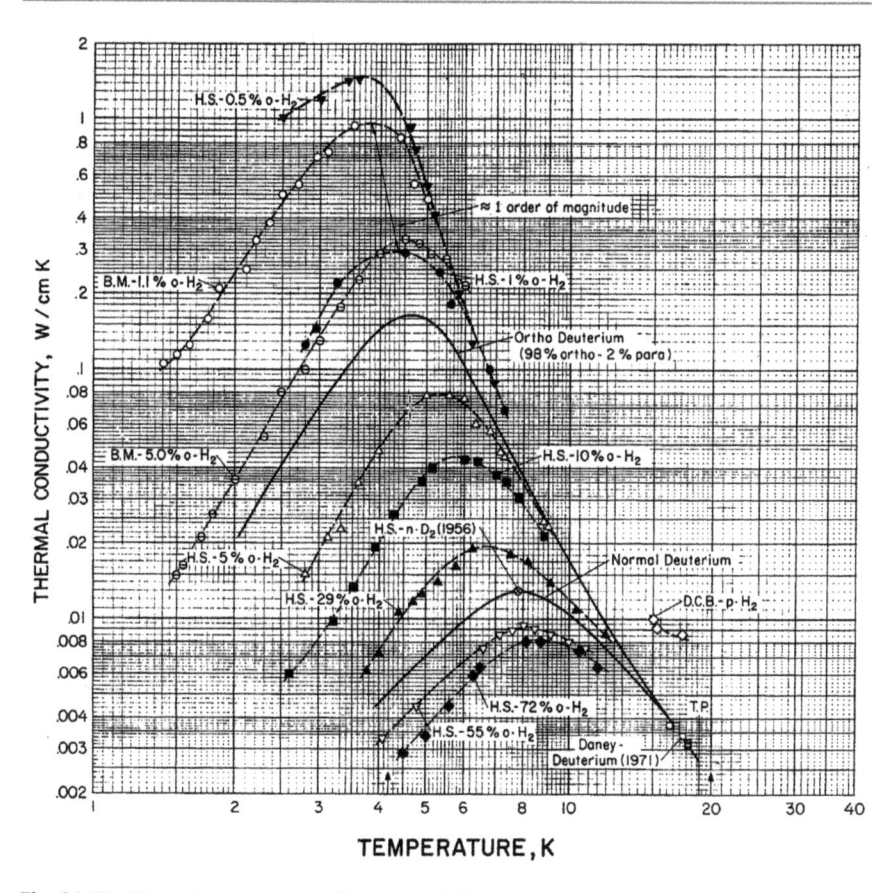

Fig. A2.28 Thermal conductivity of hydrogen [16]

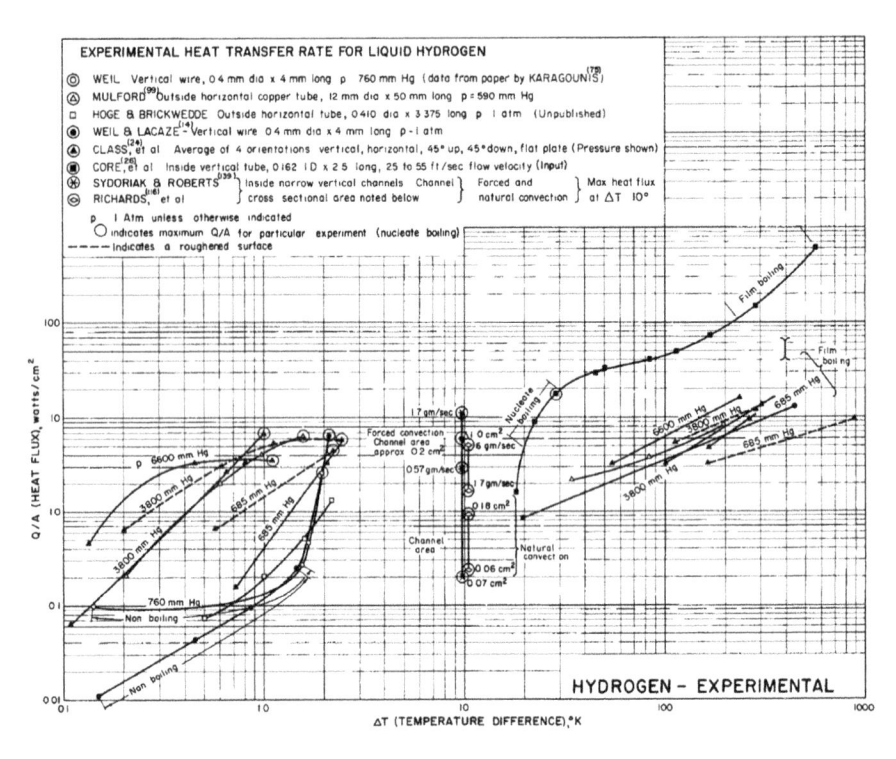

Fig. A2.29 Experimental heat transfer rate for liquid hydrogen [17]

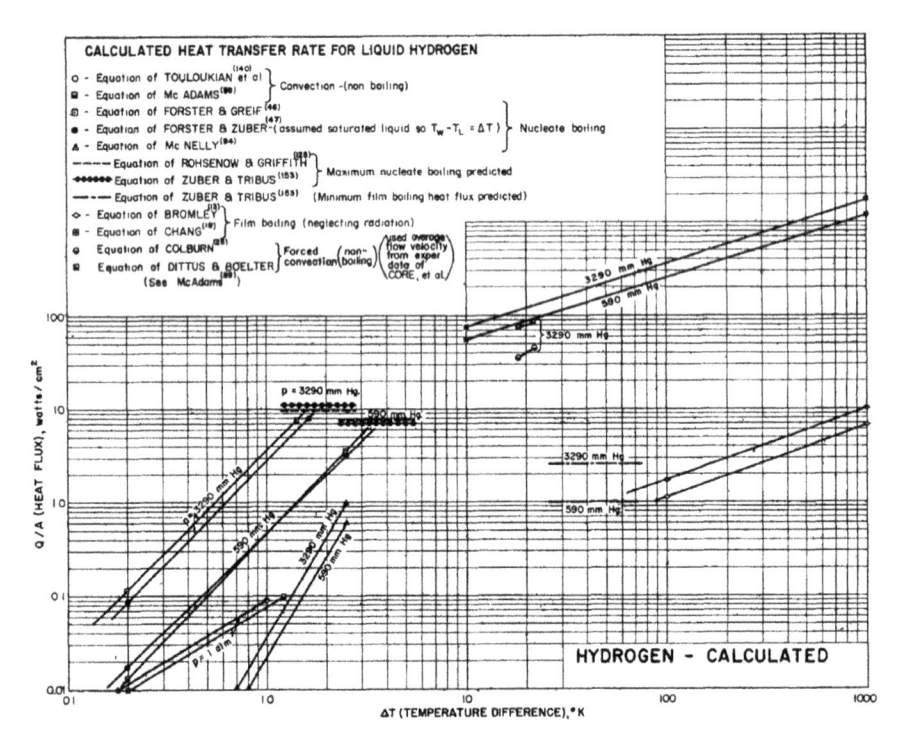

Fig. A2.30 Calculated heat transfer rate for liquid hydrogen [17]

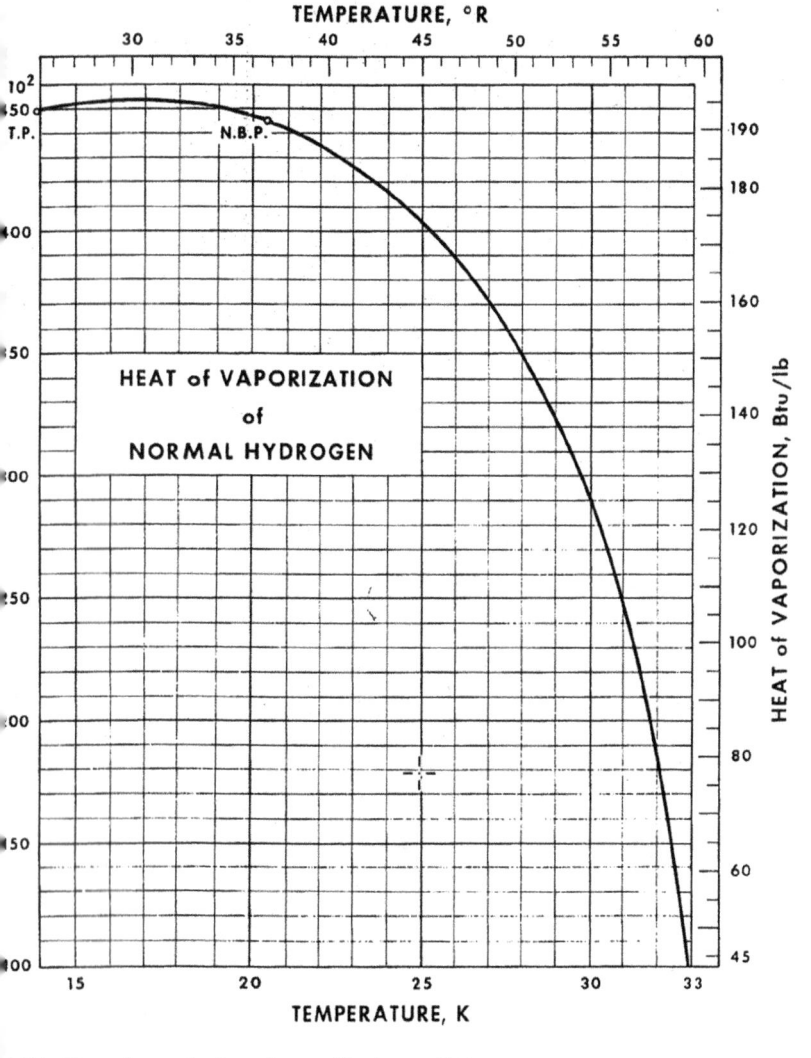

2.32 Heat of vaporization of normal hydrogen [16]

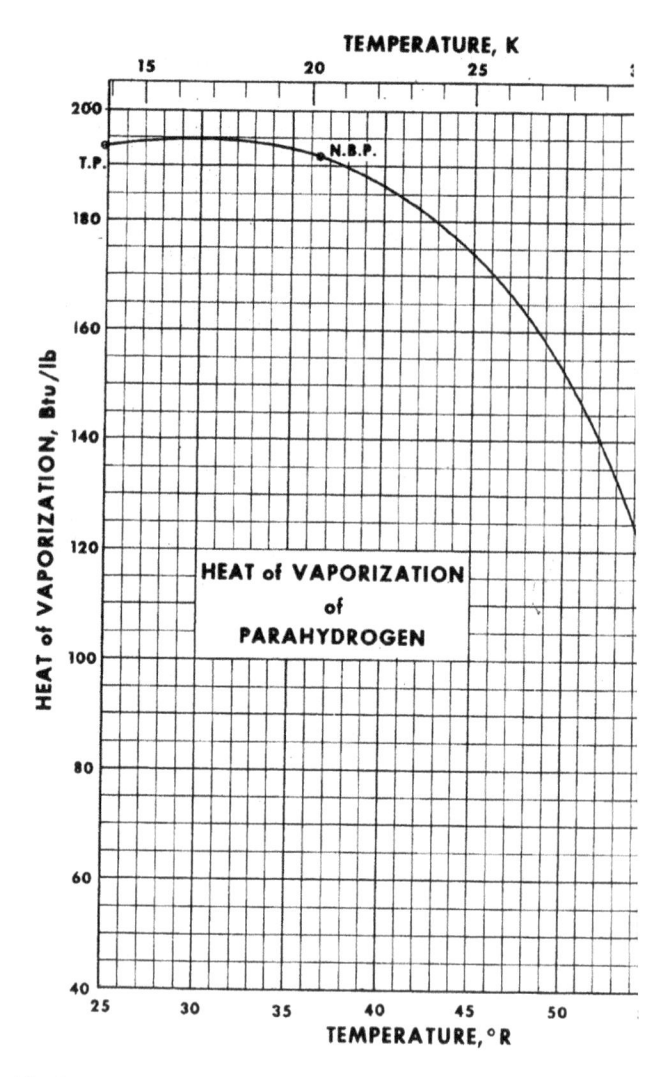

Fig. A2.31 Heat of vaporization of para hydrogen [16]

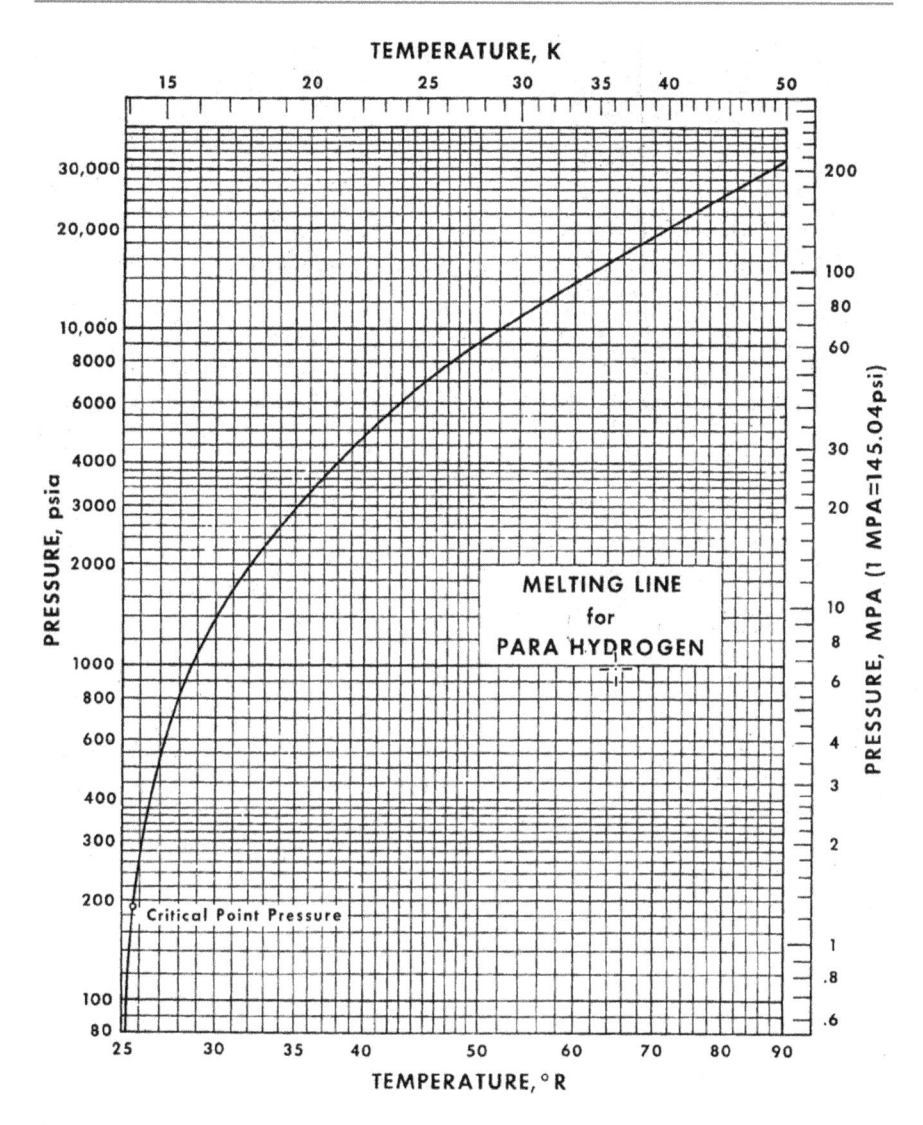

Fig. A2.33 Melting line for para hydrogen [16]

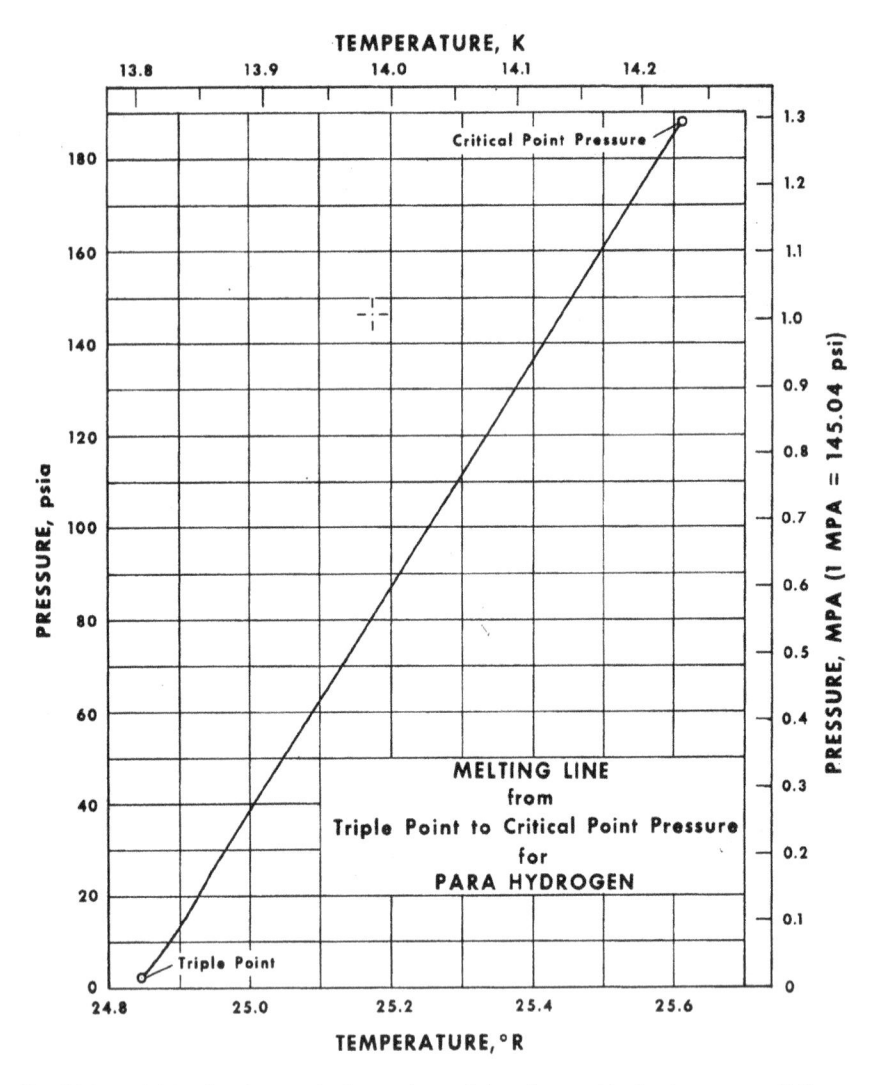

Fig. A2.34 Melting line for para hydrogen from triple point to critical point pressure [16]

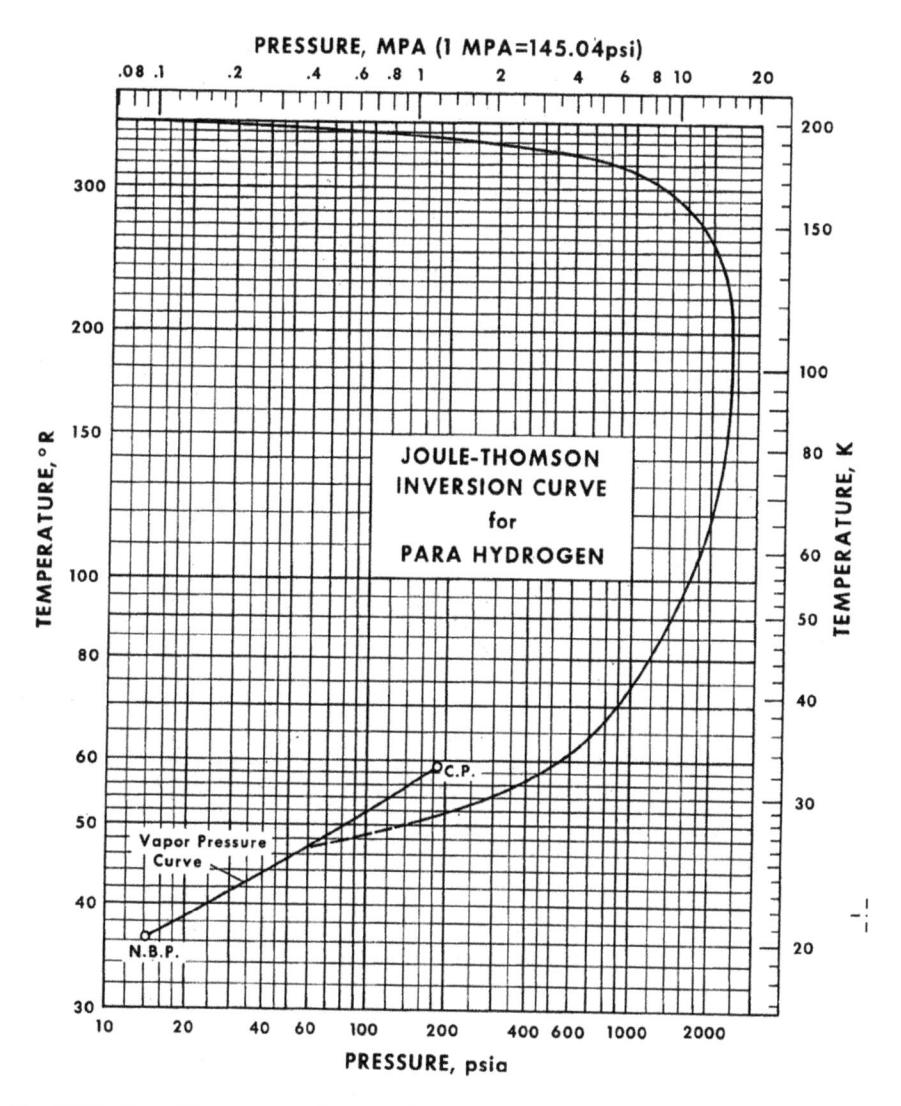

Fig. A2.35 Joule-Thomson inversion curve for para hydrogen [16]

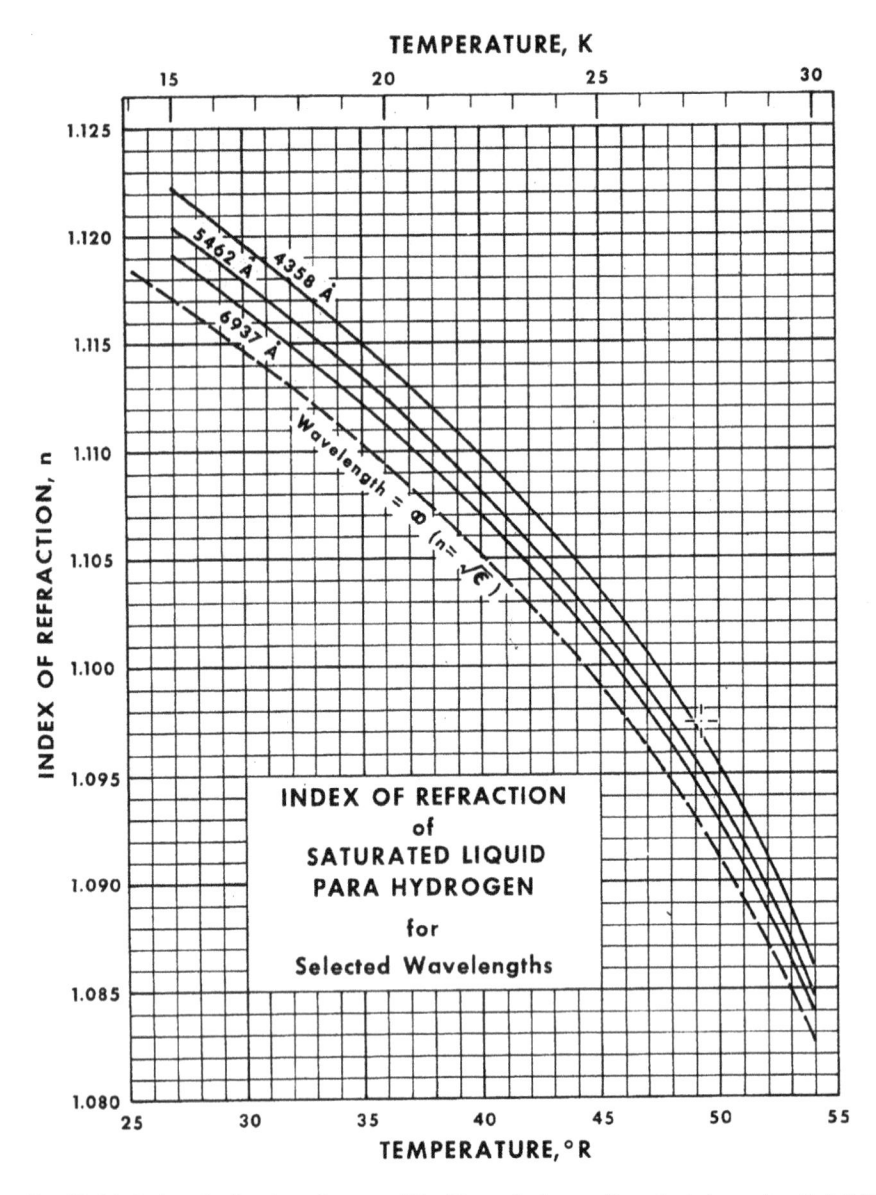

Fig. A2.36 Index of refraction of saturated liquid para hydrogen (for selected wavelengths) [16]

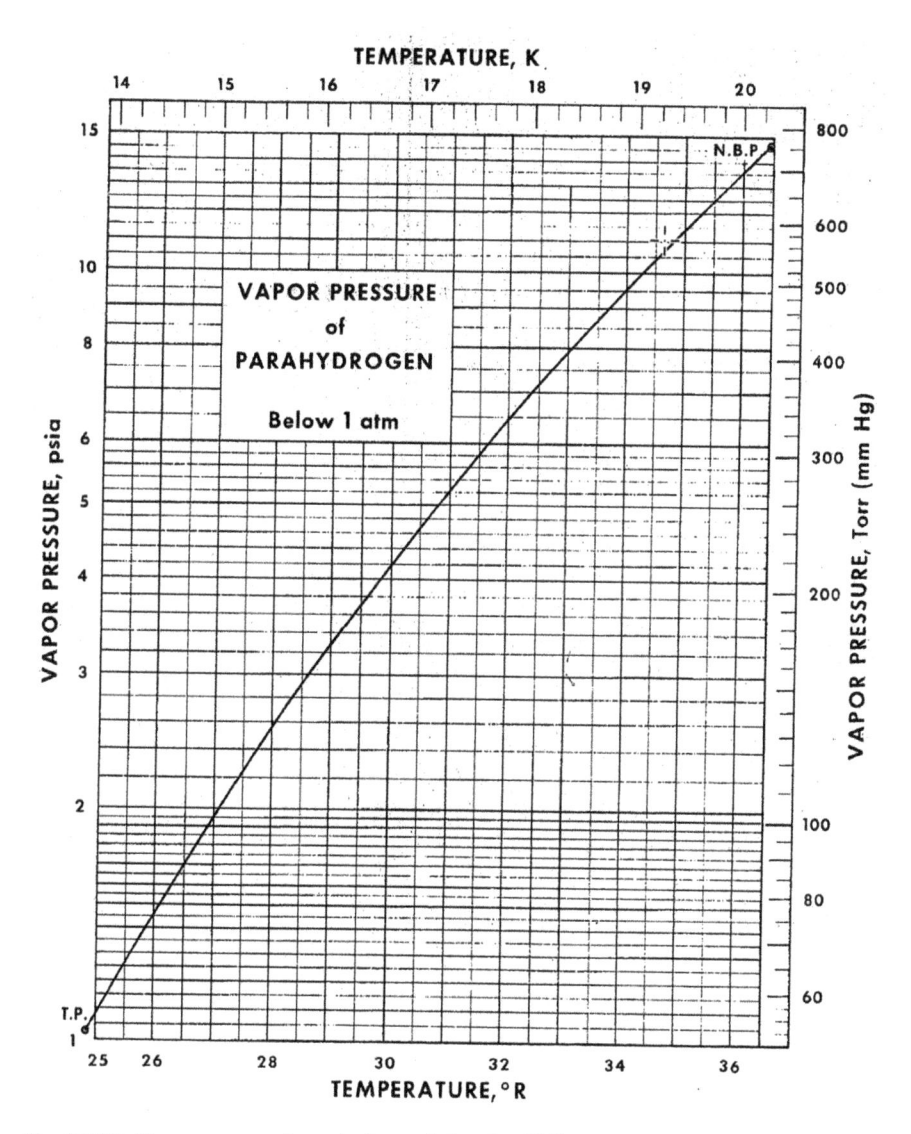

Fig. A2.37 Vapor pressure of para hydrogen below 1 at [16]

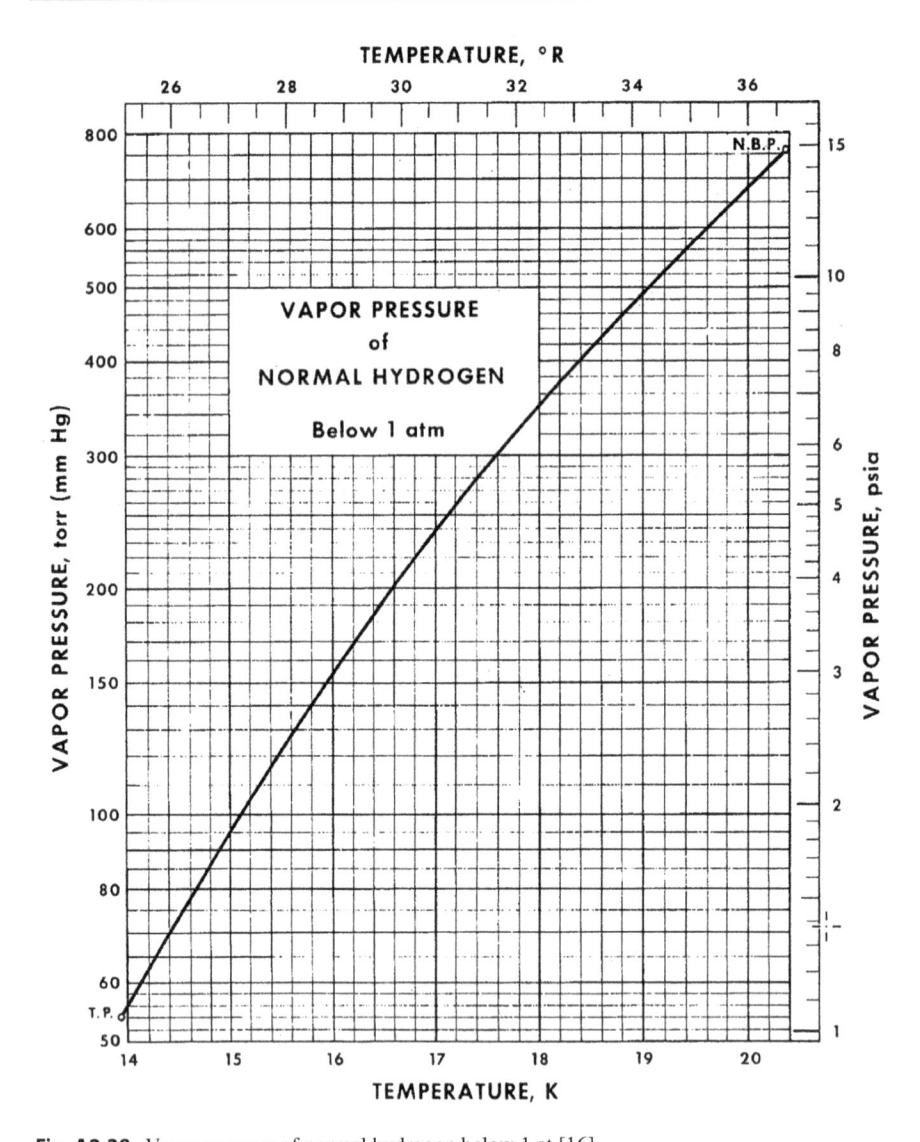

Fig. A2.38 Vapor pressure of normal hydrogen below 1 at [16]

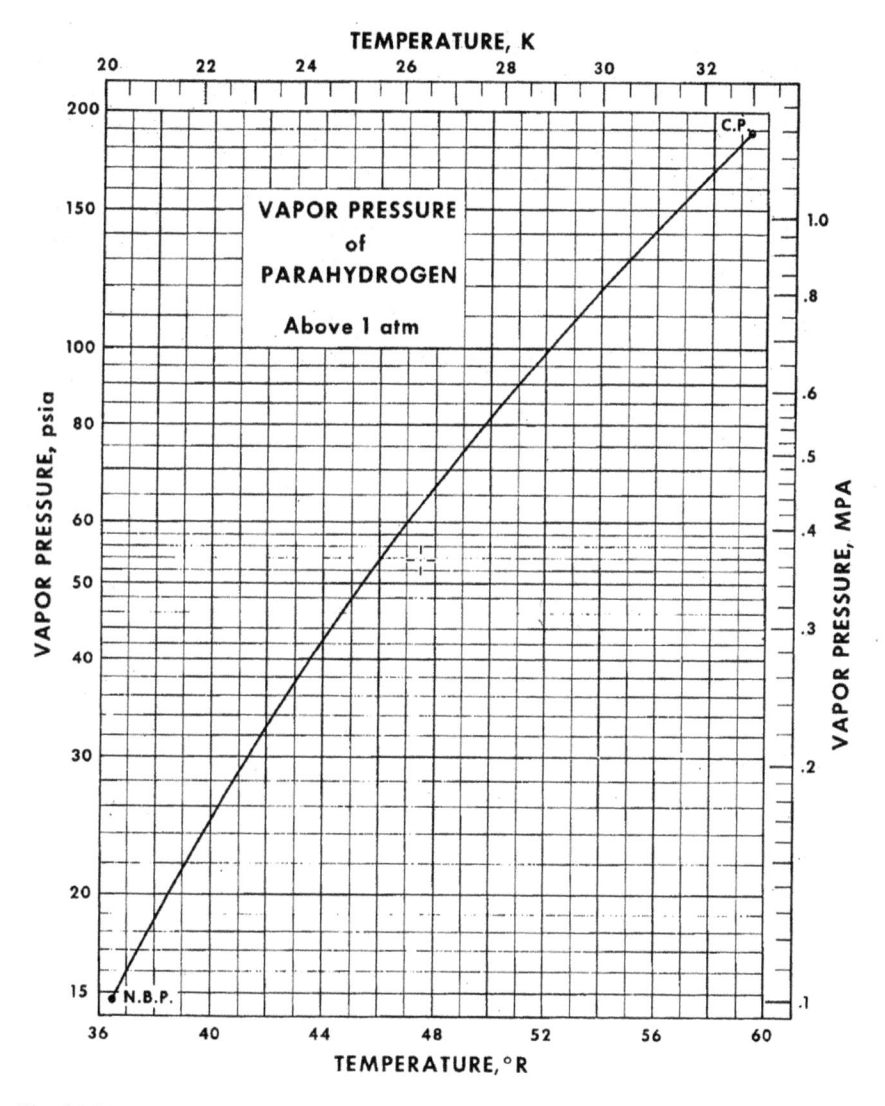

Fig. A2.39 Vapor pressure of para hydrogen above 1 at [16]

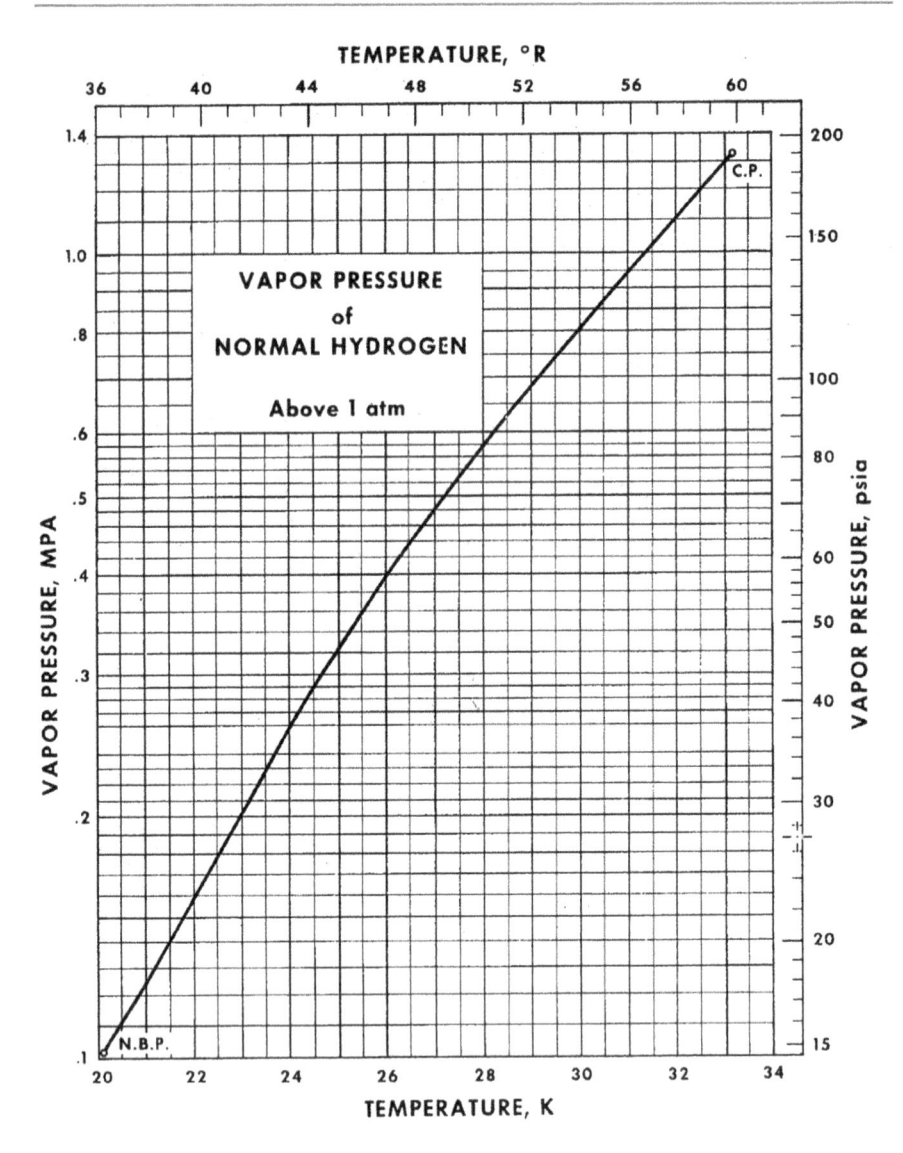

Fig. A2.40 Vapor pressure of normal hydrogen above 1 at [16]

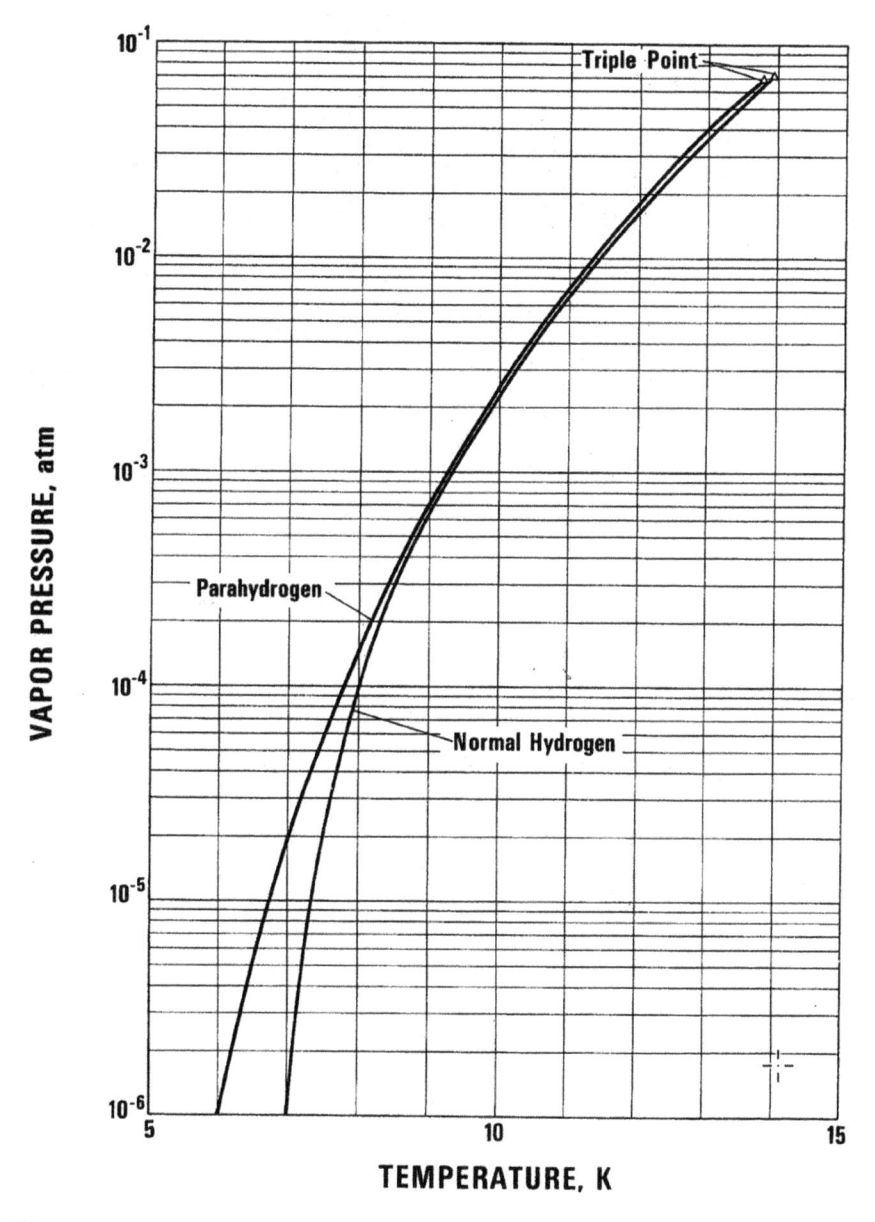

Fig. A2.41 Vapor pressure of normal and para hydrogen below 14 K [16]

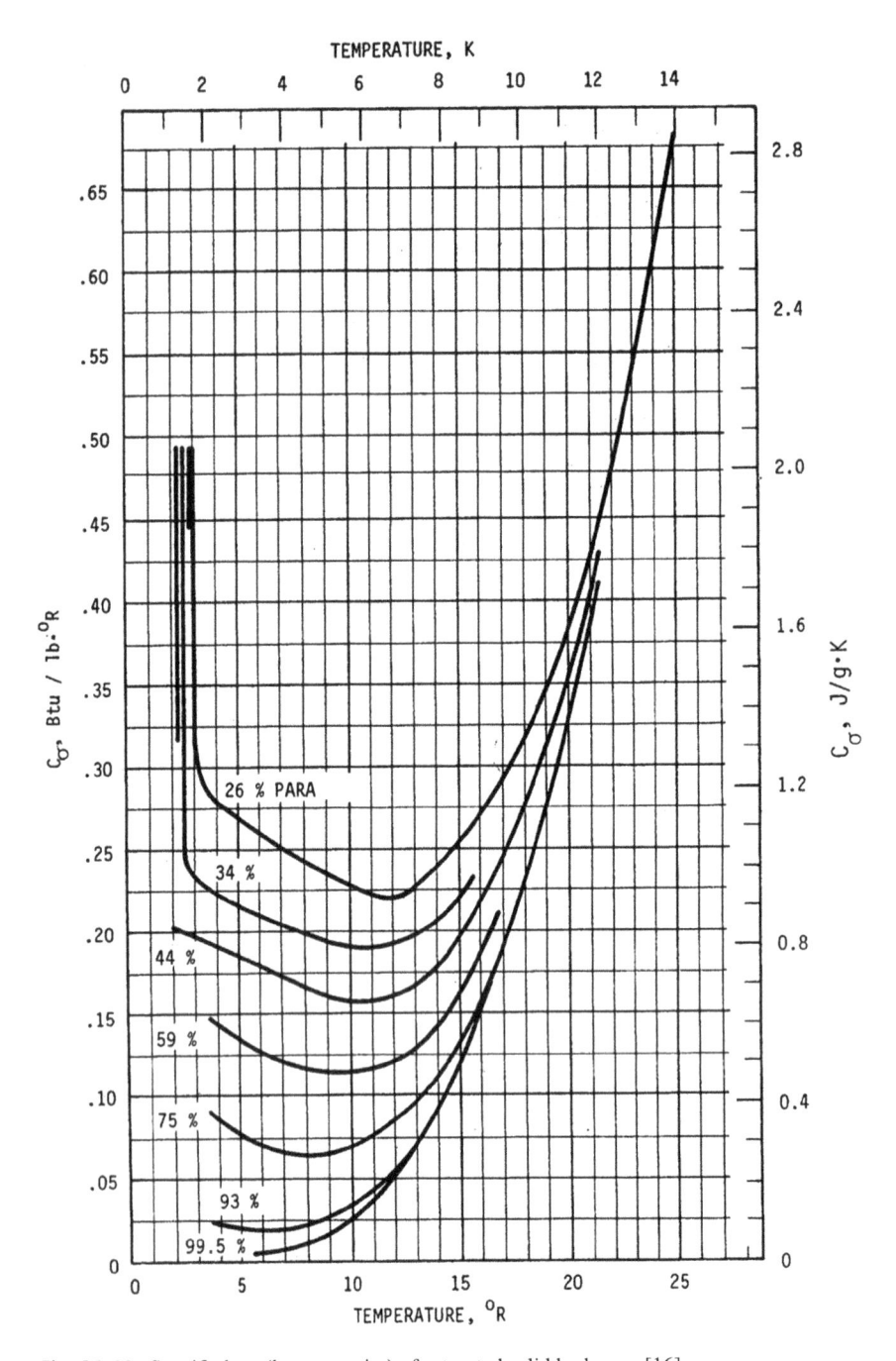

Fig. A2.42 Specific heat (heat capacity) of saturated solid hydrogen [16]

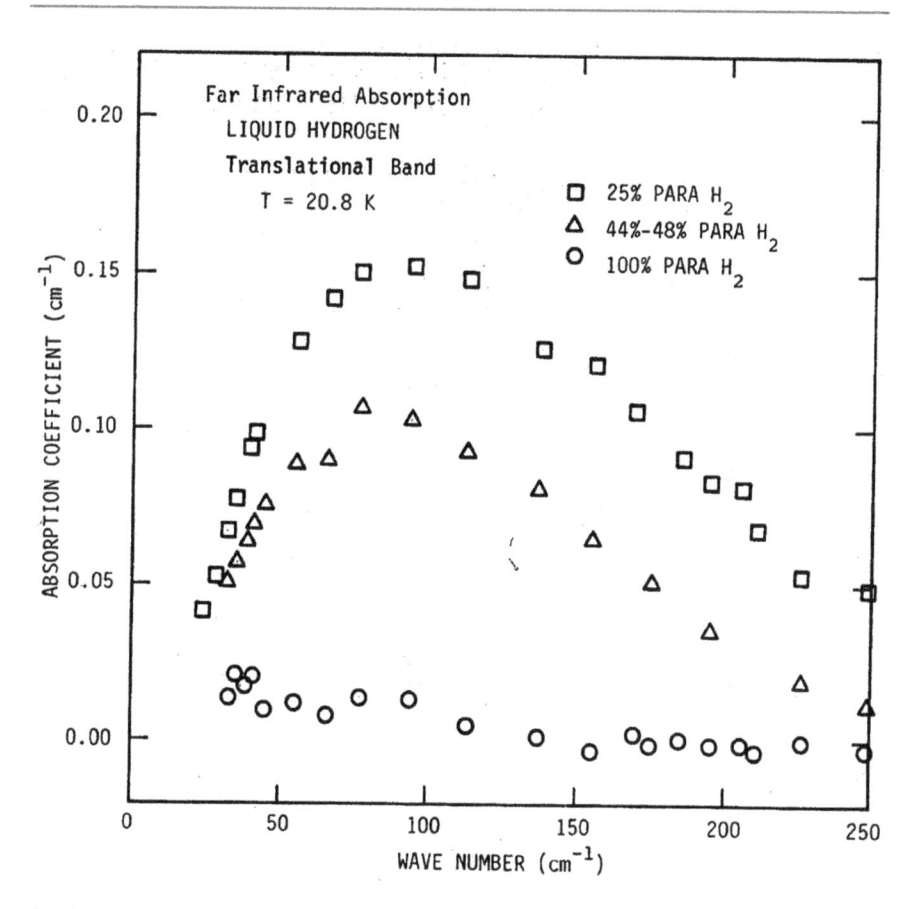

Fig. A2.43 Far infrared absorption of liquid para hydrogen as a function of wave number (translational band) [16]

Table A2.1 Hydrogen and hydrogen properties—NASA [11]

Fixed point properties of Normal hydrogen[b]

Properties	Triple point Solid	Triple point Liquid	Triple point Vapor	Normal boiling Liquid	Normal boiling Vapor	Critical point	STP	NTP
Temperature[a], °F (K)	−434.55 (13.957)	−434.55 (13.957)	−434.55 (13.957)	−422.97 (20.390)	−422.97 (20.390)	−399.93 (33.190)	32.000 (273.15)	68.000 (293.15)
Pressure (inches Hg) (mm Hg)	2.147 (54.04)	2.147 (54.04)	2.147 (54.04)	30.19 (760.0)	30.19 (760.0)	391.9 (9865)	30.19 (760.0)	30.19 (760.0)
Density, lb./ft^3 (mol/cm^3)	5.409 (0.04301)	4.817 (0.03830)	8.099 × 10^{-3} (0.0644 × 10^{-3})	4.427 (0.0352)	83.05 × 10^{-3} (0.6604 × 10^{-3})	1.879 (0.01494)	5.609 × 10^{-3} (0.0446 × 10^{-3})	5.225 × 10^{-3} (0.04155 × 10^{-3})
Specific volume, ft^3/lb. (cm^3/mol)	0.1849 (23.25)	0.2076 (26.11)	123.5 (15.530)	0.2259 (28.41)	12.05 (1514)	0.5322 (66.93)	178.3 (22.420)	191.38 (24.070)
Compressibility factor, Z = PV/RT	—	0.001621	0.9635	0.01698	0.9051	0.3191	1.00042	1.00049
Heats of fusion and vaporization, Btu/lb. (J/mol)	24.99 (117.1)	194.4 (911.3)	—	191.9 (899.1)	—	0 (0)	—	—
Specific heat, Btu/lb. °R (J/g·K) at saturation, Cσ	0.6794 (2.842)	1.642 (6.870)	−5.565 (−23.28)	2.242 (9.380)	−3.946 (−16.51)	Very large	—	—
At constant pressure, Cp	—	1.569 (6.563)	2.516 (10.526)	2.336 (9.772)	2.917 (12.20)	Very large	3.390 (14.18)	3.425 (14.33)
At constant volume, Cv	—	1.130 (4.727)	1.484 (6.211)	1.375 (5.754)	1.565 (6.548)	2.336 (9.772)	2.407 (10.07)	2.419 (10.12)
Specific heat ratio, γ = Cp/Cv	—	1.388	1.695	1.698	1.863	Large	1.408	1.416

Fixed point properties of normal hydrogen[b]

Properties	Triple point Solid	Triple point Liquid	Triple point Vapor	Normal boiling point Liquid	Normal boiling point Vapor	Critical point	STP	NTP
Enthalpy Btu/lb. (J/mol)	68.63 (321.6)	93.62 (438.7)	288.1 (1350)	117.0 (548.3)	308.9 (1447)	248.4 (1164)	1654 (7749)	1776 (8324)
Internal energy, Btu/lb. (J/mol)	67.84 (317.9)	92.83 (435.0)	263.5 (1235)	116.5 (545.7)	276.2 (1294)	–	1169 (5477)	1256 (5885)
Entropy, Btu/hr.·°R (J/mol K)	2.41 (20.3)	3.40 (28.7)	11.1 (93.6)	4.14 (34.92)	9.36 (78.94)	6.47 (54.57)	16.5 (139.6)	16.8 (141.6)
Velocity of sound, ft./s (m/s)	–	4206 (1282)	1007 (307)	3612 (1101)	1171 (357)	–	4088 (1246)	4246 (1294)
Viscosity, centipoise (μPa·s)	–	0.026 (26)	0.00074 (0.74)	0.0132 (13.2)	0.0011 (1.1)	0.0035 (3.5)	0.00839 (8.39)	0.00881 (8.81)
Thermal conductivity, k Btu/ft.·h·°R (μW/m·K)	0.52 (90)	0.042 (7.3)	0.0072 (1.24)	0.057 (9.9)	0.0098 (1.69)	Anomalously large	0.098 (17.40)	0.106 (18.38)
Prandtl no.	–	2.34	0.630	1.30	0.798	–	0.682	0.688
Dielectric constant, ε	1.287	1.253	1.00039	1.231	1.0040	1.0937	1.000271	1.000253
Index of refraction	1.134	1.119	1.000196	1.1093	1.0020	1.0458	1.000136	1.000126
Surface tension, lbf/in (N/m)	–	$0.0171 \times 10^-$ (3.00×10^{-3})	–	0.0111×10^{-3} (1.94×10^{-3})	–	0 (0)	–	–
Equivalent volume per volume of liquid at NBT	0.8184	0.9190	546.3	1.000	53.30	2.357	789.3	847.1

Note: Dashes indicate not applicable
[a]These temperatures are based on the IPTS-1968 temperature scale.
[b]McCarty et al. [15]

Table A2.2 Aircraft fuel comparison data (recompiled after NBS 690 [12])

Property	Hydrogen	Methane	Gasoline
Molecular weight	2.016	16.043	~107.0
Triple point pressure, atm	0.0695	0.1159	–
Triple point temperature, K	13.803	90.680	180 to 220
Normal boiling point (NBP) temperature, K	20.268	111.632	310 to 478
Critical pressure, atm	12.759	45.387	24.5 to 27
Critical temperature, K	32.976	190.56	540 to 569
Density at critical point, g/cm^3	0.0314	0.1604	0.23
Density of liquid at triple point, g/cm^3	0.0770	0.4516	–
Density of solid at triple point, g/cm^3	0.0865	0.4872	–
Density of vapor at triple point, g/cm^3	125.597	251.53	–
Density of liquid at NBP, g/cm^3	0.0708	0.4226	~70
Density of vapor at NBP, g/cm^3	0.00134	0.00182	~0.0045
Density of gas at NTP, g/cm^3	83.764	651.19	~4400
Density ratio: NBP liquid-to-NTP gas	845	649	~156
Heat of fusion, J/g	58.23	58.47	161
Heat of vaporization, J/g	445.59	509.88	309
Heat of sublimation, J/g	507.39	602.44	–
Heat of combustion (low), kJ/g	119.93	50.02	44.5
Heat of combustion (high), kJ/g	141.86	55.53	48
Specific heat (C_P) of NTP gas, J/g-K	14.89	2.22	1.62
Specific heat (C_p) of NBP liquid, J/g-K	9.69	3.5	2.20
Specific heat ratio (C_p/C_v) of NTP gas	1.383	1.308	1.05
Specific heat ratio (C_p/C_v) of NBP liquid	1.688	1.676	–
Viscosity of NTP gas, g/cm-s	0.0000875	0.000110	0.000052
Viscosity of NBP liquid, g/cm-s	0.000133	0.001130	0.002
Thermal conductivity of NTP gas, mW/cm-K	1.897	0.33	0.112
Thermal conductivity of NBP liquid, mW/cm-K	1.00	1.86	1.31
Surface tension of NBP liquid, N/m	0.00193	0.01294	0.0122
Dielectric constant of NTP gas	1.00026	1.00079	1.0035
Dielectric constant of NBP liquid	1.233	1.6227	1.93
Index of refraction of NTP gas	1.00012	1.0004	1.0017
Index of refraction of NBP liquid	1.110	1.2739	1.39
Adiabatic sound velocity in NTP gas, m/s	1294	448	154
Adiabatic sound velocity in NBP liquid, m/s	1093	1331	1155
Compressibility factor (Z) in NTP gas	1.0006	1.0243	1.0069
Compressibility factor (Z) in NBP liquid	0.01712	0.004145	0.00643
Gas constant (R), cm^3-at/g-K	40.7037	5.11477	0.77
Isothermal bulk modulus (α) of NBP liquid, MN/m^2	50.13	456.16	763
Volume expansivity (β) of NBP liquid, K^{-1}	0.01658	0.00346	0.0012

A3: Safety Database

Abstract This Appendix provides information on various safety aspects and resources to tap into.

Keywords Safety data; Safety parameters; Safety database; Safety standards

For "Safety standard guidelines for Hydrogen System Design, Materials Selection, Operations, Storage and Transportation" please see references [18–20]. See also [21].

A first very general advice on hydrogen handling is given below after Gregory [18]:

Hydrogen in Systems
- In liquid hydrogen systems, vents and valves can ice up and fail.
- Hydrogen systems should always be maintained at positive pressure.
- Trapped liquid hydrogen will evaporate, pressurize, and rupture lines.
  - Please be advised: LH_2 at 1.01 bar, if heated to room temperature can create pressures up to 200 MPa.
- Any enclosed or partially enclosed storage areas need to be well ventilated.

Materials to Use in Liquid Hydrogen Systems
- Austenitic steel, aluminum alloys, copper and copper-based alloys.
- Non-metals, like Dacron, Tefon, Kevlar, Mylar, and Nylon.
- Understand the effects and causes of hydrogen embrittlement.

Personnel Protective Equipment
- Trained and experienced personnel.
- Always use eye, hand, and non-sparking clothing (including face shields, gloves, and suitable overalls).
- Use non-porous shoes (Tables A3.1 and A3.2).

This section by Constantinos Minas In the following, a summary of some best practices for ground-based systems is given including an overview of the most important codes an. standards [22]:

Cryogenic Subsystem Purging
- The entire subsystem must be purged before the LH_2 tank valve is open. The temperature of LH_2 is typically between 20–30 K. If air or N_2 is left in the system, solid N_2 (solidification temperature 63 K), O_2 (solidification temperature 54.4 K) and ice will form causing blockage resulting in unsafe operating conditions.
- When purging, always flow forward and vent downstream. Be aware of check valves and valve positioning. Ensure that the entire subsystem is purged.
- Purging can be done with either He or H_2.

Table A3.1 Reference table to safety parameters for hydrogen [18]. See also Hord, NBS 690, Is Hydrogen Safe? [19]

Gaseous, liquefied, slush, and solid para-hydrogen[a]		
Equivalent vol solid @ TP/vol liquid @ NBT	0.8181	
Equivalent vol gas@ NTP[b]/vol liquid@ NBT	845.1	
Pressure required to maintain NBP liquid density in NTP GH_2 (fixed volume, no venting)	172 MPa	25,000 psi
Joule-Thomson inversion temperature	193 K	−112 °F
Heat of combustion (low)	119.93 kJ/g	51,573 Btu/lbm
Heat of combustion (high)	141.86 kJ/g	61.003 Btu/lbm
Limits of flammability in NTP air	4.0 to 75.0 vol %	
Limits of flammability in NTP oxygen	4.1 to 94.0 vol %	
Limits of detonability in NTP air	18.3 to 59.0 vol %	
Limits of detonability in NTP oxygen	15 to 90 vol %	
Stoichiometric composition in air	29.53 vol %	
Minimum energy for ignition in air	0.017 mJ	1.6×10^{-8} Btu
Autoignition temperature	858 K	1085 °F
Hot air-jet ignition temperature	943 K	1238 °F
Flame temperature in air	2318 K	3713 °F
Thermal energy radiated from flame to surroundings	17 to 25%	
Burning velocity in NTP air	265 to 325 cm/s	104 to 128 in/s
Detonation velocity in NTP air	1.48 to 2.15 km/s	4856 to 7054 ft./s
Diffusion coefficient in NTP air	0.61 cm^2/s	0.095 in^2/s
Diffusion velocity in NTP air	< 2.0 cm/s	0.79 in/s
Buoyant velocity in NTP air	1.2 to 9 m/s	3.9 to 30 ft./s
Maximum experimental safe gap in NTP air	0.008 cm	0.003 in
Quenching gap in NTP air	0.064 cm	0.025 in
Detonation induction distance in NTP air	L/D ≈ 100	
Limiting oxygen index	5.0 vol %	
Vaporization rates (steady state) of liquid pools without burning	2.5 to 5.0 cm/s	1 to 2 in/s
Burning rates of spilled liquid pools	0.5 to 1.1 mm/s	0.02 to 0.04 in/s
Energy of explosion (theoretical explosive yield)		
≈ 24 (g TNT)/(g H$_2$)	≈ 24 (lbm TNT)/(lbm H$_2$)	
0.17 (g TNT)/(kJ H$_2$)	4.0×10^{-4} (lbm TNT)/(Btu H$_2$)	
2.02 (kg TNT)/m^3 NTP GH$_2$)	0.126 (lbm TNT)/(ft^3 NTP GH$_2$)	
1.71 (g TNT)/(cm^3 NBP LH$_2$)	107.3 (lbm TNT)/(ft^3 NBP LH$_2$)	

Note: Temperature dependent ortho-para changes continue to influence properties in the solid state. Several properties (such as specific heat, thermal conductivity, and thermal diffusivity) are highly sensitive to actual ortho-para composition
[a]McCarty et al. [16]
[b]NTP = 293 K (68 °F) and 101.3 kPa (14.69 psia)

Table A3.2 List of codes and standards applied to LH_2 infrastructure site

Standard	Application
NFPA 2 hydrogen technologies code	Entire system
NFPA 30 flammable & combustible liquids	LH_2 tank
CGA H-5 standard for bulk H_2 supply systems	LH_2 tank
CGA G-5.5 vent systems	Vent stack
ASME B31.3/12	Pressure piping
ASME boiler & pressure vessel code	Ground storage
SAE J2601-1 fueling protocol for light vehicles	Dispenser
SAE J2601-2 fueling protocol for heavy duty	Dispenser
SAE J2600 compressed H_2 vehicle refueling	Nozzle
NFPA 70 electric safety code	Dispenser
IFC international fire code	System siting

Cryogenic Subsystem Pressure Test
- Testing any subsystem at maximum allowable working pressure (MAWP) and ambient temperature is not sufficient.
- Most leaks occurred when the system is cold (65–180 K).
- To ensure the system is leak-free, it must be tested with H_2, at MAWP and lowest design temperature.

High-Pressure Subsystem Purging
- When purging, always flow forward and vent downstream.
- Be aware of check valves and valve positioning.
- Ensure that the entire subsystem is purged. Purging can be done with H_2.

High-Pressure Subsystem Pressure Test
- Be aware of valve positioning and check valves during pressure test.
- Most leaks occurred when the ambient temperature drops to its minimum (−40 °C, depending on location) caused by embrittlement of elastomeric seals of valves and sensors.
- To ensure the system is leak-free, it must be tested at MAWP and lowest design temperature.

Dispenser Pressure Test
- The H_2 delivered to the dispenser is precooled to −40 °C by the chiller/heat exchanger.
- Most leaks occurred when the nozzle and elastomeric seals get cold after several fills.

- To ensure the system is leak-free, the dispenser must be tested after it reaches thermal steady state by performing multiple sequential fills.

Defueling a Horizontally Mounted Tank with Faulty Valve

In the case of the tank valve which is stuck closed, the tank needs to be defueled for valve replacement.

Procedure
- The tank is moved away from ignition source, outdoors if not already.
- The valve is loosened until it starts leaking.
- After an hour the valve and end plug can be removed.
  First lesson learned:
- Do not conclude that all the H_2 is vented.
- H_2 is still trapped in the space above tank openings.
  Second lesson learned:
- Avoid ignition sources.
- Do not use any source of heating in a tank which may have any amount of H_2.
- *A halogen light bulb will cause detonation of any H2 left in the tank.*
- **If any light is needed to inspect the tank interior surface, select one with LED bulb.**

In Summary
- Cryogenic subsystem purging with He, or H_2, is critical to avoid contamination and blockage, resulting in unsafe operating conditions.
- Cryogenic subsystem pressure test must be carried out at cryogenic temperature.
- High-pressure subsystem and pressure test must be carried out at −40 °C, when the seals are cold.
- When defueling a tank be aware of trapped H_2.
- Avoid ignition sources like halogen light.

Advanced Monitoring Methods
- Continuous monitoring of fueling can detect problems as they occur and interrupt the process.
- A H_2 sensor network can be used to determine unsafe operation, leak location, and H_2 loss quantification.
  For further details, see also reference Najjar [23].

A4: Non-aerospace Liquid Hydrogen Storage

Abstract This Appendix covers most non-aerospace liquid hydrogen storage methods.

Keywords Liquid hydrogen storage; Non-aerospace LH_2 storage; Maritime LH_2 storage; Automotive LH_2 storage; Space LH_2 storage; Medical LH_2 storage

Liquid Hydrogen Storage Tanks

The intent of storing cryogens in a commercial aircraft is of course a new venture, where we, as much as possible need to refer to prior art and work that has been done since the early 1960s. More recent research refers more distinctly to storage tanks, automotive (truck/cars) and space applications, and potential opportunities for healthcare fields.

Initially, liquid hydrogen was used as a "piggy-back" cryogen to greatly reduce the boil off of large liquid helium storage containers [24]. During that time, key information could be obtained from those early storage tanks [25] with regard to the controls necessary to prevent hydrogen ignition.

One of the biggest storage tanks for liquid hydrogen at that time was the Los Alamos liquid tank [26] (Fig. A4.1).

Note that the liquid level is measured with a capacitive probe. This is a big size; however, temperature is held constant from top to bottom below the liquid/vapor line. Temperatures are measured using Pt100-type sensors.

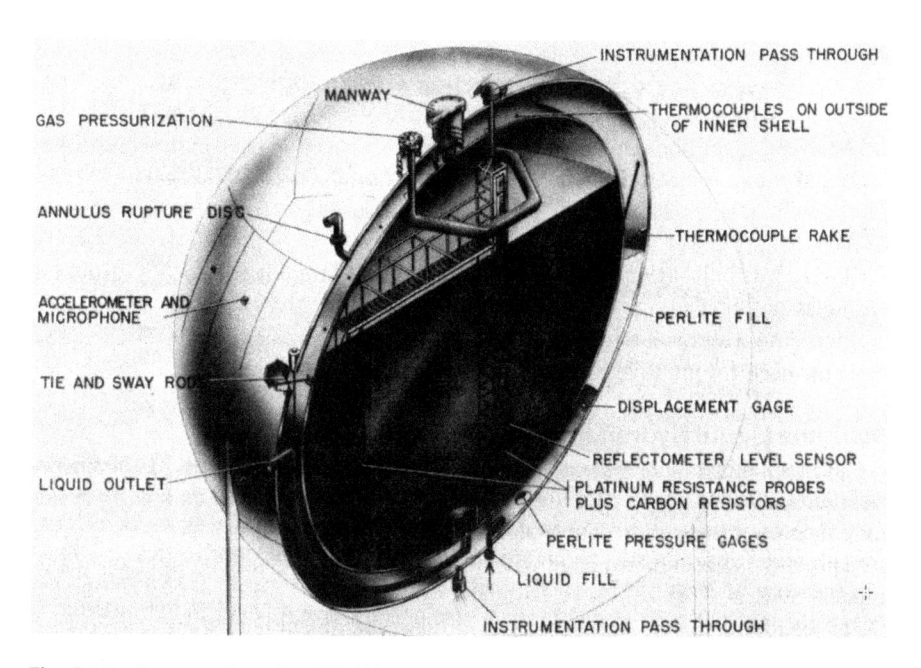

Fig. A4.1 Cutaway view of a 500,000 gal (18,927,000 liters) liquid hydrogen dewar showing instrumentation

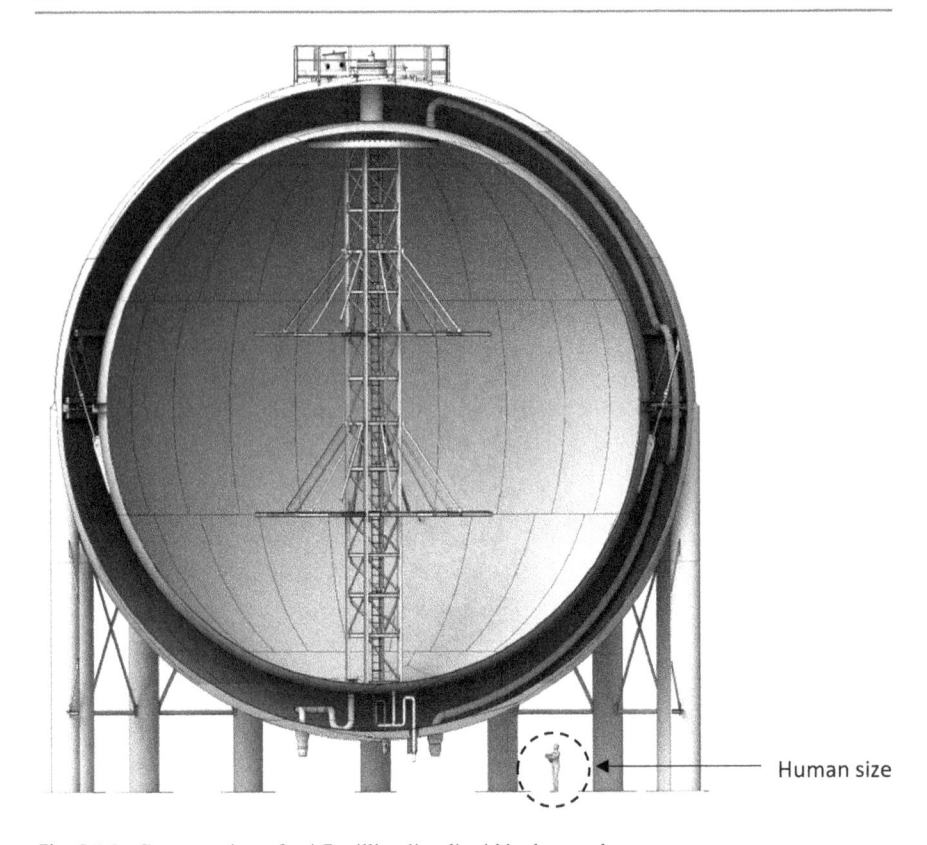

Fig. A4.2 Cutaway view of a 4.7 million liter liquid hydrogen dewar

An even bigger liquid hydrogen storage tank has recently been built by NASA in 2021 and commissioned. The usable capacity of this tank is 1,250,000 gal (4,732,000 liters) with an outer diameter of 24 m and a mean average working gauge pressure of 6 bar [27] (Fig. A4.2).

A very low boil-off rate was obtained by replacing traditional aerogel insulation with glass bubbles.

Figure A4.3 shows a typical 124,918-liter storage container used by NASA [28]. Note the need for multiple reinforcement rings (stiffeners) on the outer shell.

Maritime Liquid Hydrogen Tanks

An even smaller liquid hydrogen tank for example is produced by MAN Energy Solutions [29], out of steel for a superyacht. The liquid hydrogen tank will provide hydrogen for a fuel cell to enable zero emission propulsion (Fig. A4.4).

This tank is vacuum insulated with a total liquid hydrogen fill volume of 92,000 liters at 20 K.g

To feed the fuel cell with gaseous hydrogen at ambient temperature, vaporizers and other process equipment are installed, such as safety valves, housed in the square box at the right end (tank connection space).

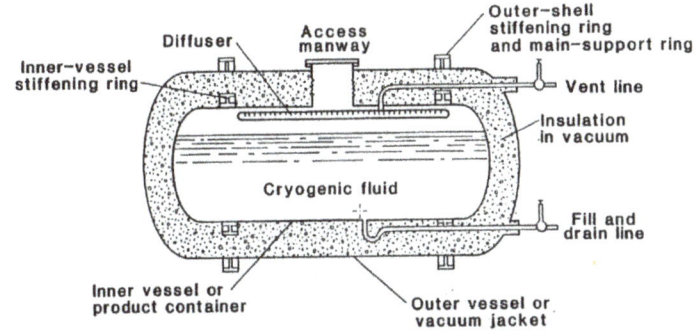

Fig. A4.3 Horizontal, cylindrical LH$_2$ storage container. (Courtesy of NASA)

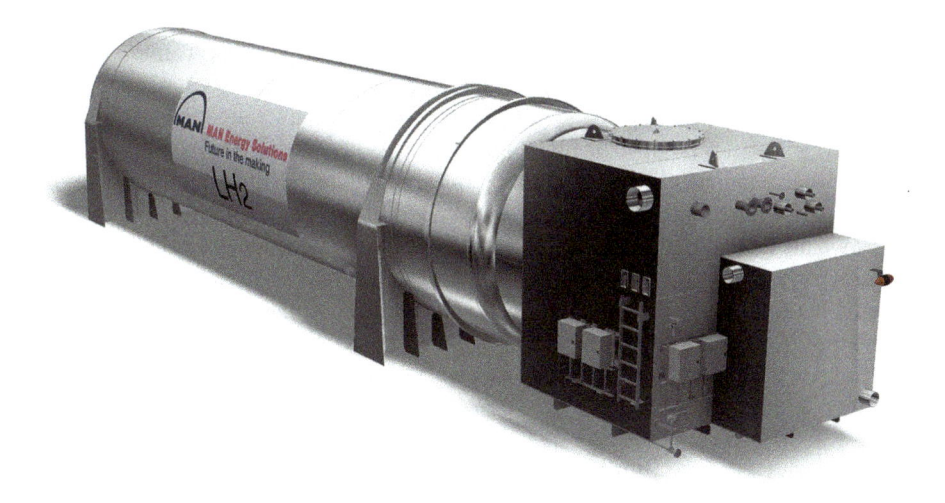

Fig. A4.4 MAN Energy Solutions LH$_2$ tank. (Courtesy of MAN Energy Solutions Sverige AB)

Liquid Hydrogen Cars

Introduction
In 1973, L.O. Williams of Martin Marietta Aerospace, based at Denver, published his visionary review article of why hydrogen-powered automobiles must use liquid hydrogen [30] when comparing liquid hydrogen with other fuels. Below is the abstract on this publication that is applicable and up to date as it is today, in each and every aspect, particularly with respect to environmental pollution and new technology:

> *A problem of particular interest and importance in the evaluation of a future hydrogen-oxygen economy is hydrogen's use as a fuel for highway vehicles.*
>
> *If, as many authors suggest, hydrogen is the only fuel that will simultaneously keep the environment clean and conserve natural resources, it must eventually be applied to the private vehicles that are a major pollution offender. Although a number of automobiles have been successfully operated on hydrogen (and this represents no large problem even with unmodified engines), the question remains of how to carry the fuel. This problem is primarily in the technology of the fuel tank—that is how to carry the hydrogen on board the vehicle safely, economically, and with a minimum of other penalties. It is unfortunate that this problem has attracted a number of suggested solutions that are less than optimum (for example, high-pressure gas and metallic hydrides) and that detract from the credibility of entire concept. If, as this author contends, the use of cryogenic liquid hydrogen is the sole, real candidate, its use will have a broad effect on industry. While requiring no basic research, it will necessitate key engineering developments in both the very large cryogenic production facilities and the small vehicle storage vessels and ancillary systems and hardware.*

Compared with other hydrogen storage methods, the following advantages for storing liquid hydrogen were postulated:

It offers the lowest cost and weight per unit energy, needs only a simple supply logistics, provides a normal refueling time, and most likely offers a low implementation cost. It also does not appear to show unsurmountable safety problems. Disadvantages are the loss of fuel when the car is not in use. May require a large tank size accompanied by engineering problems with respect to liquid safety.

Williams also continued to work on his proposals propagating "Clean energy via Cryogenic Technology" and how to build a hydrogen infrastructure [31].

As early as 1980, the Los Alamos Scientific Laboratory developed and tested a liquid hydrogen-fueled car with a 128-liter dewar tank with a range of 241 km [32]. Pointing to the future, this car was thought to be economical and practical when petroleum products would no longer be available, see also US patent 4,043,140 for further details.

Automotive Systems Burning Liquid Hydrogen
A different solution is offered by the Salzburg Aluminum Group, featuring a 2-tank design to store liquid hydrogen at an operating pressure of <10 bar. The design allows a 30% higher volumetric energy density as compared to compressed hydrogen at 700 bar. Hold time is approx. 9 days of an inventory of 100 to 150 liters for

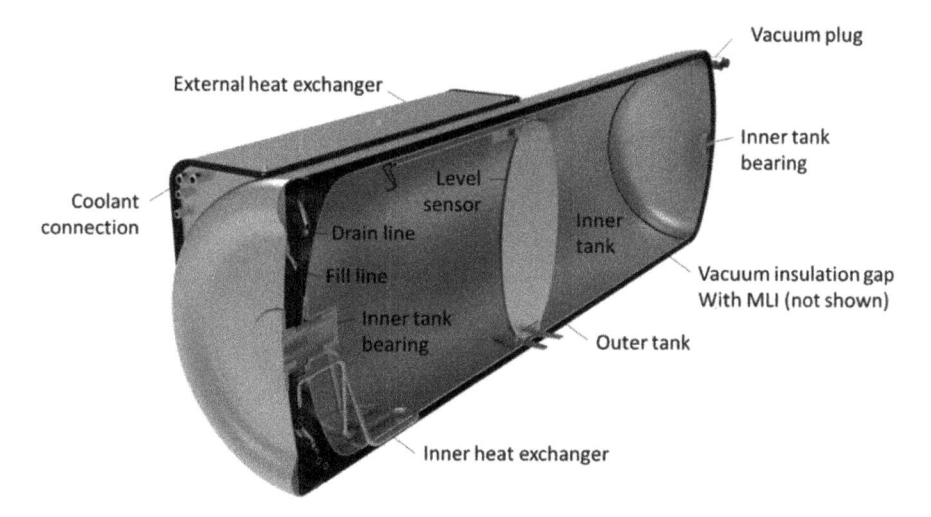

Fig. A4.5 SAG metallic liquid hydrogen tank

trucks [33]. The tank is built with an inner and outer shell using multilayer insulation in between this space (Fig. A4.5).

Automotive Systems Using Liquid Hydrogen with Fuel Cells

Although quite a few cars use compressed hydrogen at 700 bars [34] (BMW started to build the first cryo-compressed hydrogen unit in 2011), only BMW very early on realized the full potential of using liquid hydrogen as an efficient storage medium.

Magna Steyr collaborating with BMW recently showed the LH_2 tank of the BMW Hydrogen 7 tank [35]. As of May 2024 Magna Steyr however discontinued engagement in liquid hydrogen tanks (Fig. A4.6).

The cryogenics of this tank is explained by Linde as follows, mentioning downsides for vehicles [36].

Even though liquid hydrogen tanks can store more in a given volume than compressed hydrogen tanks and liquid hydrogen carries more energy density than gaseous hydrogen, there are some downside to using liquid hydrogen to power large vehicles (Fig. A4.7).

Linde acknowledges issues with high hydrogen boil-off, liquefaction, weight, volume, and tank cost. Those parameters obviously need to be addressed for any economical use of liquid hydrogen.

Safety concerns are also more of an issue with hydrogen that is in a liquid state.

The storage vessel holds about 8 kg of liquid hydrogen [37] (Fig. A4.8).

Further Reading

Groundbreaking modeling work on the dynamics of cryogenic hydrogen storage in pressure vessels has been executed by Ahluwalia [38] at Argonne National Lab with funding from US Department of Energy's office of Energy Efficiency and Renewable Energy.

Fig. A4.6 BMW Hydrogen 7 Magna Steyr composite liquid hydrogen tank

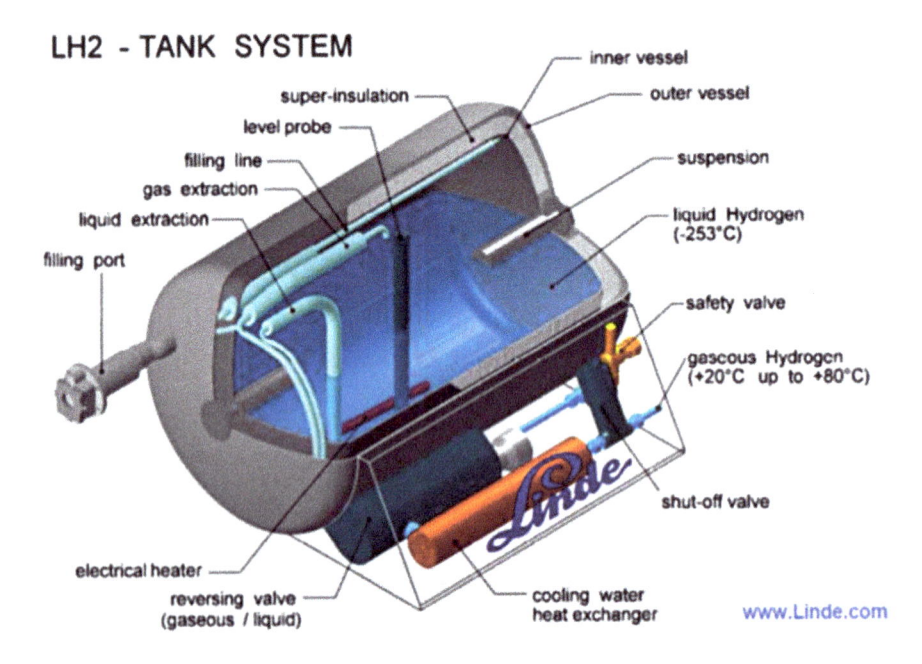

Fig. A4.7 Linde tank system

Appendices 303

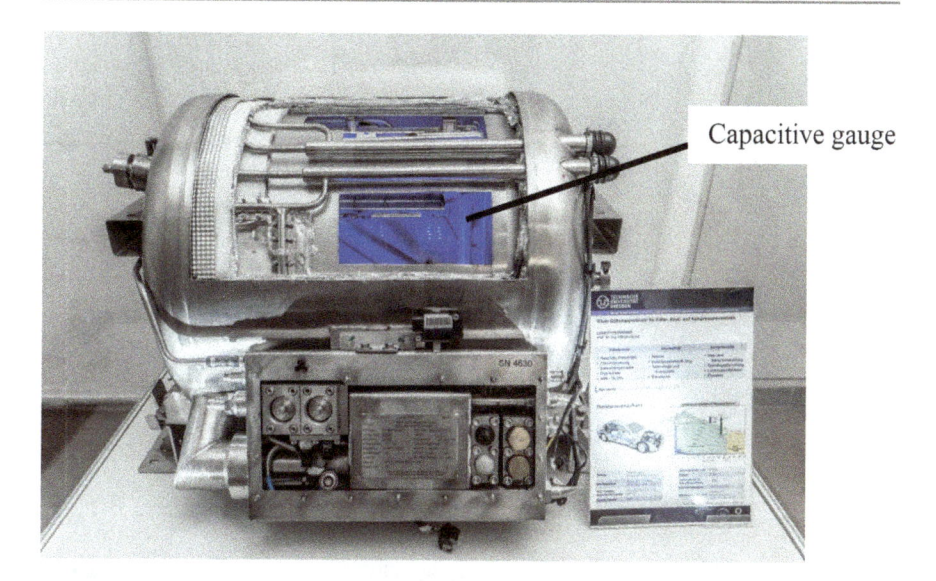

Fig. A4.8 LH$_2$ storage vessel for automotive application with a capacitive level gauge

Fig. A4.9 Launch vehicle composite cryotank development steps. (Courtesy of NASA)

Liquid Hydrogen in Space

NASA has been researching composite tank designs and build for their space missions, see Fig. A4.9.

The figure depicts the development progress in the composite LH$_2$ cryotank development for those launch vehicles.

A composite tank would give an approx. 30% weight reduction and 25% in cost as compared to an Aluminum-Lithium tank (see also CCTD project) [39].

Fig. A4.10 Liner-less composite tank test—2021. (Courtesy of Boeing)

Liquid Hydrogen in Space and for Future Aviation
Boeing Cryogenic Fuel Tank

Figure A4.10 by Guzman shows Boeing's all-composite cryogenic fuel tank undergoing pressure testing at NASA's Marshall Space Flight Center. The liner-less tank has a diameter of about 4.3 meters and is built on NASA composite technology. The vessel did undergo multiple test cycles to operating pressure [40].

For further space composite tanks and tank designs, please references [41–46].

Besides this effort companies like Composites—Gloyer-Taylor Laboratories, Inc. (GTL) actively pursue the volume production of ultra-lightweight LH_2 composite tanks based on the patented BHL™ cryotank technology developed for space launch applications. Vertical as well as horizontal versions are being tested with composite tubes attached.

Fabrum industries eyes the marine and aviation market in their development of composite tanks including some Japanese companies who very early on developed composite dewars for providing measurements in a non-magnetic environment.

Liquid Hydrogen Tanks for Medical Applications

Figures A4.11 and A4.12 shows a dual-purpose extremities scanner working either with liquid hydrogen or with liquid helium [47].

Figure A4.12 postulates a possible, liquid hydrogen-cooled MRI magnet system, using medium or high temperature superconductors [48]

Fig. A4.11 GE—MgB$_2$ scanner, 1.5 to 3 T, 2015

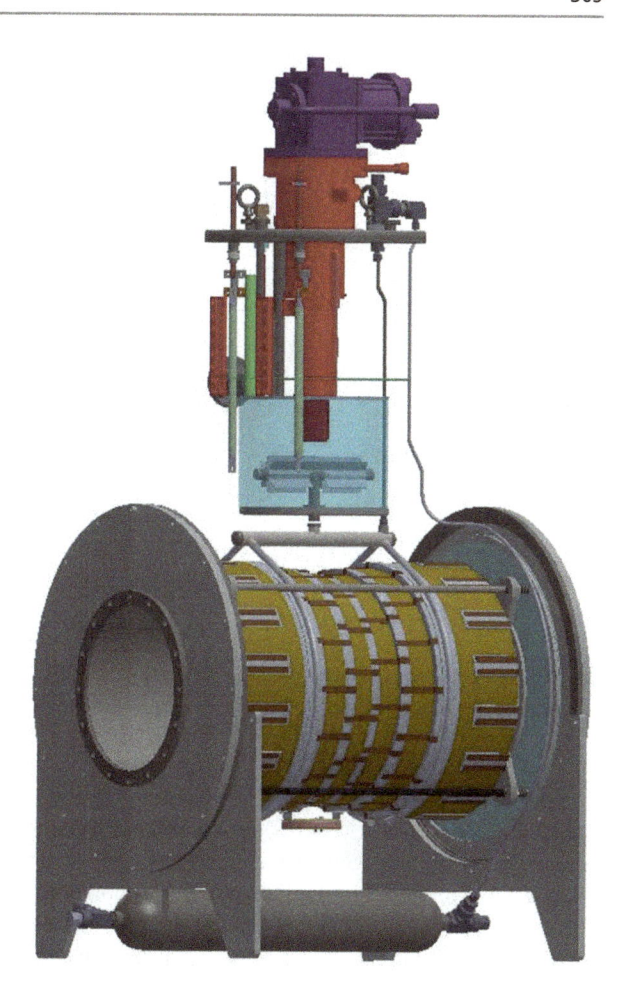

Liquid Hydrogen for Superconducting Applications

Early economic calculations in late 1960 6 revealed that large high-purity aluminum structures for use in high-field magnets suffer from two main aspects. Firstly, a hydrogen refrigerator is required to keep the magnet at 21 K, for cooldown and keeping bay of static and dynamic loads during operation and, if the magnet coil should be immersed, the added quantity of liquid hydrogen required. The initial hope at that time vaporized in that aluminum coils may be cost-efficient in the light of Kohler's rule. With magnetic fields higher than 4 Tesla, however the magneto-resistive effect was much higher than expected at 21 K with increased wire purity.

Wit the arrival of new high temperature superconductors, superconducting magnet applications become attractive using hydrogen instead of nitrogen in its liquid phase to increase the operating margin of the superconducting component. Usually, we would like to operate at a 65 K maximum temperature when using HTS with LN$_2$ cooling but allowing to get down to 25–30 K results in a big benefit on the higher critical current carrying capacity.

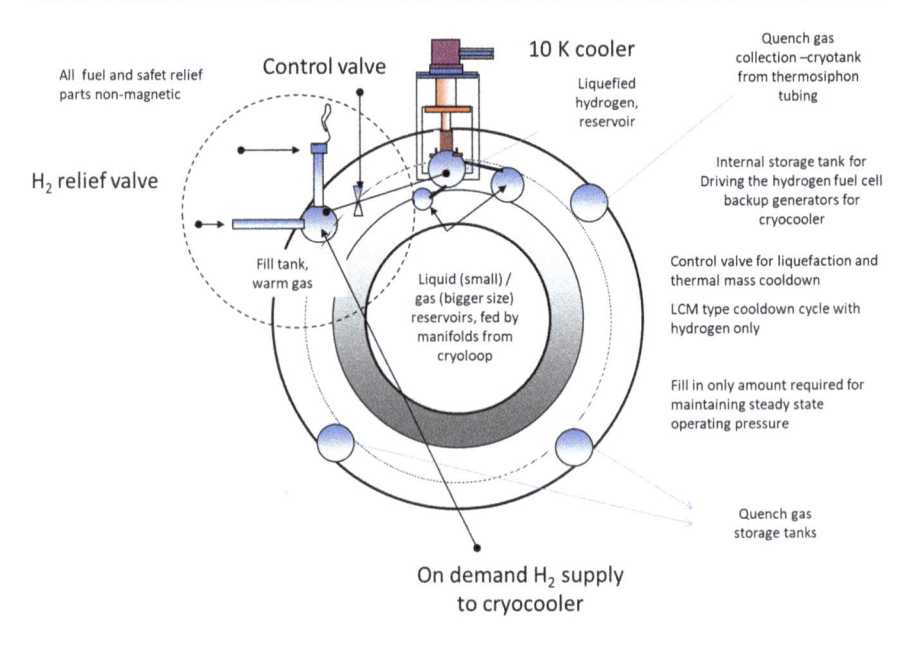

Fig. A4.12 GE—liquid hydrogen cooled MRI magnet system

One of the first applications pondered the use of a superconducting magnetic energy storage system (SMES) as an emergency power supply, embedded in an already existing hydrogen environment [49, 50].

In 2009, thoughts of establishing a micro power grid began to take shape in the form of feasibility studies combining superconducting cables, SMES, hydrogen infrastructure, fuel cells, and power converters, whereas a cable would be transferring electrical power as well as hydrogen as a refrigerant fuel [51].

Subcooled hydrogen is also being used to cool the supercritical H2 neutron moderators that surround the target. This is achieved by providing a cooling power of 20 kW at 16.5 K via gaseous helium to a He/supercritical H_2 heat exchanger [52].

Please Note
Superconducting applications provide the greatest synergy when designing for aircraft composite tanks. Many reinforced epoxy mixtures and fibers work for superconducting magnet applications. An up-to-date review of current resins for magnet impregnation is given by Feldman in [53].

A5: Internet Snippets on Cryogenics and Superconductivity for Aircraft

Abstract This Appendix covers the Airbus activities in cryogenics and superconductivity, as publicly available of 2024.

Keywords ASCEND; Electric propulsion; ASCEND electric motor; ASCEND powertrain; Sustainable air travel

The Airbus Program ASCEND
A Supercooled Breakthrough for Future Electric Propulsion Systems
Figure A5.1 [54] tank is an early depiction of a liquid hydrogen tank proposed to go on an aircraft.

ASCEND 500 kW powertrain consists of the following components:

- Superconducting distribution system, including cables and protection item.
- Cryogenically cooled motor unit.
- A superconducting motor.
- A cryogenic system (Fig. A5.2).

This is a key consideration, as increasing the power of current electrical aircraft systems from a few hundred kW to the MW required for a fully electric aircraft is no easy feat. Simply put, more power increases weight and installation complexity, and generates more heat.

Source: https://www.airbus.com/en/newsroom/stories/2021-03-cryogenics-and-superconductivity-for-aircraft-explained

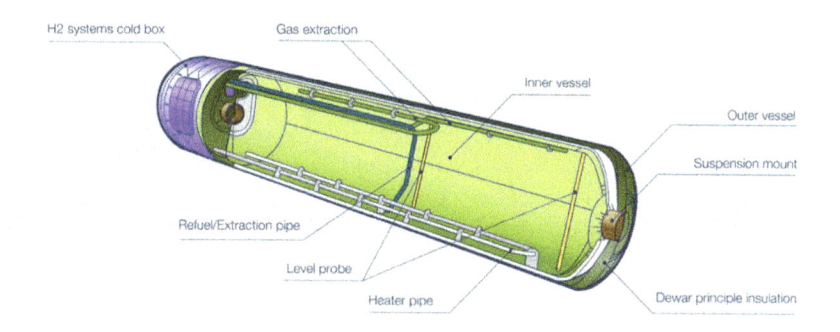

Fig. A5.1 Airbus liquid H2 tank—artist's view [54]

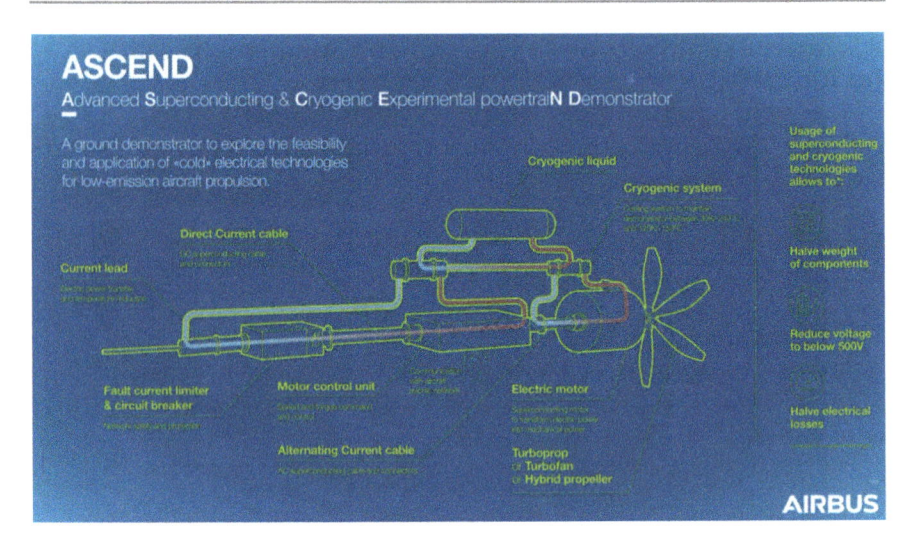

Fig. A5.2 ASCEND ground demonstrator with liquid tank, electric motor, and cryogenic components

Hydrogen-Based Aviation

King's College London's Prof David Moxey, Dr. Mark Ainslie, and Dr. Mashy Green explore some of the engineering hurdles on the path to Jet Zero.

Hydrogen-based solutions are a leading candidate in the move towards net-zero aviation, pivoting away from conventional jet fuels such as aviation kerosene, to hydrogen combustion (H_2C) using cryogenic liquid hydrogen (LH_2). Currently, no achievable alternatives for 100 per cent carbon-free long-haul flights exist other than H_2C.

Tank Development Steps

A move to H_2C will require redesigning not only combustion engines—an ongoing effort by the likes of Airbus, Rolls-Royce and others—but of the whole aircraft to be able to safely store and manage LH_2 onboard. To meet these challenges, significant research is required to better understand the fundamental behavior of liquid hydrogen.

Industry needs to embrace models which accurately simulate LH_2 and superconductors' combined electromagnetic, thermal and mechanical behaviors to avoid the need for costly physical iterations, so they can be deployed in a tightly regulated industry.

For cooling, LH_2 is being explored as a viable option for -250 °C temperatures since superconductors perform significantly better the cooler they are. Researchers are even looking at dual-purpose LH_2 systems for fueling and cooling all-electric aircraft, like the NASA-backed CHEETA project.

Source: https://www.theengineer.co.uk/content/opinion/comment-hydrogen-and-cryogenics-hold-keys-to-net-zero-aviation/

Airbus Recently Announced to Develop a Cryogenic Hydrogen Tank for Future Planes as of June 2023

Airbus, Zero Emission Development Centre (ZEDC) has already started its work on the development of the cryogenic hydrogen tank, for its next generation of aircraft. The UK facility in Filton, Bristol has already started working on the technology's development.

All Airbus ZEDCs are expected to be fully operational and ready for ground testing with the first fully functional cryogenic hydrogen tank during 2023, and with flight testing starting in 2026, said the company.

Source: https://www.cryogenicsociety.org/index.php?option=com_dailyplanetblog&view=entry&category=industry-news&id=59:airbus-facility-to-develop-cryogenic-hydrogen-tank-system-for-future-planes

ASCENDing to New Heights with Cryogenic Superconductivity

ASCEND, Airbus' superconductive powertrain demonstrator, has achieved a world first. Over the last 3 years, a small team of experts has developed and manufactured a cryogenic superconducting electric propulsion system purposely built to aerospace specifications.

In November 2023, ASCEND took its final step. The team successfully powered-on the 500 kilowatt powertrain at the system's core. It's an exciting breakthrough.

In the quest for cleaner transportation, the marriage of hydrogen-powered fuel cells and high temperature, or cryogenic, superconductivity could be a game changer.

The current density of superconducting tape is one hundred times that of a copper equivalent. Cryogenically freezing that tape enables it to carry electrical power from a fuel source to a propulsion system with practically no resistance.

That's why, nestled inside a large warehouse just outside of Toulouse, France, Airbus UpNext's **A**dvanced **S**uperconducting and **C**ryogenic **E**xperimental powertrai**N** **D**emonstrator (ASCEND) team has spent the last 3 years exploring the impact of cryogenic superconductivity on the electrical infrastructure that could power the next generation of low-carbon aircraft.

In November 2023 with the successful powering-on of a 500 kilowatt powertrain, consisting of superconducting tape, a cryogenic motor control unit and cooling system, and a superconducting motor. For context, getting the equivalent of today's city-hopping turboprops off the ground using electric power alone would require around eight megawatts (Fig. A5.3).

The ASCEND powertrain drove this superconducting electric motor during a successful power-on at Airbus' e-Aircraft System testing facility in Ottobrunn, Germany.

All Airbus ZEDCs are expected to be fully operational and ready for ground testing with the first fully functional cryogenic hydrogen tank during 2023, and with flight testing starting in 2026 (Fig. A5.4).

Source: https://www.airbus.com/en/newsroom/stories/2023-12-ascending-to-new-heights-with-cryogenic-superconductivity

Fig. A5.3 ASCEND electric motor

Fig. A5.4 ASCEND electric motor, assembly

A Cryogenic Aircraft Engine for Sustainable Air Travel

ASuMed is another program funded by the European union as part of the Horizon 2020 Research and Innovation program.

The goal of the ASuMed project was to develop an aircraft that utilizes superconductivity to achieve the power densities and efficiencies required for hybrid-electric distributed propulsion aircraft (HEDP).

ASuMED motor is based on double cryostat concept and includes two separate cryostats with two separate cooling systems. The rotor is cooled using liquid helium, while the stator is cooled with hydrogen.

Demaco and other members of the consortium demonstrated that a compact and superconducting electric motor can absorb enough energy to be used in aviation industry.

Source: https://demaco-cryogenics.com/projects/a-cryogenic-aircraft-engine-for-sustainable-air-travel/

Aircraft Hydrogen Combustion Efforts

Hydrogen burns 10 times faster than jet-fuel, causing concerns on flame stability. Additional challenges here are the need for cryogenic fuel pumps and new piping and seals to accommodate liquid hydrogen temperatures.

Possible long-range hydrogen-powered A380 aircraft will have liquid hydrogen tanks for combusting hydrogen in modified GE Passport turbofans [55].

A6: Cryogenic Reference Books

Abstract This Appendix is meant for the new generation of students or researchers who would like to learn more about Cryogenics.

Keywords Cryogenic books; Cryogenic references; Further reading; Cryogenic hydrogen history

Alphabetical *(without superconducting technology books)*

1. CEC/ICMC *Advances in Cryogenic Engineering* Conference proceedings (semi-annually since 1954).
2. Ackermann R A *Cryogenic Regenerative Heat Exchangers* Plenum Press 1997.
3. Atrey M D *Cryocoolers Theory and Applications* International Cryogenics Monograph Series Springer 2020.
4. Bailey C A *Advanced Cryogenics* Plenum Press 1971.
5. Barron R *Cryogenic Systems* Mc Graw-Hill New York 2nd edition 1985.
6. Barron F R Nellis G F *Cryogenic Heat Transfer* 2nd edition CRC press 2016.
7. Barron T H K White G K **Heat capacity and thermal expansion at low temperatures** Springer International Cryogenics Monograph Series 1995.
8. Bell *Cryogenic Engineering* N J Englewood Cliffs 1963.
9. Maytal B-Z Pfotenhauer J M *Miniature Joule-Thomson Cryocooling* Springer International Cryogenics Monograph Series 2013.
10. Clark A F Reed R P Hartwig G *Nonmetallic materials and composites at low temperature* Plenum Press 1979.
11. Codlin E M *Cryogenics and refrigeration*—a bibliographical guide Plenum Press New York 1968.
12. Conte R R *Elements de Cryogenie* Masson & Cie 1970 (in French).
13. Croft A J *Cryogenic Laboratory Equipment* Plenum Press 1970.
14. Demko J A Fesmire J E Shu Q S *Cryogenic Heat Measurement* CRC Press 2022.

15. Dana L l **Cryogenic Science and Technology** Union Carbide Corporation 1985.
16. Din F Cockett A H *Low temperature techniques* Interscience Publishers New York 1960.
17. Gilmore D G *Spacecraft thermal control handbook Volume I: Fundamental Technologies* The Aerospace Press AIAA 2002.
18. Donabedian M *Spacecraft thermal control handbook Volume II: Cryogenics* The Aerospace Press AIAA 2003.
19. Donnelly R J *Experimental Superfluidity* Chicago Lectures in Physics The University of Chicago Press 1967.
20. Edeskuty F J Stewart W F *Safety in the Handling of Cryogenic Fluids* Plenum Press 1996.
21. Ekin J W *Experimental techniques for low-temperature measurements* Oxford University Press 2006.
22. Elrod C W *Design Handbook of liquid and gaseous helium* AD 410935 US Department of Commerce 1963.
23. Enss C Hunklinger S *Low-temperature physics* Springer Berlin 2005.
24. Fastowski W G Petrowski J W Rowinski A E *Kryotechnik* Akademie Verlag 1970 (in German).
25. Filina N N Weisend II J G *Cryogenic Two-Phase flow* Cambridge University Press 1996.
26. Flynn T M *Cryogenic Engineering* Marcel Dekker 2nd edition New York 2005.
27. Frey H Haefer R A Tieftemperaturtechnologie *(Low temperature technology)* VDI Verlag (in German) 1981 (in German).
28. Frost W *Heat transfer at low temperatures* Plenum Press 1975.
29. Haefer R A *Cryopumping Theory and Practice* Oxford Science Publications 1989.
30. Hands B A *Cryogenic Engineering* Academic Press 1986.
31. Hartwig G *Polymer properties at room and cryogenic temperatures* Plenum Press 1994.
32. Hartwig G Evans D *Nonmetallic materials and composites at low temperature 2* Plenum Press 1982.
33. Hartwig G Evans D *Nonmetallic materials and composites at low temperature 3* Plenum Press 1984.
34. Haselden G G (Ed) *Cryogenic Fundamentals* Academic Press London 1971.
35. Hausen H Linde H *Tieftemperaturtechnik* Springer-Verlag Berlin 1985, 597 pp (in German).
36. Hoare F E Jackson L C Kurti N *Experimental Cryophysics* Butterworth London 1961.
37. Jacobsen R T Penoncello S G Lemmon E W *Thermodynamic Properties of Cryogenic Fluids* Springer International Cryogenics Monograph Series 2017.
38. Jensen T E et al. **Brookhaven National Laboratory Selected Cryogenic Data Notebook** BNL 10200-R DOE Vol. 1 1980.
39. Jelinek J Malek Z *Kryogenni technika* Praha SNTL Nakladatelstvi 1982 (in Czech) and Done R *The safe use of Cryogenic Technologies* IOP Publishing 2021.

40. Jha A R *Cryogenic Technology and Applications* Butterworth Heinemann 2006.
41. Kaganer M G *Thermal insulation in cryogenic engineering* Israel Program for Scientific Translations Jerusalem IPST Press 1969.
42. Kalia S F Shao-Yun *Polymers at Cryogenic temperatures* Springer 2013.
43. Keller W E *Helium-3 and Helium-4* Plenum Press 1969.
44. Kent A *Experimental Low-Temperature Physics* American Institute of Physics New York 1993.
45. Kropshot R H Birmingham B W Mann D B *Technology of Liquid Helium* National Bureau of Standards Monograph 111 US Government Printing Office 1968.
46. Leachman J W, Jacobsen R T, Lemmon E W, Penoncello S G *Thermodynamic Properties of Cryogenic Fluids* Springer International Cryogenics Monograph Series 2017.
47. Lounasmaa O V *Experimental Principles and Methods below 1 K* Academic Press 1974.
48. Malkov Fradkow *Kältetechnik* Moskau 1971 (in Russian).
49. McClintock M *Cryogenics* Reinhold Publishing Corporation 1964.
50. Miropolsky Z L Soziev R I *Fluid Dynamics and Heat transfer in superconducting equipment* Hemisphere Publishing Corporation 1987.
51. NBS technical books online, see https://pages.nist.gov/NIST-Tech-Pubs/date.html (Appendix 7).
52. Organ A J **Stirling and Pulse-tube Cryocoolers** Professional Engineering Publishing Limited 2005.
53. Pobell F *Matter and methods at low temperatures* 3rd edition Springer 2007.
54. Pohlmann W Iket GmbH *Taschenbuch der Kältetechnik* VDE Verlag 2018 (in German).
55. Putselyk S *Cryogenic Receivers* Physics research and technology Nova Science Publishers 2012.
56. Putselyk S (ed.) *Recent developments in cryogenic research* Physics research and technology Nova Science Publishers 2019.
57. Reed R P Clark A F *Materials at low temperatures* American Society of Metals Carnes Publication Services 1983.
58. Reed R P Horiuchi T *Austenitic Steels at low temperatures* Plenum Press 1983.
59. Richardson R C Smith E N *Experimental Techniques in Condensed Matter Physics at low temperatures* Westview Press Oxford 1998.
60. Rose-Innes A C *Low temperature laboratory techniques* English University Press London 1964.
61. Sawyer C Astin A V *Low-Temperature Physics* Proceedings of the NBS Semi-centennial Symposium on low temperature physics 1952 NBS Circular 519.
62. Scoczeń B T *Compensation Systems for Low Temperature Applications* Springer Verlag 2004.
63. Scott R B *Cryogenic engineering* Van Nostrand New York 1959.
64. Scott R B Denton W H Nichols C M *Technology and uses of liquid hydrogen* Pergamon Press 1964.
65. Scurlock R G *Low-Loss storage and handling of Cryogenic Liquids: The Application of Cryogenic Fluid Dynamics* Kryos Publications 2006.

66. Serafini T S Koenig J L *Cryogenic properties of polymers* Marcel Dekker New York 1968.
67. Smith A U *Current Trends in Cryobiology* Plenum Press 1970.
68. Timmerhaus K D Flynn T M *Cryogenic Process Engineering* Plenum Press 1989.
69. Timmerhaus K D Reed P R *Cryogenic Engineering,fifty years of progress* Springer International Cryogenics Monograph Series 2007.
70. Van Lammeren J A *Technik der tiefen Temperaturen* (in German) Berlin Springer 1941.
71. Van Sciver S *Helium Cryogenics* Plenum Press 1986.
72. Vance R W Duke W NM *Applied Cryogenic Engineering* Wiley New York 1962.
73. Venkatarathnam G **Cryogenic Mixed Refrigerant Processes** Springer 2008.
74. VDI Wissensforum *Cryogenics* 1977, 2010, VDI Verlag.
75. Reiss H Chapter *K6 Superinsulation* and other cryogenic topics in **VDI Heat atlas** 2nd Edition Springer 1993.
76. Ventura G Risegari L *The Art of Cryogenic Low Temperature Experimental Techniques* Elsevier 2008.
77. Walker G *Cryocoolers Part 1: Fundamentals* Plenum Press 1983.
78. Walker G *Cryocoolers Part 2: Applications* Plenum Press 1983.
79. Walker G **Miniature Refrigerators for Cryogenic Sensors and Cold Electronics** Oxford Science Publications 1989.
80. Walker I R *Reliability in Scientific Research* Cambridge University Press 2011.
81. Weinstock *Cryogenic Technology* Boston Technical Publishers 1969.
82. Weisend II J G *Cryostat Design Case Studies, Principles and Engineering* International Cryogenics Monograph Series 2016.
83. Weisend II J G *Handbook of Cryogenic Engineering* Taylor & Francis 1998.
84. Weisend II J G *He is for Helium* Cryogenic Society of America 2018.
85. White G K Meeson P J *Experimental Techniques in Low Temperature Physics* 4th ed. Clarendon Press Oxford Science Publications 2002.
86. Wigley D A *Mechanical Properties of Materials at Low Temperatures* Plenum Press 1971.
87. Wilson D *Supercold: an introduction to low temperature technology* The Bowering Press Ltd. 1979.
88. Williamson K D Edeskuty F J *Liquid Cryogens Volume I Theory and Equipment* CRC press 1983.
89. Williamson K D Edeskuty F J *Liquid Cryogens Volume II Properties and Application* CRC press 1983.
90. Zabetakis M G *Safety with Cryogenic Fluids* Plenum Press 1967.
91. Zhao Z Wang C *Cryogenic Engineering and Technologies, Principles and Applications of Cryogen-Free Systems* CRC Press 2020.
92. Zohuri B *Physics of Cryogenics* Elsevier 2017.

Upcoming

Stautner W Ansell P Haran K Minas C *Aircraft Cryogenics* Springer International Cryogenic Monograph Series 2024.

A7: NBS/NIST: Cryogenic, Relevant Key Resources

Abstract This Appendix highlights the relevant key resources provided by NBS/NIST for future use of hydrogen, to create and accelerate the build of the required hydrogen infrastructure for aircraft and other technology fields, like off-shore superconducting generators with superconducting cables for energy and liquid hydrogen transfer as well as for other superconducting applications that use high temperature superconductors, including maritime engineering.

Keywords NIST liquid hydrogen temperature database; Selected hydrogen reports; Key documents

NBS 63	Warren/Reed **Tensile and impact properties of selected materials from 20 to 300 K** 1962
NBS 101	Reed/Mikesell **Low temperature mechanical properties of copper and selected copper alloys** 1967
NBS 168	Mc Carty/Hord **Selected properties of hydrogen** (Engineering design data) 1981
NBS 519	Astin **Low-Temperature Physics** 195
NBS 1053	Fickett **Electrical properties of materials and their measurement at low temperatures** 1982
NBS 111	Kropshot/Birmingham/Mann **Technology of liquid helium** 1968
NBS 120A	Dean **A tabulation of the thermodynamic properties of normal hydrogen from low temperatures to 540 °R and from 10 to 1500 psia** 1962
NBS 122	Richards/Steward/Jacobs **A survey of the literature of heat transfer from solid surface to cryogenic fluids** 1961
NBS 13	McClintock/Gibbons **Mechanical properties of structural materials at low temperatures** 1960
NBS 131	Childs/Ericks/Powell **Thermal conductivity of solids at room temperature and below** 1973
NBS 132	Schramm/Clark/Reed **A compilation and evaluation of mechanical, thermal and electrical properties of selected polymers** 1973
NBS 21	Corruccini/Gniewek **Specific heats and enthalpies of technical solids at low temperatures** 1960
NBS 29	Corruccini/Gniewek **Thermal expansion of technical solids at low temperatures** 1961
NBS 317	Brentari/Giarratano/Smith **Boiling heat transfer for oxygen, nitrogen, hydrogen and helium** 1965
NBS 556	Powell / Blanpied **Thermal conductivity of metal alloys at low temperatures** 1954
NBS 564	Hilsenrath/Becket et al. **Tables of thermal properties of gases** 1955
NBS 596	Jacobs **Single-phase transfer or liquid gases** 1958
NBS 664	Olien/Schiffmacher **Hydrogen-future fuel-A bibliography with emphasis on cryogenic technology** 1975
NBS 690	Hord **Is hydrogen safe?** 1976
NBS 724	Roberts **Properties of selected superconductive materials**

NBSIR 74-359	Sparks/Fickett et al. **Semi-annual report on materials research in superconducting machinery** 1974
NBSIR 76837	Clark/Weston/Arp/Hust/Trapan **Characterization of a superconducting coil composite and its components** 1976
NBSIR 76-848	Reed /Hust et al. **Semi-annual report on materials research in superconducting machinery** 1976
NBSIR 78-884	Fickett/Reed et al. **Materials studies for magnetic fusion energy applications at low temperatures-I** 1978
NBSIR 80-1633	Ekin/Kasen/Read et al. **Materials studies for superconducting machinery coil composites** 1979
NBSIR 84-3007	Hust/Lankford **Thermal conductivity of aluminum, copper, iron and tungsten for temperatures from 1 K to the melting point** 1984
NIST 177	Simon/Drechsler/Reed **Properties of Copper and Copper Alloys at Cryogenic Temperatures** 1992
NISTIR 90–3935	Goodrich/Goldfarb/Bray **Development of standards for superconductors** 1990
NISTIR 3979	Reed/Purtscher/Simon et al. **Aluminum alloys for ALS cryogenic tanks: Comparative measurements of cryogenic mechanical properties of Al-Li alloys and alloy 2219** 1993
NSRDS-NBS 8	Ho/Powell/Liley **Thermal conductivity of selected materials** 1966
NSRDS-NBS 16	Ho/Powell/Liley **Thermal conductivity of selected materials Part 2** 1968

Other key material documents NBS-ARPA:

A.F. Clark **Materials Research in Support for Superconducting Machinery**, 1973.

A.F. Clark **Materials Research in Support for Superconducting Machinery II**, 1974.

R.P. Reed **Materials Research for Superconducting Machinery III**, 1974.

R.P. Reed **Materials Research for Superconducting Machinery IV**, 1975.

R.P. Reed **Materials Research for Superconducting Machinery V**, 1976.

R.P. Reed **Materials Research for Superconducting Machinery VI** 1977.

See also:

SLAC-TN-03-023 J. G. Weisend II, V. Flynn, E. Thompson, R.P. Reed **A Reference Guide for Cryogenic Properties of Materials** Stanford Linear Accelerator Center, Stanford University, Stanford, CA 94309 2003.

Battelle Columbus Laboratories, **Handbook on Materials for Superconducting Machinery** AD-A035 9,261,977.

Schwartzberg/Osgood **Cryogenic Materials Data Handbook** AD-609562 Volume I 1968 and II 1970.

The TPRC series, e.g.:

Touloukian Y S / DeWitt D P **Thermophysical properties of matter**—The TPRC series Vol. 8 **Thermal radiative properties—nonmetallic solids 1972.**

A8: Cryogenic Seals and Vacuum Technology

Abstract Critical vacuum sealing technology for cryogenic applications.

Keywords Cryogenic seals; Vacuum technology; Design examples

Many components that we currently use in the automotive industry come with a small leak rate. This is tolerable if the leakage flow is in an open environment.

Seals are either required to seal against leakage of cryogenic media and leaks into vacuum space at cryogenic temperatures [56].

When using cryogenic hydrogen in an aircraft vacuum seals need to be however hermetically leak tight. Currently, many components, like safety valves and other process instrumentation, like sensor feedthroughs and other do not comply with these safety standards and are specified with a determined leak rate by the manufacturer.

In addition, many common seals will not seal properly against composite materials.

For metal to metal material combinations copper gaskets can be used. Alternatively, copper gaskets and indium in V-shaped channels can be used together if those process elements need to be replaced for service and maintenance.

Knife edge copper gasket seals may give the best leak rate if both mating materials are the same.

O-rings in tandem with Kapton foil are also common for some cryogenic applications (see Fig. A8.1).

Fig. A8.1 O-ring/Kapton foil seal [57]

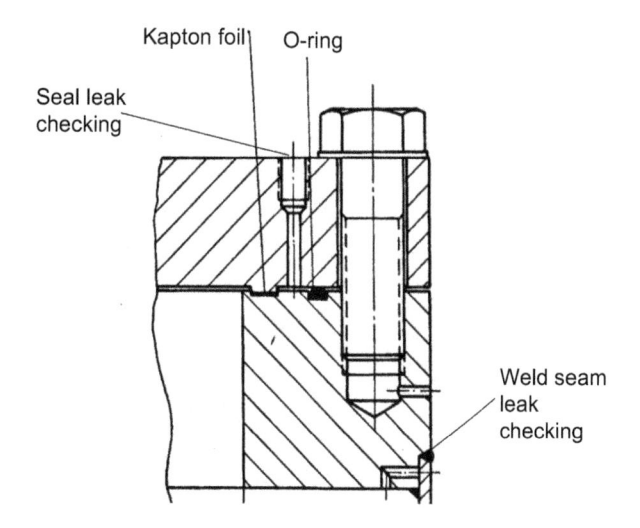

For Kapton films to seal, a foil width of 8 mm and a thickness of at least 0.13 mm is required.

The downside is the high contact pressure that needs to be applied to the Kapton foil in the range of 50 N/mm^2.

For permanent connections several composite adhesives can be used. Those are Stycast 2850 FT (blue), some Epoxy Adhesives DP460 by 3M™ and some Epibond adhesives, e.g., 121 and others.

Indium Seals

When using indium wire as a sealant as shown in Fig. A8.2, the "squeeze curve" needs to be known.

Figure A8.3 depicts an experimental curve that shows the pressure required for an indium wire with a diameter of 1 mm per cm length [57].

For vacuum seals see also [56]

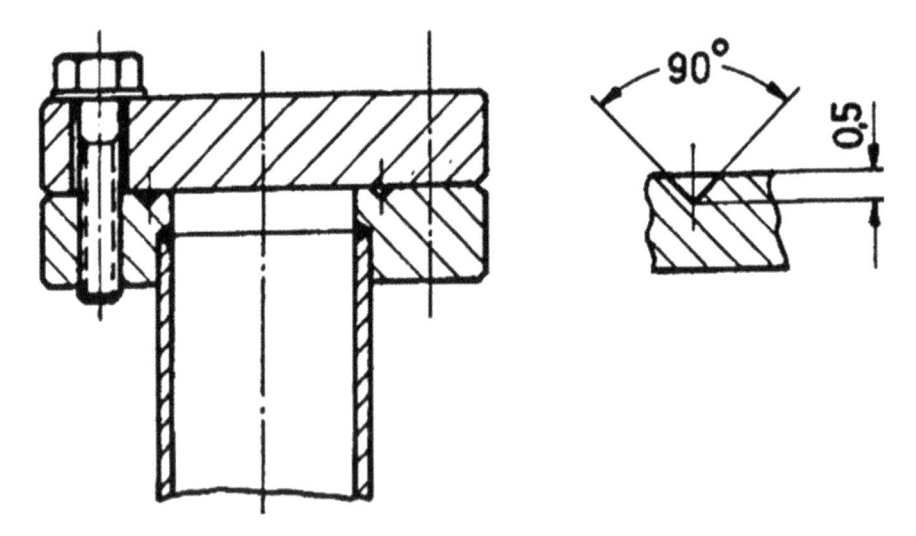

Fig. A8.2 Indium seal in V-groove (dimensions in mm) [57]

Fig. A8.3 Indium squeeze curve (dimensions in mm)

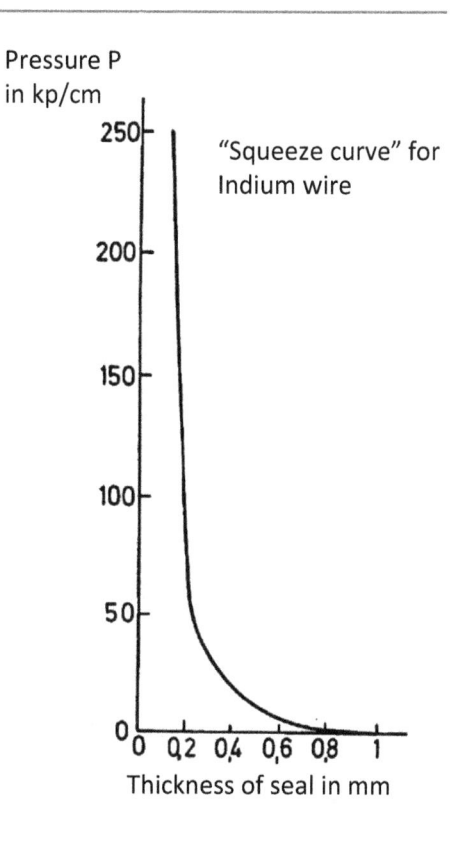

"Squeeze curve" for Indium wire

Pressure P in kp/cm

Thickness of seal in mm

References

References for Appendix 1

1. J. A. Lee, "Hydrogen Embrittlement", NASA/TM-2016–218602, 2016
2. F.D. Gregory, "Safety standard for hydrogen and hydrogen systems" NASA NSS 1740.16, 1997
3. Chandler, W. T. and R. J. Walter "Testing to Determine the Effect of High Pressure Hydrogen Environments on the Mechanical Properties of Metals" in *Hydrogen Embrittlement Testing*, ASTM, 1974
4. C. San Marchi, B P Somerday "Technical reference for hydrogen compatibility of materials" SANDIA report SAND2012-7321, 2012
5. J.W. Ekin, "Experimental techniques for low-temperature measurement", Oxford University Press, 2006
6. R.R. Conte *Elements de Cryogenie* Masson & Cie 1970 (in French)
7. J.L. Christian, J.F. Watson, "Properties of 7000 series aluminum alloys at cryogenic temperatures", Advances in Cryogenic Engineering, vol. 6, pp 604–621 1961
8. H.J. Hucek, K. E. Wilkes, K. R. Hanby, and J. K. Thompson, "Handbook on Materials for Superconducting Machinery, Second Supplement, VICIC-HB 04S2, 1976
9. E.H. Schmidt "Fatigue Properties of Sheet, Bar, and Cast Metallic Materials for Cryogenic Applications", Rocketdyne Engineering, Technical Report R-7564, Contract NAS8-18734, 1968
10. R.F. Muraca, J.S. Whittik, "Materials Data Handbook", Aluminum Alloy 7075, 2nd edition, 1972

Further Readings for Appendix 1

J. E. Campbell, *Review of Current Data on the Tensile Properties of Metals at Very Low Temperatures* Battelle Memorial Institute, DMIC Report No. 148 1961

F. R. Schwartzberg et al. *Cryogenic Materials Data Handbook, Volume I and Volume II* National Technical Information Service, Springfield, Virginia 1970

R.H. Kropschot, R. M. McClintock, D. A. Van Gundy, *Mechanical Properties of Some Engineering Materials Between 20 K and 300 K* Advances in Cryogenic Engineering, Vol. 2, Plenum Press, New York, pp. 93–99 1960

J. L. Zambrow, M. G. Fontana, *Mechanical properties including fatigue of aircraft alloys at very low temperatures* Trans. ASM 41 pp. 480–518 1949

W.M. Dubbs, *Properties of Engineering materials for use in rotating machinery at cryogenic temperatures Part 1 and 2*

W. Stautner, *Compendium of cryogenic material properties for Al 6061 gas tanks* GE Research, internal report 2016

D Mann *LNG materials and fluids: a user's manual of property data in graphic format* OSTI ID: 6271374 1977

References for Appendix 2

11. F.D. Gregory, "Safety standard for hydrogen and hydrogen systems" NASA NSS 1740.16, 1997
12. J. Hord, NBS Technical Note 690, "Is Hydrogen Safe?", US Department of Commerce, 1976
13. R.T. Jacobsen S.G. Penoncello E.W. Lemmon *Thermodynamic properties of cryogenic fluids* Plenum Press 1997
14. J.W. Leachman, R.T. Jacobsen, E.W. Lemmon, S.G. Penoncello *Thermodynamic Properties of Cryogenic Fluids* Springer International Cryogenics Monograph Series 2017
15. GasPak, GASPAK® of Cryodata Inc., Horizon Technologies
16. R.D. McCarty, J. Hord, H.M. Roder **Selected properties of Hydrogen** (Engineering Design Data) NBS Monograph 168 1981
17. R. J. Richards, W. G. Steward, R. B. Jacobs **A survey of literature on heat transfer from solid surfaces to cryogenic fluids** NBS Technical Note 1961

References for Appendix 3

18. F.D. Gregory *Safety standard for hydrogen and hydrogen systems* NASA NSS 1740.16, 1997
19. J. Hord, NBS Technical Note 690 **Is Hydrogen Safe?** US Department of Commerce, 1976
20. F.J. Edeskuty W.F. Stewart *Safety in the Handling of Cryogenic Fluids* Plenum Press 1996
21. Molkov V *Fundamentals of Hydrogen Safety Engineering 1 and 2* courses by HySafer Centre, UK, free ebooks at www.bookboon.com 2012
22. C. Minas w Stautner *Best Safety Practices in LH2 Infrastructure and Fueling* CEC/ICMC 23 C2Po2B-10 2023
23. Y.S.H. Najjar *Hydrogen safety: The road toward green technology* International Journal of hydrogen energy 38 pp 10716-10728 2013

References for Appendix 4

24. C.W. Elrod, "Design handbook for liquid and gaseous helium handling equipment", 1963, ASDTD061-226.1-505
25. F. Edeskuty, W. F. Stewart, "Safety in Handling of Cryogenic Fluids", Plenum Press, 1996, page 56
26. K.D. Williamson, F.J. Edeskuty, "Liquid cryogens, Volume 1 Theory and Equipment", CRC Press, 1983, page 152
27. J.E. Fesmire, A. M. Swanger, et al "Energy efficient large-scale storage of liquid hydrogen" CEC/ICEC 2021
28. W. Johnson, T. Tomsik, J. Moder, "Fundamental of Cryogenics", 25th Annual TFAWS, 2014
29. MAN energy solutions, https://www.man-es.com/company/press-releases/press-details/2020/11/10/man-cryo-announces-series-of-hydrogen-projects
30. Williams L O, Hydrogen powered automobiles must use liquid hydrogen, Cryogenics pp 693–698 1973

References

31. Williams L O, Clean Energy via Cryogenic Technology Advances in Cryogenic Engineering Vol 18 pp 50–524 1973
32. Research News, Cryogenics page 606 1980
33. https://www.sag.at/en/development/hydrogen/
34. T. Brunner, BMW Hydrogen Workshop, "Cryo-compressed Hydrogen Storage", 2011 and W. Peschka, "Liquid Hydrogen: Fuel of the Future", Springer-Verlag Wien New York, 1992
35. D. Schultheiß, "Permeation barrier for lightweight hydrogen tanks", PhD Thesis, 2007
36. https://www.hydrogencarsnow.com/index.php/liquid-h2-fuel/
37. T. Funke, C. Haberstroh, "Capacitive density measurement for supercritical hydrogen", *IOP Conf. Ser.: Mater. Sci. Eng.* 278 2017 012071
38. R.K. Ahluwalia, J.K. Peng "Dynamics of cryogenic hydrogen storage in insulated pressure vessels for automotive applications", International journal of hydrogen energy 33 pp 4622-4633 2008
39. T. F. Johnson, D. W. Sleight, R. A. Martin, "Structures and Design Phase I Summary for the NASA composite cryotank technology demonstration project" (CCTD) 54th AIAA/ASME/ASCE/AHS/ASC Structures, Structural Dynamics, and Materials Conference, 2013 https://doi.org/10.2514/6.2013-1825
40. Source: https://boeing.mediaroom.com/news-releases-statements?item=130996
41. Aleck, B., "Fiberglass-Overwrapped 2219-T87 Aluminum Alloy Low-Pressure Cryogenic Tankage", Society of Aerospace Material and Process Engineers National Techincal Conference, Space Shuttle Materials. Vol. 3, 1971, pp. 131-11:4.
42. Alfring, R. J., Morris, E. E. and Landes, R. I., "Cycle-Testing of Boron Filament-Wound Tanks", NASA CR-72899, National Aeronautics and Space Administration, Lewis Research Center, August 1971 (N71-38023).
43. Barber, J. R., "Design and Fabrication of Shadow Shield Systems for Thermal Protection of Cryogenic Propellants", NASA CR-72595, National Aeronautics and Space Administration, Lewis Research Center, Cleveland, Ohio, November 1969 (N70-25098).
44. Baucom, R. M., "Tensile Behavior of Boron Filament-Reinforced Epoxy Rings and Belts", NASA TN D-5053, Langley Research Center, Hampton, Virginia, March 1969 (N69-19918).
45. Bullard, B. R., "Cryogenic Tank Support Evaluation", NASA CR-72546, NASA Lewis Research Center", Cleveland, Ohio, December 1969 (N70-13085).
46. https://leehamnews.com/2022/03/25/bjorns-corner-sustainable-air-transport-part-12-hydrogen-storage/#more-38857
47. W. Stautner, M. Xu, S. Mine and K. Amm, "Hydrogen Cooling Options for MgB_2-based Superconducting Systems", AIP Conf. Proc., vol. 82, 2014, pp 82–90.
48. W. Stautner, Cryogenic Cooling System US Pat. appl. 20130104570A1
49. H Hirabayashi, Y Makida, S Nomura, T Shintomi "Feasibility of hydrogen cooled superconducting magnets", IEEE Transactions on Applied Superconductivity vol 16 No 2, pp 1435–1438, 2006
50. H Hirabayashi, Y Makida, S Nomura, T Shintomi "Liquid hydrogen cooled superconducting magnet and energy storage", IEEE Transactions on Applied Superconductivity vol 18 No 2, pp 766–769, 2008
51. Nakayama, T Yagai, M Tsuda, T Hamajima "Micro PowerGrid System with SMES and superconducting cable modules cooled by liquid hydrogen", IEEE Transactions on Applied Superconductivity vol 19 No 3, pp 2062–2065, 2009
52. G. Weisend II, P. Arnold, J Fydrych, W. Hees, J. Jurns et al "Cryogenics at The European Spallation Source" ICEC/ICMC 2014 Conference July 2014
53. J. Feldman, W. Stautner, C. Kovacs, N. Miljkovic, K. S. Haran "Review of Materials for HTS magnet Impregnation" Supercond. Sci. Technol. 37 033001 2024

References for Appendix 5

54. https://leehamnews.com/2022/03/25/bjorns-corner-sustainable-air-transport-part-12-hydrogen-storage/#more-38857
55. Emission-Free Flying on the Horizon, ColdFacts Vol 38 number 2 2022

References for Appendix 8

56. A. Roth **Vacuum Sealing Techniques** American Vacuum Society Classics AIP Press 1994
57. H. Katheder, private communication, 1976

Index

A
Acceleration, 42–48, 52, 53, 163
　lateral, 47, 48
　profile, 45, 46
Accommodation Coeff, 37
Advanced technology, 159
Aerogel insulation, 298
Aeronautics, 15
Aerospace, 12, 67, 147, 166, 205, 300, 309
Aft, 106, 110, 113, 115, 120, 225
　tanks, 111, 225
Agreement, 87, 169
Air, 3, 12, 15, 26, 27, 29, 34, 59, 71, 167, 168, 176, 178, 192, 223, 227, 230, 293, 310
　air foil structure, 110
　flow, variable, 178
　pressure calculations, 58
Airbus, 4, 12, 25, 44, 45, 112–114, 116, 307–309
　cryoplane, 112, 114
　ZEROe design, 116
Aircraft, 1, 7, 19, 63, 105, 123, 151, 173, 185, 201, 213, 233, 243
　all-electric, 3, 4, 22–24, 105, 118–119, 147, 233, 308
　　system, 4, 222
　altitude, 59
　altitude change, 58
　applications, 16, 28, 34, 89, 154, 176, 218
　commercial aircraft, 16, 214, 297
　configuration, 119, 120
　cryogenic circuit, 167
　cryogenic structure, 123
　design analysis, 112
　designs, transport, 12
　detection system, 178–180
　dormant, idling, 10
　efficiency, 110
　electric, 15–16, 235, 307
　engine core, 176–178
　fuel, 12–13, 111, 177, 229, 292
　fuel distribution, 177
　fueling pump, 146
　fuel tanks, 105, 111, 120, 304
　fuel transport, 111
　hybrid electric, 115, 117, 310
　hydrogen driven, 155
　hydrogen fueled, 4, 22, 111
　hydrogen tank, 4, 19–60, 63–64, 69, 73, 86, 89, 93, 105, 110, 114, 115, 120, 123–149, 155, 167, 298–299, 301, 307, 309
　industry, 1
　leak location, 179
　LH$_2$ powered, 134
　lightning, 168
　liquid hydrogen, 46
　propulsion motors, 185, 190
　safety, 23, 217, 229
　storage tanks, 16, 24, 25, 33, 34, 58, 63, 67, 87, 113, 118, 131–133, 173, 174, 214, 220, 224, 297–298
　subsonic transport, 110, 111
　systems, powered, 2
　transport, 12, 99, 110, 111, 120
　turnaround, 220, 227, 228
　wide-body, 177
　XLR hybrid-electric, 115–118
Aircraft engine
　core, 176–178
Air flow
　variable, 178
Airframe, 4, 13, 16, 23, 25, 44, 86, 100, 105–121, 155, 233
Airgap Bpk, 191

Air ingress
  accumulates, 167
Air permeability, 192
Airport, 2, 25, 145, 146, 213–230
  electrical loads, 222
  local distribution, 219–224
  premises, 218
Airport terminals, 214, 221
Air transportation, 15
Alternative
  fuels, 13, 14
Altitude, 12, 15, 25, 58, 59
  changes LH_2, 57
Aluminum, 8–10, 13, 21, 37, 46, 65, 69, 70, 73, 83, 84, 86, 89–91, 93–99, 129, 131, 134, 144, 203, 248, 249, 254, 305
  alloy, 37, 65, 69, 93, 94, 99, 243, 247–249, 253, 255, 293, 316
  conductor, 134
  layer, 83, 91, 95, 96, 98, 144
  lined carbon, 97
  property data, 249, 250
  S-N curves, 256, 258
  structure, 20, 21, 305
  tank, 8, 9
Aluminum-Lithium tank, 303
Aluminum structure, high-strength, 20, 21
Aluminum, wire purity, 10
Applied internal pressure, 93, 96
Armature, 205, 208
Armature windings, 22, 23, 182, 183, 187–197, 201, 202
Atomic radius, 80–81
Automotive industry, 11, 43, 63, 85, 127, 145, 146, 317
Average working gauge, 298
Aviation, 7, 13–16, 21, 99, 118, 168, 213, 216–219, 229, 304, 308
  Administration, 168
  energy system, 219
  environmentally sustainable, 16
  Rulemaking Advisory, 171
Aviation, industry, sustainable, 16
Axial flux, 158
Axial flux class, 158

B
Backup system, 222
Baffle, 19, 25, 37, 41–43, 46–51, 64, 235
  cross wing, 48–51
  cross-wing design, 48–51
  design, 46, 48–51
  ring, 48, 50
  semi circular, 48, 50
  wing, 48–51
Ball bearing
  stainless steel, 206
Barometric altitude formula, 59
Barrier, 28, 33, 34, 69, 80, 81, 83, 86, 152
Baseline, 46–49, 229
  tank design, 46
Base plate, 85
Batteries, 23, 24, 234
Bayonet connection, 162
Bearing, 8, 38, 86, 88, 91, 156–159, 201–211
  axial, 158
  ball, 38, 201–211
  cryogenic, magnetic, 156–159
  heat pipe, 208–210
  magnetic, 156–159
  material shrinkage, 129
  nonmetallic liners, 8
  outer race, 208
  shaped structures, 38
Bellow, 126, 156
Berylco, 207, 209, 210
Bimetallic penetrations, 129
Black Coal Gasification, 215
Blanket, 30, 31, 35, 166
  thickness, 30
BMW, 7, 25, 53, 63, 99, 128, 145, 173, 178, 301, 302
Body forces, 45
Boeing, 13–15, 25, 63, 64, 109–113, 170, 304
Boiling
  explosive, 168
Boil-off
  excessively high, 54
  rate, 9, 26, 35, 54, 298
Brown Lignite Gasification, 215
Brushes, 203, 204
BSCCO, 190–195
BSCCO wire, 184, 192, 193
Buckling, 86, 88, 90, 92, 95, 100
  pressure, 88, 92, 95
Bushing, 37, 38

C
Cable cryostat (CC), 134, 137–140, 234
Capacitance, 134
  decrease, 134
Capacitive
  level meter, 133
Carbon fiber, 8, 35, 37, 64, 71, 73, 76, 78, 84, 86, 89–91, 93, 95–98, 100, 233

Index

Carbon fiber composite (CFC), 8, 9, 63, 65, 89, 91–93
Carbon fiber/epoxy composite, 78
Carbon-number, 213
Carnot, 188, 189
Cavitation, 12, 43, 156, 162–164, 234, 235
   avoidance, 162
   conditions, 163
CC, *see* Cable cryostat (CC)
CCTD project, 63, 303
Center for High-Efficiency Electrical Technologies for Aircraft (CHEETA), 13, 15, 16, 41, 46, 48, 86–87, 89–96, 98, 99, 118–121, 123, 139, 148, 155, 167, 201, 236, 308
   all-electric aircraft, 118–121
   research effort, 118
   stress analysis, 89–96
   tank types, 93
CERN, 37
CFC, *see* Carbon fiber composite (CFC)
CFC laminate, 91
CFD
   dedicated analysis, 128
   typical, 179
CFD analysis, 128, 178
CFD analysis, transient, 178
CFD calculations, 209
CFRP, 65, 68–71, 83
CH_2, *see* H_2 concentration (CH_2)
Chamber, 40, 43, 88, 92, 125, 162, 183, 209
CHATT project, 63
CHEETA, *see* Center for High-Efficiency Electrical Technologies for Aircraft (CHEETA)
Cheetah, 13, 15
CHEETA program, 120, 123, 148
Chemical classification per weight, 74
Chilldown, 59, 69, 70, 79–80, 87, 93, 107, 110, 143–144, 226, 228, 235
Circulation system, 185, 186, 188
Cisterns, 55
Clean air, 176
Clean break disconnects, 226
Climate change, 1, 2
Climb phase, 45
Cloth epoxy matrix, 81
Cloth winding, reinforced, 75
CNRS PM-SC NbTi, 191
Coefficient of performance (COP), 188–189, 194–197
CO_2e/kg, 215

CO_2 emissions, 1, 15, 16
Coil, 10, 73, 156, 183, 185, 186, 201, 202, 205, 208, 209, 305, 316
Cold, 4, 10, 23, 24, 36, 88, 126, 127, 149, 160–164, 168, 183, 184, 201, 234, 295, 296
Cold trap, 164
Collapse, 162
Combustion, 3, 4, 12–14, 176, 229, 292, 294, 308, 311
   properties, 13
Compliance, 155
Composite, 8, 20, 63, 109, 124, 152, 177, 233
   fully wrapped, 8
   hydrogen, affecting, 70
   LH_2 cryotank, 63, 303
   materials, 86, 152, 156, 235, 317
   sandwich, 169
   shell, 68–70, 88
   tank, 8, 43, 65, 67, 68, 71–74, 76, 78–83, 87, 88, 109, 124, 126, 131, 143, 144, 168, 169, 233, 236, 243, 303, 304, 306
   tank shell, 78
   tank test, 304
   wall, 33, 79, 85, 131, 143
   wall enabling, 85
Composite/steel design, 129
Compressibility, 264, 290, 292
Compression, 37, 92, 151, 160, 162
Compression chamber, 162
Compressive pressure, 206
Compressive Strength MPa, 92
Concentration, spatial distribution, 179
Conditions, steady-state, 55
Conduction cooling, 183, 185, 188
Conductor, 134, 159, 201
Configuration, 29, 35, 37, 67, 69, 91, 99, 105, 107, 109, 110, 119–121, 126, 129, 164, 165, 176, 183, 184, 202, 204, 218, 228
Continuous steam-iron process, 2
Control, 11, 20, 21, 35, 54, 71, 83, 88, 107, 120, 123, 127, 128, 132, 134, 141, 142, 145, 148, 156, 164, 184, 187, 214, 297, 309
   elements, 128, 132, 134, 141, 145
   valves, 142
Convection heat, internal, 58
Convective flows, external, 58
Coolant
   circulating, 185
Coolant channel, 135

COOLCAT, 30
Cooldown, 10, 20, 37, 71, 73, 79, 125, 127, 132, 162, 203, 207, 305
Cooling, 3, 4, 9, 10, 22–24, 79, 123, 135, 139, 144, 151, 158, 162, 181–197, 201–203, 206, 210, 233–235, 305, 306, 308, 309, 311
  cryogenic bath, 183, 186, 188, 197
  cryogenic system, 196
  effect, 139
  forced-flow, 151, 183–185, 187, 188, 197
  liquid helium, 135
  MRI magnet, 54, 304, 306
  neon thermosiphon, 184
  open-loop, 186–187, 196
  superconducting components, 3, 187
  technologies, 181–197
  thermosiphon, 183–184, 187, 188, 197
Cooling capacity, 184, 185, 187, 188
Cooling duct design, 188
Cooling power, 10, 188, 194, 196, 203, 306
  required, 194
Cooling power, higher, 188, 203
Cooling requirement, 194, 196
Cooling, rotor, 184, 203
Cooling, source, 183, 184
Cooling, system, 181–197, 309, 311
COP, *see* Coefficient of performance (COP)
COP expected, 189
Copper, 68, 70, 73, 75, 78–80, 130, 203, 204, 206–208, 210, 244, 247, 293, 309, 315–317
  balls, 207
  coating, 70
  powder, 78
  straps, 207
Core pressure
  expected, 169
Cost, 11, 105, 107, 108, 137, 160, 184, 187, 216, 218–220, 222, 227, 300, 301, 303, 305
Coupling loss, 191, 193, 194
Crack formation, 129, 217
Critical state model, 191
Crossplies, 68, 80, 84
Crude oil, 213
Cruising, increasing, 59
Cryo aircraft, 22, 118, 233, 234
  CHEETA, 118, 119, 233, 234
  minimum tank, 96–100
Cryocomp, 78
Cryocooler, 10, 54, 127, 165, 183–186, 188–190, 196, 197, 201–211, 235, 311, 313, 314

  on-board, 185–186, 188, 197
  rotating, 185, 186, 201–211
  rotating, test, 202, 203
  spinning, 204
  test facility, 202, 203
Cryogenic circulation, helium, 151
Cryogenic liquid, outlook, 16, 234
Cryogenics, 3, 7, 19, 63, 105, 123, 151, 177, 181, 201, 213, 233
  aircraft liquid, 58
  aircraft motor, 205
  circuit, 123, 132
  deflection, measuring, 85
  effect, 72
  engineering, 22, 311–314
  fluids, 151, 152, 183, 186, 312–315
  Hydrogen-Electric Aviation, 118
  insulation performance, 141
  LH_2 tanks, 293
  liquid evaporation, 55
  liquid helium, 54
  liquid, reservoirs, 155
  liquids, 16, 19–60, 69, 128, 139, 145, 155, 162, 234, 300, 308, 313
    introduction, 42–43
    transport, 139
  materials, 66, 67, 85, 100, 243, 254, 316, 321
    common, 254
  power electronics, 24, 233–237
  service, 141, 235
  state, 219
  tank, 37, 38, 45, 54, 71, 106, 124, 143, 148, 316
  temperature
    higher, 23, 38, 64–66, 69, 73, 77, 85, 89, 108, 135, 141, 185, 186, 188, 197, 233, 256, 296, 316, 317
  transfer, 12, 24, 106, 107, 134–140, 148
  vacuum environment, 202
  vessels, 55, 141, 148
Cryogenic temperature, higher, 66, 135
Cryogenic vessel, single, 148
Cryogens, 3, 4, 7, 9, 10, 16, 19, 33, 49, 52, 54, 58, 63–65, 67–69, 76, 89, 123, 126, 129, 134, 137, 143, 152, 155, 164–167, 187, 188, 201, 297, 314
  storing, 297
Cryoplane, 11, 112, 114, 155
Cryoplane design, 112
Cryopumping, 169, 312
Cryostat, 12, 54, 55, 84, 125, 148, 167, 311
  design, 148, 314

Index

Cryotank, 34, 52, 57, 63, 88, 129, 235, 303, 304
   applications, 63, 99
Cryotankage, 13, 25–60, 63–100, 105–121, 123, 166, 167
   heat sources, 25–34, 43
   operating conditions, 87–100
Cycling, 28, 71–73, 83, 155–159
Cylinder, single, 160

D

Damping
   sloshing motion, 49
Damping elements, 58
Data readout, 85
dB/dt, 183, 189–191, 193–197
Decarbonization, 134, 216
Defueling, 59, 69, 110, 146–147, 296
Dense fluid layer, 58
Density layer, higher, 58
Design, 3, 10, 19, 63, 105, 123, 151, 178, 181, 201, 220, 233
   cylindrical, 64, 170
   flaws, 123, 125
   improvement, 203
   principles, generic, 202
   simplified, 34–60, 87, 187
   solutions, 64
   study, 109, 111, 112, 114–116
Designs, Titanium alloy, 37, 68
Detection, 29, 33, 169, 173–180, 235
   ratio, 178, 179
Development plan, 217
Dewar
   liquid cryogen, 166
Diagnostics, 106, 125, 169
Dielectric constants, 133, 134, 263, 264, 291, 292
Diffusion barrier, 33, 34, 69
Dimpled ribbon, 165
Direction, lateral, 44, 48, 49
Direct operating cost (DOC), 11, 108
Discharge, 161–163
   valve, 132, 162
Discharge pressure, 147, 152, 234
Discharge temperature, 162, 163
Disk, sintered, 165
Dispensing, 213, 218
Distance, 2, 168, 214, 294
Distribution, 16, 24, 91, 173, 177–180, 184, 213–230, 233, 307
Distribution, network, 215–217, 223
DOC, *see* Direct operating cost (DOC)

Dominant loss components, 192, 193
Downstream, 43, 162, 163, 293, 295
Drag, 16, 110, 112, 118
Drive train
   bare-bone, 202
   high-level, 202
Drive train, integration, 202
Dual-purpose extremities scanner, 304
Duration, 4, 21, 42, 100, 120, 162, 228, 229
Dynamical systems, 181

E

E-aircraft, 308
Economy
   hydrogen future, 219, 300
Eddy-current loss, 191, 193
Effective resistivity, 193, 194
Efficiencies, 12, 15, 16, 24, 49, 107, 109, 110, 151, 156, 157, 165, 169, 177, 181, 183, 184, 187–189, 194, 196, 197, 201, 222, 234, 301, 310
Elasticity, 76, 78, 85, 92
Electric
   propulsors, 115–121
Electric aircraft, 15–16, 22–24, 235, 306
Electrical grids, 217
Electricity, 16, 215–218
   demand, 218
   generation, 215, 216
   grid, 215
Electric power
   transfer, 203, 306
Electrolysis, 2, 16, 215, 218, 219
Electrolyte, 176
Electrolyzers, 2, 25, 142, 217–219
   large-scale systems, 217–219
Electromagnetic performance, 181, 182, 189
Emission, 1, 3, 12, 15, 16, 152, 154, 173–180, 201, 215, 216, 218, 298, 309
   level, 16
Emissions, indirect, 215
ENEA PM-SC REBCO, 191
Energizing, solenoid, 156
Energy, 1, 12, 42, 78, 105, 134, 151, 181, 213, 235
   application, 215, 316
   demand, 219, 220
   density, 13, 14, 16, 196, 219, 227, 230, 300, 301
   equivalent, 219
   future carrier, 217
   life cycle, 218
   Solutions, 196, 298, 299

Energy (*cont.*)
  storage, 16, 306
  transitions, 219
Engine, 12, 16, 25, 88, 108, 113, 300, 308
  core, 176–178, 180
  core, compartment, 176, 177
Engineering, 23, 123, 131, 141, 148, 176, 260–292, 300, 308, 311–315
Enhanced electromagnetic performance, 189
Enthalpy, 9, 59, 78, 82, 291, 315
Environment, 1, 3, 25, 26, 29, 30, 34–39, 42, 56, 67, 84, 88, 89, 105, 124, 126, 129, 137, 143, 157, 166, 187, 202, 205, 210, 234, 243, 300, 304, 306, 317
  Embrittlement HEE, 243
  microgravity, 42, 166
Environmental impact, 15, 196
Epoxy, 27, 64–67, 70, 71, 73, 76–81, 84–86, 89, 129–131, 137, 144, 233, 236, 306, 318
  fiberglass-laminates, 65–67
  filled, 130
  matrix fiber, 81
  penetrations, 129, 131
  reinforced, 73, 78, 80, 81, 84, 85, 129–131, 137, 236, 306
  resins, 71, 73, 76–80, 84, 89
  resin samples, 71
  resins, reinforced, 73, 80, 84
Eulerian, 44
Evaporation, 49, 50, 52
Explosive mixture, 179
External, polyurethane foam, 27
External, port leakage, 166
External, vibration modes, 58
Extraction, 123, 146, 147, 213, 218
Extremities scanner, 304

F
Failure, 29, 31, 65, 69, 77, 88, 89, 119, 120, 131, 143, 148, 163, 166–169, 174, 202
Fan, 202
Fatigue, 20, 29, 37, 58, 65–67, 70, 71, 76–77, 83, 86, 93, 129, 204, 233, 235, 250, 252
  crack growth, 77
  epoxy resins, 77, 79
  life, 205
  strength, 66, 67, 252, 258
Federal Aviation Administration, 168
Feedstock, 13, 215

Feedthrough, tight thermal, 130
Fiber, 20, 21, 27, 34, 37, 38, 64, 65, 73, 75–78, 80, 81, 84–87, 89, 91, 93–98, 233, 235, 236, 305
  orientations, 78, 81
Fiber glass, typical, 80
Fiber orientation, matrix, 78, 81
Fiber wrap, 20, 21, 34
  Al permeation, 34
Field coils, 201, 205, 208, 209
Field winding, 23, 24, 158, 182–188, 197, 202
Filters, 162, 165, 207
Findings, 70, 77, 85, 127, 148
Flashback arresters, 141
Fleet, 220, 221
Flight cycles, 27, 28
Flight profile, 21–23, 44, 58
Flow, 9, 11, 12, 22–24, 58, 70, 105, 120, 123, 125–128, 131–133, 139, 145, 147, 148, 151, 152, 155, 157, 158, 160, 162, 164, 165, 168, 173, 174, 178, 183–185, 208, 223, 224, 227–229, 233–235, 295, 317
  circuit, 22, 23, 105, 123, 128, 131–133, 234
  pulsating frequency, 127
Flow circuit
  extended, 128
Flow control system, 148
Flow meter
  self-calibrating, 145
Flow rate, 11, 120, 131, 139, 145, 147, 151, 152, 155, 157, 158, 160, 162, 164, 168, 173, 184, 227–229, 234
  coolant, 184
  typical, 155
Fluid management, 187
Fluid, slosh problem, 47
Foam, 26–29, 46, 64, 89, 108, 125, 164–166
  additively printed, 165
Foil, 31–33, 35, 36, 40, 64, 68–70, 83, 86, 317
  tin, 83
Foil, thick, non-flammable, 31
Fraction, 11, 19, 69, 85–88, 93, 96–100, 106–108, 110, 144, 164, 165, 233
Fracture, toughness, 253–259
Fuel, 1, 7, 21, 63, 105, 123, 151, 173, 196, 202, 213, 233
  cell, 2–4, 7, 16, 20–24, 123, 132, 142, 155, 158, 176, 196, 197, 202, 233–235, 298, 301, 306, 309
  displaced, 152
  distribution system, 173, 177
  fossil, 213, 215, 216

leakage, 73, 87
station, 151
storage, higher, 86
system, 155, 173, 229
tank, 2, 16, 26, 46, 48, 63, 77, 105, 107, 111, 120, 141, 168, 235, 236, 300, 304
type, 1, 12
Fuel cells, PEM, 155
Fueling
station, 25, 146
system interface, 141
Fuel redundancy, 119, 148
Fuel sloshing suppression, 49
Fuel transfer, structure, 21
Fuselage, 13, 23, 58, 71, 105, 106, 108, 110–115, 117, 118, 120, 123
extended, 112
twin, 113, 115

G
G10, 37, 38, 78, 80, 82, 89, 131, 144
integral enthalpy, 82
Gap
magnetic, 156
Gas, 1, 2, 7–9, 20, 21, 24, 27–29, 31, 34–37, 40, 43, 60, 67, 70, 71, 83, 84, 107, 108, 120, 123, 126, 129, 139, 141, 152, 164, 165, 169, 176, 201, 208, 210, 215, 217, 218, 222–224, 292, 300, 315
leakage, 71
permeation, 70, 84
phase, 60
storage tanks, 34, 67
Gas conduction, residual, 31, 34–36
Gas, high pressure, 67, 152, 300
Gasification, 2, 215, 218
GasPak, 59, 60, 262
Gate, 220–224, 228–230
GE, *see* General Electric Company (GE)
General Electric Company (GE), 10, 12, 90, 137, 156, 166, 169, 186, 211, 305, 306, 311
Generators, 184–186, 190, 197, 315
Getters, 35, 40
GFRP tank, cylindrical, 64
GH$_2$, 27, 169, 178, 218, 220, 222, 223, 226, 229, 294
Gifford-McMahon, 185
Glass, 40, 65, 73, 75, 76, 78, 80, 81, 84, 85, 87, 89, 93, 96, 137, 220, 298
bubbles, 220, 298

Glass cloth, epoxy, 81, 89
Glass fiber, 27, 37, 38, 64, 75, 76, 84, 85, 87, 93, 96
rods, 37
Global environmental problems, 1
Goal, 2, 21, 73, 87, 96, 201, 206, 310
Gravity, 30–32, 58, 120, 164, 183
assisted phase, 164
Gray Natural gas, 215
Greases, agricultural residues, 14
Greenhouse gas, 215
Greenhouse gas, emissions, 215, 218
Ground support equipment, 217, 224, 226
GRP
glass composite, 65
tubes, 71, 131
Guide, 38, 130, 149, 316

H
Hardener, 75, 76
H$_2$ concentration (CH$_2$), 173, 174, 179
Head, correct choice, 108
Heat, 9–11, 13, 14, 22–24, 30, 31, 33–38, 43, 50, 52, 53, 55, 58, 59, 64, 78–80, 88, 107, 123, 125–127, 137, 139, 152, 156, 162, 167, 176, 182–184, 186, 188, 196, 202, 203, 205–210, 215, 220, 230, 233, 235, 243, 262, 265, 266, 277, 278, 288, 290, 292, 294, 306, 307
flux, 33–35, 129, 167
latent, high, 196
Heater system, single, 148
Heat exchangers, 24, 123, 179, 183, 184, 208, 220, 235, 243, 295, 306, 311
Heat flux, density, 35, 167
Heat load, 9, 10, 30, 31, 34–38, 53, 88, 126, 137, 139, 182, 202
higher, 10
parasitic, 137, 139
Heat pipe, 127, 208–210
spoke, 209, 210
Heat pipes, cooldown, 127
Heat ratio, 290, 292
Heat sources, 25–34, 43, 53
Heat transfer, 79, 165, 184, 186, 201–211, 260, 276, 315
capability, 208
design, 203
enhanced, 184
high, 186, 187, 208
rates, 209, 276
rotating, 201–211

Heavy-duty trucks, 213, 219
HEE, see Hydrogen environment embrittlement (HEE)
Helium, 3, 9, 10, 25, 29, 31, 41, 42, 54–56, 64, 67, 71, 80, 81, 88, 107, 108, 125, 127, 129, 130, 133–135, 145, 151, 156, 157, 165, 167, 168, 183, 185, 208–210, 244, 246, 297, 304, 306, 311, 315
   reservoirs, 133
High-Efficiency Electrical Technologies, 15, 118
Hinetics, 202, 203
Homopolar motor, 182
Homopolar REBCO, 191
Honeycomb, 28, 29, 64, 166
Hose connections, 223, 224, 226
HTS, see Superconductor, high temperature
Hydrant pits, 223
Hydrant, systems, 214, 220, 221, 224
Hydrocarbon, 16, 120, 213, 214, 229
Hydrogen
   compressed hydrogen gas, 8
   compressed hydrogen tanks, 7, 301
   dedicated pipelines, 217
   economy, 1, 2, 219, 234, 300
   economy, integrated, 2
   energy, 16
   PEM fuel, 155
   physical properties, 12, 165, 229, 230
   storage, 7–16, 21, 63, 64, 99, 110, 112, 113, 117, 217, 229, 300
      options, 7, 8
      tank, 24, 131–134, 297–298
   storage systems, 2, 16, 133
   system design, 293
   tank, 4, 7, 11, 19–60, 63–64, 68–70, 73, 85–87, 89, 93, 97, 105, 106, 109, 120, 123–149, 169, 233–235, 297, 301, 304, 309
   tank pressure, 42, 53, 58, 59
   tanks, lightweight, 233
   tank test, 304
   vapor, 11, 34, 147, 168, 263
   vapor, high-pressure, 147
   ZBO designs, 10
Hydrogen-air mixture, 223
Hydrogen-based economy, 1, 2, 219, 234
Hydrogen embrittlement, 217, 293
Hydrogen environment embrittlement (HEE)
   index, 69, 243, 244, 246
   reasons, 141
Hydrogen permeation, 70, 217
Hydrogen production, 16, 215–218

Hydrogen production methods, 215–217
Hydrogen properties T-S, 261, 262
Hydrogen pump
   suction, 146
Hydrogen sloshing
   modelling, 44
Hydrogen system, purging, 222
Hysteresis, 189
Hysteresis loss, 191, 192, 194

I
Ice formation, 168, 169
Inclined tubes, 126–131
Induction, 182
Infrastructure, 2, 4, 22, 24, 173, 175, 176, 185, 213–230, 233–235, 295, 300, 306, 309, 315
Inspection, 21, 26, 43, 87, 168
Instabilities, 50, 128, 130
Installation, 120, 175, 216, 222
Insulation, 11, 13, 23, 26–36, 69, 93, 106–108, 137, 139, 162, 164, 167, 168, 203, 220, 298, 301
   thickness, 11, 28
Insulation, outer, foam, 164
Insulation, polyurethane foam, 108
Intercept, 38
Interfaces, 59, 129, 141, 155, 226
Interference fit, 130
Intermittent convection layer, 50
Isolation capability, 148
ITER, 151

J
JAERI, 151
Jelly type tape, 165
Jet-A, 1, 3, 13, 14, 16, 106, 110
   fuel wing, 43
Jet-A, LH$_2$ COx, 16
Jet fuel, 4, 16, 213, 214, 219, 308, 311
Jet-fuel, conventional, 214, 219, 308
Johnston, H.L., 139

K
Kerosene, 13, 16, 213, 308
Key components, 233
Key technology areas, 233
Kinetic energy, 47, 48
Knee-shaped configuration, 126
kW-class prototype REBCO, 194

L

Lagrangian, 44
Laminate, 64–68, 71, 83, 88–92, 169
  Cytec CYCOM, 64
Landing, 29, 120
  gear, 120
Launch, 26, 63, 64
  vehicles, 64, 113, 115, 220, 303
Launch vehicles
  orbit, 115
Layered silicate nanocomposites, 69
LCH$_4$, 51
Leak, 22, 31, 35, 68, 71, 84, 88, 107, 120, 126, 130, 139, 141, 152, 163, 168, 173–180, 217, 221, 223, 235, 295, 317
  potential locations, 173, 175
  rate, 141, 235, 317
Level gauge, capacitive, 303
Level meter, design, 133
LH$_2$ filling, 223
LH$_2$ pipelines, 219, 221
LH$_2$ pump, 153, 158–163, 175, 177, 179, 234
  outlet, 177, 179
  shutdown, 175
LH$_2$ refueling, utilize, 221, 228, 229
LH$_2$ tank, 53, 73, 107, 113, 114, 116, 132, 159–162, 177, 179, 293, 295, 299, 301
LH$_2$ transportation
  direct, 219
Lightening
  protection, 168
  protection requirements, 168
Linear compressor, 203
Liner, 8, 28, 33, 40, 41, 58, 64, 67–71, 73, 75, 83–86, 90–93, 100, 143, 144, 233, 243, 301
  metallic, 8, 33, 70, 83, 144
Liquefaction, 8–10, 27, 188, 213, 217–219, 221, 224, 301
Liquefaction facility, 222
Liquid, 2, 7, 19, 63, 105, 123, 151, 173, 183, 201, 217, 233
  cryogen tank, 76
  helium, 10, 25, 41, 42, 54–56, 64, 88, 125, 130, 135, 151, 156, 157, 167, 183, 210, 297, 304, 311, 315
  helium bath, 41
  helium tank, 25, 56, 88, 135
Liquid fraction, 164
Liquid helium, nonmagnetic, 130

Liquid hydrogen (LH$_2$), 4, 11, 19, 63, 107, 131, 158, 175, 213, 233
  burst disk, 148
  cars, 300–303
  dewar, 297, 298, 300
  emission product, 16
  flow transfer, 140
  fuel sloshing, 47
  level, 148
  solenoid valve, 141, 142
  storage, 108, 220, 224, 297–306
    cylindrical, 299
  tank, 73, 107, 113, 116, 132, 160–162, 179, 293, 295, 298–299, 301
    placement, 114
    pressure, 53
Liquid, reliable nonmagnetic, 130
Loading, lateral, 45, 46, 48, 51
Loads
  high compression, 37
  thermal, 20, 25, 37, 41, 49, 56, 157
Loss density, high, 194, 196

M

Mach, 16, 112
Magnetic field, 189, 192–194, 201, 305
  applied, 192
Magnetic flux density, 192
Magnets, 3, 10, 37, 54, 55, 73, 76, 77, 100, 125, 182, 190, 236, 304–306
  NMR, 54, 55
Magnet system
  NMR, 54
Maintained, oxygen free, 29
Maintenance
  intervals, 156, 157
Maintenance area, 221, 224
Mass transportation, 2
Material change, 206–207
Material data, 243–252
Material phase transformations, 205
Materials, 16, 19, 65, 108, 129, 152, 182, 204, 217, 233
  selection, 23, 65, 243–252, 293
Materials insulation, non-integral/modular, 99, 106
Materials, nonmetallic, 311, 312
Matrices, replaceable, 165
Maximum gravimetric storage, 177
Maximum internal pressure, 93, 94
Maximum rated CPM, 152
Maximum stress, 93, 96, 252

Measurement, 41, 50, 52, 54, 71, 77, 123, 304, 315, 316
Mechanical
  analysis, 77
  performance, 73
  properties, 65, 66, 71, 86, 235, 273, 315, 316
Medical application, 236, 304
Member
  mechanical support, 37
Mesh, 44, 165, 208
  size, 165
Mesh type, configuration, 165
Metal foam, 165
Metal hydride, 7
Metal oxide semiconductor (MOS), 176
Metals, 3, 7, 25, 28, 33, 37, 39, 65, 70, 73, 77, 84, 89, 124, 129, 131, 132, 162, 165, 176, 193, 194, 204, 243, 244, 317
MgB$_2$, 134, 139, 158, 182, 190–196, 305
Microcrack, 64, 65, 71, 76
  formation, 73
Microcracking, 65, 73, 169
Micron/mTorr, 35, 36
Minimum weight, 4, 86, 109, 110
Mission profile, 20, 21, 25, 42, 58, 59, 106, 112, 146, 234
  parameters, 19
MLI, 31, 32, 34, 35, 88, 166, 167
Modulus, 65, 74, 76, 78, 86, 89, 92, 253, 273, 292
Modulus (GPa), 65, 92
Modulus MN/m^2, Fracture stress MN/m^2, 74
Monitoring, continuous, 175, 296
Mono-filament structure, 193, 194
Motor, 4, 7, 22–24, 115, 123, 132, 151, 152, 157, 158, 160–163, 181, 183–186, 188, 190, 201–211, 234, 235, 307–311
  windings, 23, 24, 158, 163, 202
MRI
  cryo vessel, 125
  hydrogen cooled, 304, 306
  magnet system, 10, 37, 125, 304, 306
MRI, heavy weight, 54
Multi-filament, 194
Multi-stage, 160
Multitank
  operation, 147–149
Multitank, logistics, 148–149

N
NASA, 10, 12, 26, 39, 42, 63, 64, 99, 144, 170, 186, 220, 260, 290, 298, 299, 303, 304, 308
National hydrogen association (NHA), 237
Natural gas, 215, 217
Natural gas, steam, 2
Nautical mile mission, 177
NbTi, 191–195
Net positive suction head (NPSH), 12, 162, 163
NFPA, 137, 175, 214, 229, 295
NHA, *see* National hydrogen association (NHA)
Nm/kg, 182
Non-insulation (NI), 158
Novel solution, 177, 260
NPSH, *see* Net positive suction head (NPSH)
NsDs-pump, 160
  chart, 159

O
Offshore installations, 216
Off-site, 214, 218, 219
Off-site, electrolysis, 219
Off-site liquefaction, 218
On-airport storage, 214
On-site, 218, 219, 221
On-site hydrogen liquefaction, 221
On-site liquefaction strategy, 218, 219
On-site production, 217, 218
Operating conditions, 4, 23, 25, 64, 87–96, 110, 131, 155, 166, 201, 202, 204, 236, 293, 296
Operating loads, 73
Operating pressure, 7, 8, 19, 88, 107, 300, 304
Operating temperature, 7, 24, 71, 143, 176, 182, 188–190, 194, 195, 202
Operational usage data, 44
Optimal configuration, 109
Optimal sensor locations, 180
Optimal storage configuration, 110
Orientation dependency, 188
O-ring failure, 174
Oscillation energy, 56
Oscillations, 55, 162
Outer jacketed tube, 125
Outer, shell, 33, 38, 41, 54, 97, 126, 167, 298, 301
Outer, skin, 112
Outer, surface, 35, 92, 168, 203

Index

Outer, vessel, 13
Outgassing, 25, 28, 34, 35, 39–41, 166
Owens Corning Fiberglass, 75, 137

P

Paper mill drums, 54
Passenger, 106, 110–112, 114, 120, 228
Pathways, 86, 120, 213, 215, 216, 218
PEEK, 68, 77, 163, 205, 233
Penetrations, 31, 40, 41, 43, 69, 76, 80, 123–131, 143
  bimetallic, 129
  epoxy, 129, 131
Permanent magnet, 183, 190
Permeability, 64, 69–72, 81, 83, 84, 192
Permeation, 33, 34, 70, 80–84, 166, 217
  barrier, 34, 80, 83
Phase separator, 43, 147, 164–166, 235
Photovoltaic solar, 216
Physical parameters, 60, 260
Physical properties, 73, 214, 224, 229
Pipe
  gallery, 222, 223
Pipeline, 107, 213, 214, 217–223
  infrastructure, 221, 223
Pipe, related techniques, 75
Piping, 84–85, 123, 124, 128, 129, 132, 137, 162, 235, 295, 311
  arrangements, 124, 129
Piston, 151, 155, 156, 159, 160, 162–164, 203
  positive displacement, 160
  reverses direction, 162
  rings, 163, 164
Piston, dry, 163
Piston pump, performance, 155
Piston rings, disassembled, 164
Piston ring, wear, 163
Placement, envisaged, 112
PM-SC, 182, 190, 191
PM-SC BSCCO, 191
PM-SC REBCO, 191
Polymer, 69, 71, 169, 235, 312–315
Polyurethane foam, 27, 108
Porosity
  adjustable, 165
Port
  single inlet, 148
Power, 203
Power degradation, 203
Power density, high, 189
Power generation, 2, 16, 216, 222
Power, recondensing requirements, 10
Pre-impregnated fabrics, 71

Prepreg, 64, 70
Pressure, 7, 19, 64, 105, 127, 151, 177, 206, 234
  altitude changes, 58
  composite, 68, 69, 88–89, 155
  cryogenic vapor, 177
  expected core, 169
  gauge, 132
  high, 8, 9, 67, 86, 105, 126–129, 147, 151, 152, 159, 162, 177, 208, 295, 296, 300
  high CFC, 8, 9
  maximum, 155
  outlet, 155
  rating, 91, 233
  relief valve, 148
  saturated pressure, 60, 148
  vessel, 7, 8, 65, 67–69, 85–89, 108, 155, 295, 301
  vessel design, 155
  working, 7, 13, 152
  working gauge, 298
Pressure head, 157
Pressure, hydrogen, high, 243, 244, 246
Pressure, inlet, 157
Pressure, internal, 73, 86, 87, 90, 91, 93, 94, 96, 209
Pressure, rated for system, 132
Pressure test, 170, 295, 296, 304
Pressurization, 11, 68, 106, 169
Pressurized, multiple times, 155
Process
  flow control, 148
Process elements, 132, 149, 317
Process engineering components, 141
Process flow, typical, 131–133
Production
  life cycle, 216
Production, pathway, 216, 218
Production, technologies, 215
Properties
  temperature-dependent, 194, 294
Propulsion, 15, 16, 84, 105, 181, 185, 190, 297, 307, 309
Propulsion, motor, 185, 190
Propulsor, 3, 7, 24, 115–121, 132, 201
Protection, 25–29, 34, 168, 307
PTR, *see* Pulse tube coolers (PTRs)
Pulse tube coolers
  (PTRs), 127, 203
Pump, emergency, 155
Pumping, 11, 12, 41, 48, 49, 132, 134, 135, 155–159, 162, 234
  cycle, 162

Pumps, 2, 11, 21, 107, 132, 151, 175, 184, 205, 224, 233
  aircraft, typical, 154
  booster, 155
  cavitation, 162, 163
  centrifugal, 151, 155
  circumferential radial, 159
  common, 152
  cycling experiments, 155
  design, 151
  failure, 119, 148
  high torque, 158
  impeller, 158
  interfaces, 155
  LH_2, reciprocating, 160
  motor, 162
  operation, 156, 162, 163
  operation, efficient, 162
  performance, 155
  piston, 151, 155, 160
  process diagram, 161
  replacement, 151
  suction inlet, 146
  vibration level, 162
Pump, SGV, 161
Purge gas, 29, 108, 169
Purging, 29, 107, 222–224, 226–228, 293, 295, 296
Purging, hose chill, 226

R
Radius sphere, 206
Railcars, 219
Rail guide, 38
REBCO, 190–196
REBCO wires, 190, 192, 193
Recondensing, 10, 23
Reduced bondline strength, 169
Refinery, 213
Refueler, 214
Refueling, 143–144, 158, 173, 175, 214, 217, 220, 221, 223–230, 295, 300
  process, 223, 226, 228
Refueling stations, 173, 220
  remote-site, 220
Refuel rates, 229
Regeneration, 41
Regenerator
  cryogenic, 165
Reinforcement rings, multiple, 298
Reliability, 105, 126, 183, 185, 187, 197, 234, 314
Renewable energy, 215, 216, 218, 301

Resin, 27, 71–80, 84, 85, 89, 137, 233, 236, 306
Resin, filled, 73, 75–76, 78, 80
Resin, semi-flexible, 73
Resins, unfilled, 73, 75–78, 80
Resistivity, 193, 194, 234
Responses, typical, 174, 175
Rocket, 26, 56, 57, 99
Rods/suspensions, 37–39
Roof-mounted installation, 216
Room temperature, 7, 37, 40, 41, 65, 67, 71, 72, 78–80, 84, 85, 91, 125, 127, 128, 143, 209, 235, 236, 258, 293, 315
  storage, 7
Rotating
  coupling, 187
  cryocooler, 185, 186, 201–211
  shaft, 185, 203, 206
Rotor, 163, 183, 184, 186, 188, 190, 202, 203, 206, 208, 209, 311
Rotor, cooling concept, 203
Round-shaped wire, 192, 193
Rupture, 88, 120, 167, 293

S
Safety, 4, 20, 21, 23, 26, 29, 31, 33, 58, 63, 64, 88–91, 93, 94, 107, 109, 112, 120, 123, 141–143, 151–170, 173–175, 179, 187, 214, 217, 222–226, 229–230, 234, 235, 293–296, 298, 300, 301, 312, 314, 317
  database, 293–296
  standard guidelines, 293
Safety system, 173–175
  infrastructure, 175
Saturated region, leaving, 60
Selection, 19, 23, 65, 177, 183, 189, 243–259, 293
Semi circular, 48, 50
Semiconductor, 176
Sensing element, 176–178
Sensor, 29, 85, 87, 133–134, 143, 163, 166, 169, 173–180, 235, 295–297, 314, 317
  concentration, 175
  independent, 177
  locations, 174, 180
  module, 177
Sensor location, optimal, 180
Shaft, rotating, 185, 203, 206
Shaker, 55–57
Shear modulus, 92, 273

Index

Shell, 28, 32–34, 38, 41, 43, 47, 58, 65, 68–70, 73, 77, 78, 80, 88–90, 92, 93, 97, 100, 108, 109, 126, 128, 134, 137, 144, 166–168, 235, 298, 301
    design, 68, 109
Shell, inner, 90, 126
Shrinkage, 73, 75, 76, 129, 209, 210
Shutdown, 174, 175
    compressor, 175
    total refueling, 175
Signal distortion, NMR, 54
Simulation model, 209
Slip ring, 201–211
    common, 203
    copper, rotating, 204
Sloshing, 20, 25, 41–49, 53, 58, 107, 168, 235
Sloshing, phenomenon, 46
Slush, 10–12, 294
    transfer
        efficiency, 12
Snake design, 126, 127
Sorption kits, inline, 140
Space
    applications, 11, 64, 67, 166, 297
    microgravity applications, 67
    missions, 10, 26, 205, 303
    shuttle, 27, 28, 227, 229
Spatial distribution, 179
Spheres
    packed, 165
Stainless steel, 13, 28, 37, 86, 129, 164, 203, 206–210, 247
Standard barometric altitude, 59
Standard operating procedures (SOP), 224, 227
Standards, 9, 11, 31, 42, 59, 109, 120, 124, 126, 141, 163, 166, 168, 214, 224, 227, 229–230, 234, 237, 293, 295, 313, 316, 317
Stator, 183, 188, 190, 201, 202, 206, 208, 209, 311
Steam methane, 215, 218
Steel
    selected, 243, 246
    stainless, 13, 28, 37, 86, 129, 164, 203, 206–210, 247
    wool, coarse, 164
Stiffeners, 92, 95, 298
Storage, 2, 4, 7–16, 21, 24, 25, 29, 32, 33, 58, 59, 63, 64, 67, 86, 87, 97–99, 105, 107, 110, 112, 113, 117, 118, 131–134, 148, 166, 173, 174, 177, 187, 213, 214, 217, 219–221, 224, 229, 234, 293, 295–301, 303, 306, 313

container, 297–299
density, 7, 97, 98, 177
solutions, 110
system, high-volume, 16
tank, 16, 24, 25, 32–34, 58, 63, 67, 87, 113, 118, 131–134, 173, 174, 214, 220, 224, 297–298
tank placement, 113
vessel, 220, 300, 301, 303
Storage, gaseous, 7
Storage, on-site, 221
Straps, 38, 110, 207
Stratification, 43, 53, 58, 126, 168–169
Strength, 66, 67, 69, 75, 84, 86, 89–92, 99, 169, 193, 205, 233, 244–249, 252, 253, 258, 259, 273
    high, 65, 69, 75, 86, 99, 205, 207, 208, 233
    aluminum, 20, 21, 37, 69, 91
Strength, MPa, 92, 249
Stress, 20, 29, 43, 46, 65, 66, 69, 71, 73, 74, 76, 77, 87–97, 106, 125, 129, 131, 137, 143, 144, 162, 235, 252, 253, 257, 273
    concentrations, 43, 46, 73, 77, 93, 129, 143, 144, 235, 253
    distribution, 91
Stroke, 152, 162
Structural analysis, 90, 91
Structural component, 54, 210
Structural support
    Heim column, 37, 38
    member, 37, 39, 109
Structure
    additively manufactured, 165
Strut
    thin carbon, 38
Suction, 12, 132, 146, 147, 161, 162, 164
Suction adaptor, 161, 162
Superconducting, 3, 7, 22, 73, 115, 125, 151, 181, 201, 234
    applications, 4, 151, 181, 236, 305, 306, 315
    coil, 73, 183, 185, 186, 316
    components, sustainability, 118–121
    fully, 24, 115, 118, 183, 190
    generator, 186, 197
    level probes, 134
    LN_2 pump, 158
    machine, fully, 115, 183, 190
    machine model, 115
    machines, 115, 182–187, 189, 190, 197
    magnet technology, 73
    motors, 7, 22–24, 115, 132, 157, 181, 183, 184, 186, 205, 210, 211, 307, 309
    propulsors, 115–121

Superconducting (*cont.*)
  rotating machine, 181–197
  states, 185, 188, 197
  stationary, 156, 182, 183, 187–196
  wire, 10, 182, 183, 187, 191–193
  wire materials, 182
Superconducting, partial, 182, 183
Superconducting pump, 155–158
Superconducting pump, advanced, 156
Superconductor, 24, 134, 156, 158, 181, 182, 187, 192–194, 196, 202, 304, 305, 308, 315, 316
  filaments, 193
  high temperature, 24, 137, 156, 202, 304, 305, 315
  medium temperature MgB_2, 134
  region, 192, 193
Superfluid pump
  specification, 157
Supply chain, 213, 218–219
Support members, 37, 39, 109
Suspension, 25, 37, 39, 43, 54, 58, 85, 100, 203, 210, 235
Suspension, axial, 37
Suspension system, conventional, 39
Synchronous motor, 198
System
  single thermosiphoning, 148
System, warm-up, 223, 224

T
Take-off, 45, 46, 59
Tank, 2, 7, 19, 63, 105, 123, 151, 174, 214, 233
  aluminum, reinforced, 8, 9
  chilldown, 143–144
  closed, 20, 21, 53
  composite, linerless, 70
  conformal, 58, 86, 88, 105–107
  cryogenic design, 148
  depletion, 168
  designs, 4, 11, 19–60, 63–64, 67–70, 79, 80, 83, 85, 87, 89, 97, 99, 106–121, 123, 144, 147, 148, 220, 233, 300, 304
  ellipsoidal, 108
  failure, 143, 169
  farm, 219, 220
  faults, 87–96, 167
  fiber, 97
  high-pressure receiver, 159, 162
  inner surface, 125, 202
  installation, 120
  instrumentation, 133–134
  insulation, 26, 69, 93, 166, 168
  internal convection, 58
  liquid stratification, 168
  lobed, single, 108
  mass, 22, 48, 49, 88, 96–100, 107, 108, 144, 233
  mass weight, 22, 88, 96–100, 233
  mass weight fraction, 88, 96–100, 106, 233
  material, 82
  onboard storage, 174
  pressure change, 58
  pressure rise, 43, 53
  shape, 13, 19, 23, 44, 46, 105–121
  size, 26, 58, 128, 169, 300
  structure, 4, 10, 22, 23, 45, 65, 71, 87, 107, 108, 168, 169
  surface area, 54
  suspension, 25, 37, 58
  temperature
    homogeneity, 77
  wall, 19, 23, 29, 33, 34, 42, 64–87, 107, 168, 169, 174
  wall design, 34
Tank design, non-conformal, 58, 106
Tank fueling, first, 143
Tank, inner, 13, 33, 35, 49, 69, 70, 79, 83, 87, 88, 91–92, 166–168, 220
Tank lay-up optimized, 73
Tank, level, 50, 51
Tank mass, inner, 48
Tank mass, minimum, 96–100, 233
Tank, receiver, high-pressure, 159, 162
Tank type, 71, 83, 91, 93, 97
  wall, 91
TDC, *see* Top-dead center (TDC)
Temperature, 3, 7, 23, 64, 107, 125, 156, 176, 182, 201, 224, 233
  crank drive, 163
  distribution, 184
  domain, 37
  reduction, 59
  sensor, 163
  superconductor rotor, 202
Temperature, fluctuating
  window, 202
Temperature range, ambient, 163
Temperature, steady state, 139
Tensile strength MPa, 92
Terminal servicing equipment, 214
Terrier Sandhawk rocket, 57
Test implementation, typical, 10

Index 339

Thermal
  conductivity, 30, 34, 35, 37, 77, 78, 80, 81, 176, 206, 207, 271, 275, 294
  connector, details, 130
  contraction, 73, 75, 76, 131, 209
  cycling, 28, 71, 83
  diffusivity, 80, 294
  expansion, 77, 133, 315
  resistance, 206, 207
  response time, 80, 144
  shrinkage, 75
Thermal conductivity
  effective, 30, 34
Thermal contraction
  integral, 75, 76
Thermal gradients, high, 205
Thermal protection scheme, 25–28
Thermal radiation
  mitigation, 29–32
Thermal short, 125, 126
Thermistor, 163, 176
Thermomechanical device, 203
Titanium, alloy, 37, 68
Tolerable limits, 93
Top-dead center (TDC), 162
Topology, 4, 181–183
Torque, 156, 158, 182, 187–189, 203
Torque, average, 182
Toxicity, 13, 14
Toyota Mirai, 173, 174, 178
Trailers, 12, 42, 168, 214
Transfer design options, 203
Transport, 4, 12, 42, 44, 55, 99, 110, 111, 120, 164, 167, 191, 192, 205, 214, 218, 219
Transportation, 2, 15, 181, 213, 214, 217–221, 309
Transportation, scenarios, 219
Transport loss, 191, 192
Trimming, 148
Truck, 2, 11, 25, 46, 59, 213, 214, 218–221, 224, 226, 297, 301
T-S Equilibrium diagrams, 261, 262
Tube, 37, 54, 71, 75, 89, 105, 109, 123–132, 137, 144, 162, 164, 165, 168, 185, 201, 203, 209, 210, 256, 304
  angle
    definition, 127
    horizontal, 127
  multifunction tube
    horizontal, 125
  orientation, 128
  shrinking, 125, 131
Tube, multiple circuit, 128

Tube, thin-walled, stainless, 164
Tubing
  design, 127
  structure, 127, 129
Tubing design, workaround, 127
Tubing, multiple connected, 128
Turbofan, engine, 12, 155
Turboprop, engine, 155
Turret, horizontal, 125

U
Universe, 3, 215

V
Vacuum, 25, 69, 105, 124, 152, 201, 220, 235, 298
  achievable quality, 29
  chamber, 40, 125
  compromised, 166
  compromised conditions, 166
  conditions, 20, 30, 33–35, 134
  enclosure, 88, 100
  high, 30, 34, 105, 107
  pressure range, 34
  quality, 29, 31, 34, 36, 37, 39, 41, 87, 166
  range, 35
  shells, 28, 41, 69, 90, 128, 134, 166, 168
    linerless, 41
  space, 34, 40, 41, 69–71, 83, 87–89, 129, 210, 317
  inner, 41, 69, 83, 89
Vacuum-jacketed pipeline (VJP), 221
Valve, 20, 21, 132, 141–143, 162, 223, 293, 295, 296
  discharge valve, 132, 162
  downstream vent valve, 162
  relief valve, 132, 141, 148
  supply valve, 162
  supply valve interface, 174
Valve, socket foot, 151
Vapor
  bubbles, 164, 235
Vaporization, 9, 11, 24, 49, 54, 58, 196, 229, 277, 278, 290, 292, 294
Vapor pressure, 162, 283–287
Variable speed control, 156
Vehicle, 52, 141, 158, 169, 173, 178, 222, 224, 228, 230, 295, 300, 301
  operation, 174
  safety system, 174
  startup, 174
Velcro, 31, 32

Velos Orbiter, 113, 116
Vent, 53, 166–168, 293, 295
  multiple, 168
Venting, standard practice, 11
Vent tube
  single, 168
Vent valve, downstream, 162
Vertical turret, common, 125
Vessel design
  rating, 155
Vessel, inner, 13, 56
Vibration, 19, 20, 25, 46, 47, 50, 52–56, 58, 162, 204, 205, 207
  bulk-head baffles, 19
  canceling elements, 54
  cause of losses, 46
  floor, 54
  induced heat, 53
  liquid bulk, 42, 53, 123–131
Vibrostand, 55, 56
VJP, *see* Vacuum-jacketed pipeline (VJP)
Volumetric
  density, 96, 97
  efficiency, 109, 110, 177
  energy density, 300
Volumetric density
  comparison, 96
Volumetric efficiency, 109, 110, 177
Volumetric energy
  higher, 300
von Mises stress, 91

W

Wall, 22, 23, 26, 29, 33, 83, 85, 91, 107, 130, 131, 143, 210
  crack, 174
  fiberglass dewar, 130
Wall thickness, 7, 54, 75, 79, 80, 97, 144
Warning, 128, 175, 235
Water, electrolysis, 2, 215, 218

Water, thermolysis, 215
W/cm^3, 194–196
Wear
  highest, 156
Wear rates
  constant, 205
Webs, spar, 109
Weight, 4, 7, 11, 13, 31, 37, 38, 46, 49, 65, 66, 70, 74, 75, 78, 89, 93, 110, 129, 137, 160, 182, 189, 197, 234, 236, 300, 301, 307
  fraction, 86–88, 93, 96–100, 110
  fuel, 7
  reduced, composite, 68
  reduction, 70, 303
  total, 38, 110
Wind energy, 216
Winding, 23, 24, 31, 73, 75–77, 84, 158, 163, 186
Winding angle, 75, 76
Winding, enclosure, 183, 184
Wind turbine generators, 185, 190
Wing, 106, 109–111, 115, 116, 224
  storage, 118
  tapered
    cylindrical, 118
Wire, 176, 192–194, 196, 246, 305, 318
  MgB$_2$, 190, 194
  properties, 194
  tape-shaped, 193, 194
  temperature, 191
  twist pitch, 193, 194

X

X-33 investigation report, 169

Y

Yield strength, 91, 245, 248–249, 252